# Designing Sustainable Working Lives and Environments

Work is central to people's lives and the course of their life. The opportunities and chances an individual can have in their life are significantly connected to work. Individuals' work is also crucial for organisations, companies and for the whole of society. There is a constant need to make changes and readjustments of working life since these can deeply affect the individual and their employability. To make working lives more healthy, sustainable and attractive, being aware of the measures and changes that can be achieved in practice is of crucial importance. This book bridges the gap between the theories and explanatory models offered in research and actual work environments and workplaces.

This book constitutes a theoretical framework that visualises the complexity of working life and increases the knowledge and awareness of individuals, companies, organisations and society regarding different factors and patterns. It aims to support individual reflections and joint discussions into daily operations on the individual, organisational and societal level. This book contains practical tools to use in daily working life that analyse possible risks in the work environment when planning measures and actions for health promotion. These practical tools are derived from the four spheres for action and employability in the SwAge model. Developed by the author, the SwAge model (Sustainable Working Life for All Ages) is a theoretical, explanatory model that explains the complexity of creating a healthy and sustainable working life for all ages. By using the SwAge model as a comprehensible framework, the reader will be able to visualise the complexity of factors that affect and influence whether people are able to and want to participate in working life and in the work environment, thereby contributing to increased employability.

*Designing Sustainable Working Lives and Environments* is an essential read for students, researchers, work environment engineers, ergonomics and human factor specialists, occupational health and safety practitioners, business managers, HR staff, leadership decision-makers and labour union professionals.

**Kerstin Nilsson** is Professor in Public Health Epidemiology and Associate Professor in Work Sciences, in the Division of Occupational and Environmental Medicine at Lund University & in Division of Public Health at Kristianstad University, Sweden. Her previous affiliations include the Swedish National Institute of Working Life and

Swedish University of Agricultural Sciences, from where she received her PhD in Work Science. She also has a doctorate in Public Health and Epidemiology from Lund University. She is the developer of the SwAge model (Sustainable Working Life for All Ages).

The SwAge™ model presented in this book has been developed by Professor Kerstin Nilsson. For more information, please visit https://swage.org/en or contact kerstin.nilsson@med.lu.se.

# Designing Sustainable Working Lives and Environments
## Work, Health and Leadership in Theory and Practice

Kerstin Nilsson

Designed cover image: © Kerstin Nilsson

The book is translated from Swedish by Emma Nilsson, Translator and Bachelor of Science majoring in Business Administration

First edition published 2024
by CRC Press
2385 NW Executive Center Drive, Suite 320, Boca Raton FL 33431

and by CRC Press
4 Park Square, Milton Park, Abingdon, Oxon, OX14 4RN

*CRC Press is an imprint of Taylor & Francis Group, LLC*

© 2024 Kerstin Nilsson

Reasonable efforts have been made to publish reliable data and information, but the author and publisher cannot assume responsibility for the validity of all materials or the consequences of their use. The authors and publishers have attempted to trace the copyright holders of all material reproduced in this publication and apologize to copyright holders if permission to publish in this form has not been obtained. If any copyright material has not been acknowledged please write and let us know so we may rectify in any future reprint.

The Open Access version of this book, available at www.taylorfrancis.com, has been made available under a Creative Commons Attribution (CC-BY) 4.0 license.

Any third party material in this book is not included in the OA Creative Commons license, unless indicated otherwise in a credit line to the material. Please direct any permissions enquiries to the original rightsholder.

*Trademark notice*: Product or corporate names may be trademarks or registered trademarks and are used only for identification and explanation without intent to infringe.

*Library of Congress Cataloging-in-Publication Data*
Names: Nilsson, Kerstin, author.
Title: Designing sustainable working lives and environments : work, health and leadership in theory and practice / Kerstin Nilsson.
Description: First edition. | Boca Raton, FL : Taylor & Francis, 2024. | Includes bibliographical references and index.
Identifiers: LCCN 2023046554 (print) | LCCN 2023046555 (ebook) | ISBN 9781032590486 (hardback) | ISBN 9781032616674 (paperback) | ISBN 9781032616681 (ebook)
Subjects: LCSH: Quality of life. | Work-life balance. | Industrial hygiene.
Classification: LCC HD4904 .N555 2024 (print) | LCC HD4904 (ebook) | DDC 331.01—dc23/eng/20240104
LC record available at https://lccn.loc.gov/2023046554
LC ebook record available at https://lccn.loc.gov/2023046555

ISBN: 9781032590486 (hbk)
ISBN: 9781032616674 (pbk)
ISBN: 9781032616681 (ebk)

DOI: 10.1201/9781032616681

Typeset in Times
by codeMantra

# Contents

Preface ................................................................................................... xiii
Introduction ............................................................................................ xv
    Why Is a Book on a Healthy and Sustainable Working Life for
       All Ages Using the SwAge™ Model Needed? ......................................... xvi
    The Structure of the Book and the SwAge™ Model ..................................... xvi

## PART I   Determinant Areas for a Sustainable Working Life and Employability

    Influence and Determinant Areas for the Ability
       and Willingness to Work .......................................................................... 1
    To Stay or to Leave? ................................................................................... 3
    What Determines a Healthy and Sustainable Working
       Life for All Ages? ..................................................................................... 5
    Consider and Reflect .................................................................................. 7

## SPHERE A.   Health Effects of the Work Environment

**Area 1. Diagnoses, Self-rated Health and Functional Diversity** ....................... 11
    Different Dimensions of Health ................................................................ 11
    Health on the Individual Level ................................................................. 14
    Bodily Functions that Matter to Work ..................................................... 15
    The Dual Continuum Model of Health .................................................... 23
    The Work Environment as a Predictor of Ill-health ................................. 24
    Measuring Health and Ill-health in Working Life .................................... 27
    Consider and Reflect ................................................................................ 27

**Area 2. The Physical Work Environment** ......................................................... 29
    The Physical Design of the Workplace .................................................... 30
    Physical Workload, Heavy Lifting and Unilateral Movements ................ 30
    The Sensory System and the Work Environment .................................... 32
    Chemical Health Risks in the Work Environment ................................... 33
    Climate and Work Ability ........................................................................ 35
    Safety Risks and Occupational Injuries ................................................... 35
    Gender Differences between Physical Demands in Working Life .......... 37
    Consider and Reflect ................................................................................ 37

**Area 3. Mental Work Environment** .................................................................................. 39

    Stress that Matter to Work ............................................................................. 40
    Stressors Related to Working Life ................................................................. 43
    Coping with Work Stress ............................................................................... 44
    Bullying in the Workplace ............................................................................. 53
    Threats and Violence ...................................................................................... 54
    Gender Differences between Physical Demands
        in Working Life ....................................................................................... 55
    Consider and Reflect ..................................................................................... 56

**Area 4. Working Hours, Work Pace and Time for Recuperation** ..................... 58

    Working Hours and Work Schedule .............................................................. 58
    Working Time Regulations ............................................................................ 59
    Work Pace and Breaks during Working Hours ............................................. 59
    Exhaustion and Sleep Disorders .................................................................... 60
    Time for Recuperation .................................................................................... 61
    Working Hours, Work Pace and Time for Recuperation
        Related to Age and Gender ..................................................................... 63
    Consider and Reflect ..................................................................................... 64

**Organisational Perspective and Action Proposals that Matter
to the Health Effects of the Work Environment** ........................................................ 65

    Organisational/Meso Level and Health ......................................................... 65
    Health Promotion and Health Prevention ..................................................... 66
    Health Promotion and Organisational Productivity ..................................... 69
    The Health Prevention Process: Identify – Prioritise –
        Rectify – Reorient ................................................................................... 71
    Measures for Physical, Mental and Temporal Work
        Environment Issues and Health ............................................................. 72
    The Design of the Physical Work Environment ........................................... 75
    The Design of the Mental Work Environment ............................................. 77
    Organisation of Work to Facilitate Recuperation .......................................... 79
    Occupational Health Care .............................................................................. 81
    Consider and Reflect ..................................................................................... 82

**Societal Perspective and Action Proposals that Matter to Health Impacts
of the Work Environment** ............................................................................................... 85

    Societal/Macro Level and Health .................................................................. 85
    The Significance of Society for Healthy Choices ......................................... 86
    The Significance of Society for the Knowledge of and Compliance
        with Laws and Directives ....................................................................... 87
    Societal Perspectives on Fragmentation, Demands and Control
        in the Work Situation .............................................................................. 88
    Societal Support in Choices and Changes of Profession
        and Work Environment ........................................................................... 89

Contact-based Professions and Mental Demands ............................................. 91
Working Time Models .................................................................................... 91
The Significance of Occupational Health Care to Decrease the Negative
   Health Effects of the Work Environment ................................................... 92
Consider and Reflect ....................................................................................... 92

## SPHERE B.   The Financial Situation

**Area 5. Personal Finance and Economics** ........................................................... 97

Finance and Employability ............................................................................... 98
Economics and Financial Incentives ............................................................... 101
Consider and Reflect ..................................................................................... 102

**Organisational Perspective and Action Proposals that Matter to the
Financial Situation** ........................................................................................ 104

Occupational Injuries Affect on Finance ........................................................ 104
The Significance of Leadership for Good Health
   and Employability ..................................................................................... 105
Attitudes in Organisations and Employability ................................................ 106
Organisational Costs of Sickness Absence ..................................................... 107
Consider and Reflect ..................................................................................... 108

**Societal Perspective and Action Proposals that Matter
to the Financial Situation** ............................................................................. 110

Societal Perspective and Action Proposals that Matter to
   Personal Finance ....................................................................................... 110
Financial Perspective on Societal Support for Employability ......................... 110
Social Security Systems ................................................................................. 111
Employment Development in Society in Relation to Social
   Security and Pension Systems ................................................................... 112
Consider and Reflect ..................................................................................... 112

## SPHERE C.   Social Inclusion, Support
             and Sense of Community

**Area 6. The Personal Social Environment** ....................................................... 115

Social Inclusion outside Working Life ........................................................... 115
Boundaries between Work and Personal Life ................................................. 116
How Partner and Family Affect Labour
   Force Participation .................................................................................... 117
How the Personal Social Environment Affects Work ..................................... 118

Work-life Balance ................................................................................. 119
Consider and Reflect .............................................................................. 119

**Area 7. The Social Work Environment** ................................................ 121

Identity and Social Contexts .................................................................. 121
Social and Instrumental Support and Inclusion in the Workplace ............... 132
The Social Effects of Informal Organisations in Working Life .................. 134
Disregard, Bullying and Discrimination .................................................. 136
Norms, Norm Breakers and Power .......................................................... 140
Consider and Reflect .............................................................................. 142

**Organisational Perspective and Action Proposals that Matter to Social Inclusion, Support and Sense of Community** ......................................... 143

Organising Groups ................................................................................. 143
Leadership for Employeeship – Social Innovation and Support through Need-, User-, and Person-centred Development ...................................... 144
Situational and Age-conscious Leadership .............................................. 146
Communication as a Tool to Prevent Risks of Disregard and Discrimination in the Social Work Environment ................................... 149
Inventory of the Workplace Culture ........................................................ 152
Conflict Management in the Social Work Environment ........................... 154
Consider and Reflect .............................................................................. 156

**Societal Perspective and Action Proposals that Matter to Social Inclusion, Support and Sense of Community** ......................................... 158

Society as a Supportive Environment for Social Inclusion and Participation ...................................................................... 158
How Society Affects Disregard and Discrimination in Working Life ........ 159
How Society Affects Work-life Balance .................................................. 160
Consider and Reflect .............................................................................. 161

# SPHERE D.  The Execution of Work Tasks and Activities

**Area 8. Work Satisfaction, Motivation, Stimulation and the Core of Work** ....................................................................................... 165

Meaningfulness and Stimulation or Futility in the Work Situation ............ 165
Building a Wall or Building a Windmill .................................................. 168
The Individual's Experience of Comprehensibility, Manageability and Meaningfulness in the Work Situation ............................................ 170
Satisfaction of Needs Is Fundamental to Work Motivation ...................... 172
Effort-reward Balance/Imbalance ........................................................... 174
Motivation through Empowerment and Nudging .................................... 175

Occupational Identity ............................................................................. 177
Consider and Reflect ............................................................................. 178

**Area 9. Knowledge, Skills and Competence Development.............................. 180**

Perception and Cognition ...................................................................... 180
Intelligence ............................................................................................. 183
Memory .................................................................................................. 184
Knowledge .............................................................................................. 187
Learning and Employability .................................................................. 190
Consider and Reflect ............................................................................. 192

**Organisational Perspective and Action Proposals that Matter
to the Execution of Work Tasks and Activities ........................................ 194**

Design of the Execution of Work Tasks – A Historical Overview
  of Efficiency and Meaningfulness in Working Life .............................. 194
Satisfaction in Working Life ................................................................. 196
Employability in a Changing Working Life – Resilience .......................... 197
Employeeship – Increasing Individuals' Experiences of Being
  Part of a Greater Whole ...................................................................... 200
Organisation of Work in Relation to Employees'
  Abilities and Motivation ....................................................................... 203
The Learning Organisation .................................................................... 204
Learning Processes in Organisations and Companies ............................. 206
Methods for Competence Development and Transfer of
  Tacit Knowledge .................................................................................. 207
Consider and Reflect ............................................................................. 209

**Societal Perspective and Action Proposals that Matter to the Execution
of Work Tasks and Activities ................................................................... 211**

How Society Affects Individuals' Experience of Their Work
  Tasks and Activities .............................................................................. 211
Possibilities of Updating Knowledge .................................................... 212
Possibilities of Changing and Adjusting Work Content
  to the Individuals' Conditions .............................................................. 213
Consider and Reflect ............................................................................. 214

# PART II  *Age in Relation to a Sustainable Working Life*

The Demographic Situation .................................................................. 215
Age Definitions and Perspectives on Age in Working Life .................. 218

**Biological Age in Working Life** .................................................................. 221

    Individual Perspectives on Biological Age ................................................. 221
    Organisational Perspectives on Biological Age ......................................... 224
    Societal Perspectives on Biological Age ..................................................... 224

**Chronological Age in Working Life** .............................................................. 225

    Individual Perspectives on Chronological Age .......................................... 225
    Organisational Perspectives on Chronological Age ................................... 225
    Societal Perspectives on Chronological Age .............................................. 226

**Social Age in Working Life** ............................................................................ 228

    Individual Perspectives on Social Age ........................................................ 228
    Organisational Perspectives on Social Age ................................................ 230
    Societal Perspectives on Social Age ............................................................ 231

**Cognitive Age in Working Life** ..................................................................... 233

    Individual Perspectives on Cognitive Age .................................................. 233
    Organisational Perspectives on Cognitive Age .......................................... 234
    Societal Perspectives on Cognitive Age ..................................................... 234

**All Ages in Working Life** ................................................................................ 236

    Individual Level – Micro Level ..................................................................... 236
    Organisational Level – Meso Level .............................................................. 237
    Societal Level – Macro Level ........................................................................ 238
    Consider and Reflect ...................................................................................... 238

## PART III   Practical Application Based on the SwAge™ Model

**Practice Application of the SwAge™ Model** ............................................... 241

**Case for Reflection Supported by the SwAge™ Model** ............................. 243

**Nine Quick Ways to Increase Sustainability in the Organisation's Environment** ................................................................................................ 250

Contents                                                                                      xi

**The Survey Tool for Investigation and Reflection of Work Ability,
Employability and Work Situation** ......................................................... 253

    Questionnaire: Sustainable Working Life for All Ages ............................... 253

**The Tool for Employer/Manager-Employee Work Situation and Career
Development Conversations** ..................................................................... 256

    The Dialogue Tool for Career Development Conversations
        with Employeess ................................................................................ 257
    Before Conducting Career Development Conversations ............................ 257
    Conduct Your Career Development Conversations Based on Different
        Perspectives ...................................................................................... 258
    Documenting the Career Development Conversation ................................ 258
    Individual Goals and Competence Development ....................................... 260

**A Tool for Systematic Workplace Management and Action Plan for a
Sustainable Working Life for All Ages** ................................................... 261

    How to Conduct the Analysis ................................................................... 262
    The Matrix Tool for Systematic Workplace Management and Action
        Plan for a Sustainable Working Life ................................................. 266

**Examples of Measures Taken in Different Workplaces** ................................ 274

**Tools for Evaluations and Follow-Up of Actions and Measures** ..................... 278

    Process Evaluations ................................................................................. 278
    Impact Evaluations .................................................................................. 280
    Economic Evaluations ............................................................................. 280

**References** .................................................................................................. 287
**Index** ........................................................................................................... 301

# Preface

*In short, this book constitutes a well-organised toolbox, with theories and practical tools based on scientific research and explanations in the SwAge™ model (Sustainable Working Life for All Ages), to increase individuals' employability, and organisational/enterprise and societal measure initiative to create a healthy, sustainable and attractive working life for people of all ages.*

Work is a central part of peoples' lives. The opportunities and chances an individual gets in their life are significantly connected to work. There is a constantly increased need to make changes and readjustments in working life, these affect the individual regardless of whether they are employed, run their own business, are on sickness absence or are unemployed. For example, the global demographic structure is currently changing, resulting in a larger proportion of senior people in society. Organisations and companies face new challenges to maintain a healthy and sustainable working life with an increased number of senior employees. Therefore, additional knowledge on what these challenges and changes may entail is needed, as well as practical tools to meet the new and changed demands that are constantly put on individuals, organisations, companies and society.

The purpose of this book is to bridge the gap between the theories and explanatory models offered in research, and the actual work environments and workplaces where many of us spend large parts of our lives. In order to make working life healthy, sustainable and attractive, awareness regarding measures and changes we collectively can achieve in practice, as well as how to get started, is of crucial importance.

This book constitutes a theoretical framework, visualises the complexity of working life and increases the knowledge and awareness of individuals, companies, organisations and society regarding different factors and patterns in working life. These factors make up determinant areas that influence and affect whether an individual is able to and wants to work, in other words, opportunities as well as risks and issues in working life. Furthermore, the determinant areas relate to the considerations an individual makes when deciding whether to remain in or to leave working life. Moreover, this book addresses how ageing must be taken into account in order for people to be able to work for an entire working life, as well as to make working life sustainable.

The aim of this book is to convey theoretical knowledge, to be of support to individual reflections and joint discussions, and to be a comprehensive tool in the daily operations on the individual, organisational and societal level. The purpose of this book is to deepen knowledge and awareness to be able to observe and identify risk factors for ill-health, occupational injuries and issues in the work environment. Furthermore, the purpose is to introduce the need of possible measures of prevention and promotion with the purpose of developing work and the work environment to become more healthy and sustainable for people of all ages.

This book contains practical tools to use in the daily working life, for example in employee performance reviews, in order to analyse possible risks in the work environment, and when planning measures and actions of health promotion in working life. These practical tools are derived from the four spheres for action and employability in the SwAge™ model, which includes nine different determinant areas for a healthy and sustainable working life, which previously have been published academically.

The SwAge™ model (Sustainable Working Life for All Ages) is a theoretical, explanatory model which provides an overview and pedagogically explains the complexity of creating a healthy and sustainable working life for all ages. The SwAge™ model offers – and is used as – a comprehensible framework and tool for individuals, in workplaces, organisations and authorities nationally and internationally. Using the SwAge™ model as a comprehensible framework makes it easier to visualise the complexity of factors which affect and influence whether people are able to and want to participate in working life and in the work environment, thereby contributing to increased employability.

Lund April 2023 Kerstin Nilsson

# Introduction

Several factors in working life affect and influence individuals' possibility, ability and willingness to work. In order to enable the reader's understanding of the complexity of working life, this book is based on the SwAge™ model (sustainable working life for all ages) in order to visualise and organise the complexity, and to make the connections comprehensible. The SwAge™ model is a theoretical model of determinant areas for a sustainable working life for all ages (Figure 1), based on four spheres for action and employability: (A) health effects of the work environment; (B) financial incentives; (C) social inclusion, support and sense of community and (D) execution of work tasks and activities.

Employability and the ability and willingness to participate in working life are strongly connected to health and how the work environment impacts health. Work is a substantial part of our lives, we spend a large part of our life time and waking hours at work; therefore, work is significant to our health and well-being. We want to experience health and well-being every day. Health is an everyday approach. However, it is also a goal for people who do not experience sound health and well-being.

A sufficiently sound health is a prerequisite for employability. However, working life affects our health, our bodies and our mental health based on the demands, tear, or strengthening influences and development our work tasks and work environment offer.

Another prerequisite for the ability and willingness to work is to be able to afford living expenses, which employment contributes to fund. Furthermore, working life can contribute to a sense of coherence, and the experience of inclusion in a group, participation and security. However, working life can also contribute to the experience of exclusion and disregard, or even discrimination. Additionally, social inclusion, support and participation are significantly connected to whether people want to and are able to be a part of work units and teams in the workplace, human beings are herd animals after all. Moreover, since working life is in a state of constant change, we need to remain employable given the demands and ability to execute the work tasks and activities which our occupation requires. We spend a large proportion of our waking hours executing work tasks and activities, these can be a source of stimulation and joy, a challenge and a way to learn new skills and develop; however, work tasks and activities can also be a source of boredom and stagnation.

We change and age during our (working) lives. Our occupation affects our ageing, and on the other hand, our ageing affects which occupations we can cope with and how we execute our work tasks. Age and ageing have previously been ignored in research regarding health and working life. However, different age definitions must be related to work in order for working life to become sustainable for all ages. The age of retirement has been postponed in many societies, based on global demographic changes currently taking place. It is important to take preventive measures in order to make working life sustainable for all ages throughout, and not only when issues and ailments occur at the end of working life.

## WHY IS A BOOK ON A HEALTHY AND SUSTAINABLE WORKING LIFE FOR ALL AGES USING THE SWAGE™ MODEL NEEDED?

Working life has been an area of research for hundreds of years. In order to comprehend the complexity of a healthy and sustainable working life and how different factors affect and influence each other, an interdisciplinary and intersectional overview covering all factors is needed. That is, different areas of research need to be connected in one single model. Previously, there has been a lack of explanatory models which offers an overview and weighs in the entire complex image of determinant areas for a sustainable working life. Rather, one or a few factors have been studied at a time. For example, medical practitioners and technicians may primarily observe health effects caused by the work environment regarding functional diversity, physical and mental work environment and the possibility of recuperation. Economists and legal advisers focus on economics, financial incentives and regulations. Sociologists and organisational researchers may focus on leadership, social relations, support, inclusion and work life balance. Furthermore, pedagogical researchers and psychologists may primarily observe the experience of meaningfulness, motivation, stimulation, knowledge, development and the execution of work tasks and activities. The joint, complex overview is covered in the SwAge™ model, which this book is based on.

## THE STRUCTURE OF THE BOOK AND THE SWAGE™ MODEL

The first part of the book handles several areas and factors from the foundation of a healthy and sustainable working life on different levels. As previously mentioned, four spheres for action affect the individual's employability as described in SwAge™ model: (A) health effects of the work environment; (B) financial incentives; (C) social inclusion, support and sense of community and (D) execution of work tasks and activities. The first part of the book is divided into these four spheres, containing the nine determinant areas that affect whether an individual has the ability and willingness to work in a healthy and sustainable working life, that is (1) health, (2) the physical work environment, (3) the mental work environment, (4) working hours, work pace and time for recuperation, (5) financial incentives, (6) personal social environment, (7) social work environment, (8) meaningfulness, stimulation and motivation through work tasks as well as (9) knowledge, skills, competence development (please note the micro level, the yellow circle in Figure 1). Each of the nine determinant areas will be presented in detail further on.

The possibility to work does not solely depend on the individual. It depends on decisions made on the organisational level regarding production, work environment and organisational culture. Furthermore, prerequisites are created on this level regarding working hours, access to competence development, climate and attitudes towards employees of different ages, discrimination etc. (please note the meso level, the middle circle in Figure 1). The workplaces of organisations and companies constitute this level, where conditions for employees' ability and willingness to work are created. The external framework, the conditions of working life and the possibility to work are created on the societal level (please note the macro level, the outer circle in Figure 1). Decisions are made on the societal level, for example regarding

# Introduction

rules, regulations, social welfare systems and the age of retirement. These conditions control and affect the individual's opportunities, incentives and prerequisites to work. Furthermore, the laws of society control the organisations' and companies' conditions and possibilities for production, as well as incentives to motivate organisations and companies to take measures for their employees. Therefore, each sphere for action does not only relate to the individual level, but also to the organisational and the societal level.

The second part of the book describes different age definitions, and how these are connected to a sustainable working life (you can see the connections between different age definitions and other factors in Figure 1).

To conclude, in the third part of the book, we find examples of practical tools for a healthy and sustainable working life based on the SwAge™ model. This book can be read from cover to cover, or be used as a reference book.

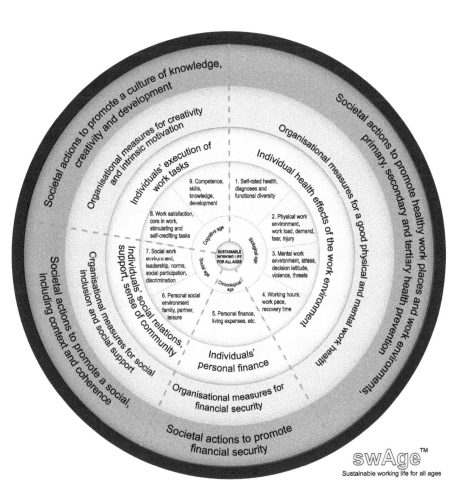

**FIGURE 1** The SwAge™ model (sustainable working life for all ages). A theoretical model of influence and determinant areas that affect a sustainable working life [1].

# Part I

# Determinant Areas for a Sustainable Working Life and Employability

Several factors in working life affect and regulate individuals' ability and willingness to work, as well as their employability. The possibility to work is set and constituted by influence and determinant areas which work on three levels, that is the individual level, organisational level and societal level. Due to the SwAge™-model those influence and determinant areas can be visualised, categorised and remedied systematically (see Figure 1).

## INFLUENCE AND DETERMINANT AREAS FOR THE ABILITY AND WILLINGNESS TO WORK

To begin with, it is important to reflect on the difference between the *ability* to work and the *willingness* to work. The fact that an individual is able to work does not necessarily mean that they want to work. Likewise, the fact that an individual wants to work does not necessarily mean that they are able to work. Which factors affect and relate to our ability and willingness to work? By analysing this question, research has shown how the nine determinant areas for a sustainable working life differ between the *ability* to work and the *willingness* to work [1–8].

Analyses show that of the nine determinant areas for a sustainable working life, the individual's self-rated health, diagnoses and functional diversity, as well as financial incentives and the individual's personal social life, in other words, the individual's life and conditions outside work, are factors of significance to the individual's *ability* and *willingness* to work. These three factors show statistically significant connections to both the *ability* to and the *willingness* to work [3–5,8–10] (Figure 2). The fundamental reason to work is that we need a way to fund our living expenses, in other words, to earn money for food, housing, lifestyle, etc. However, our health and well-being must be sufficient in order for us to be able to cope with working life and avoid developing ill-health. Furthermore, we are dependent on our family situation and relationships when we work. For example, if we need to take care of young children or relatives, and our personal social situation is dependent on and affects our working life. Additionally, analyses show that what primarily affect our ability to work are factors which can be described as the "hardware" of working life (Figure 2). Therefore, of the nine determinant areas, it is our physical work environment, our mental work environment, working hours, work pace and time for recuperation as well as our knowledge, competence and competence development that affect our ability to work. The physical work environment, with factors such as physical workload, ergonomics, technical aids, climate and toxic substances, affect whether we are able to work. Factors in the mental work environment, for example stress levels, demand and control, must be reasonable for us to be able to work, additionally, threats and violence must not occur. Working hours, work pace, work schedule, time for recuperation and breaks during and between work shifts must be appropriate and sufficient. Furthermore, we need the knowledge, competence and have access to competence development in order to be able to work. However, analyses show that what primarily affects the *willingness* to work is what can be described as the "software" of working life (Figure 2).

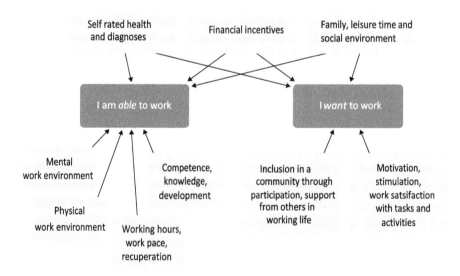

**FIGURE 2** Factors that affect whether individuals have the ability and willingness to work, as well as their employability in the workplace [3].

Accordingly, of the nine determinant areas, the willingness to work relates to the social work environment, to satisfaction with work tasks and activities, and to motivation. We need to experience inclusion in a social group in the workplace, and must not experience discrimination or disregard in decisions or in the organisation of work. We need good leadership, sufficient support and help from managers and co-workers in order to execute our work tasks. Furthermore, in order to experience motivation and a willingness to work, we must perceive our work tasks as interesting – satisfaction with the core of our work and that the tasks we execute matter to the organisation. We can experience work satisfaction by being an important cog in the organisation.

## TO STAY OR TO LEAVE?

When an individual makes decisions in life, they weigh the pros against the cons of the different options and their expected outcomes, in other words, they may ask themselves "what's in it for me?". The final decision is usually made based on the possibility to achieve the best-expected outcome of the available options. Likewise, the decision to stay in or to leave a workplace is made by considering one's options [1,2,6]. There are essentially four (Figure 1) considerations an individual more or less consciously makes when deciding whether to stay in or to leave their workplace, for example for a new workplace or for retirement. These four considerations are:

A. *Health effects of the work environment*, in other words, an individual's health, functional diversity and well-being in relation to their physical work environment, mental work environment, working hours, work pace and the possibility of sufficient recuperation when working in or when leaving the workplace.
B. *Financial incentives*, in other words, an individual's personal finance when working in or when leaving working life.
C. *Social inclusion, support and sense of community,* in an individual's personal social environment in relation to their social work environment. In other words, how and where an individual finds social support and wants to spend their time, in their working life or in their personal life.
D. *The execution of work tasks and activities*, in other words, the possibility of experiencing motivation and stimulation with work tasks, and whether the individual can utilise their skills, knowledge, and competence as well as the opportunity to develop at or outside their job.

Individuals more or less consciously summarise these four considerations and their reflections to determine their own individual opportunities for the best possible outcome. Thereafter, the individual decides whether to stay in or whether to leave the workplace. The four considerations will be described in greater detail further on in the introduction of each sphere for action, i.e. (A) health effects of the work environment, (B) financial incentives, (C) social inclusion, support and sense of community, (D) the execution of work tasks and activities.

Decision makers and politicians have an important task in enabling employability and a sustainable working life for all ages, to facilitate employees' possibilities of remaining in working life for a longer time and to increase the employers' production possibilities. Society needs to enable individuals to achieve their expectations in working life, based on the four considerations which determine an individual's willingness and ability to work. Therefore, organisations and companies need incentives from society, authorities and decision makers to be able and willing to take action and measures related to the four spheres of determination when needed, in order to enable a healthy and sustainable working life for individual employees. By enabling organisations and companies to make working life, the work situation and the work environment sustainable, society indirectly contributes to increased opportunities for individuals to be able and willing to work (please note the arrows between the different levels in Figure 1).

Furthermore, society directly influences individuals and their possibilities of making the "right" choices to remain employable by influencing attitudes, expectations and decisions. Moreover, individuals are influenced by the ambition for status and power, the possibility of experiencing self-determination and achieving a higher position, which contributes to better well-being compared to experiencing powerlessness and lower status than the surroundings and the rest of society. Individuals generally experience well-being when expectations and outcomes are balanced. Furthermore, individuals' (life) choices are set by their expectations. The individual's past, present and future affect these expectations. That is, what their life and situation have been like, what it currently is like and what they want their life to be like in the future. These expectations are primarily shaped by the individual's own attitudes. However, these attitudes are influenced by general attitudes, norms and culture in the society the individual spends their life, as well as by other people in their surroundings, or by people they look up to. Furthermore, the individual's (life) choices are affected by how easy or difficult it is for the individual to achieve their expectations in life.

Let us use a metaphor of well-being in order to reflect on how expectations affect and influence individuals' decisions and actions. Many people decorate their homes during traditional holidays. For example, during the Western Christmas holiday, a lot of people who celebrate Christmas traditionally decorate their homes with Christmas trees and ornaments. This creates expectations of experiencing well-being and Christmas spirit in a home decorated for this particular holiday. Experiencing the Christmas spirit by bringing a fir tree into one's home is not a rational causal link in itself, it is created by a general attitude in Western, traditionally Christian, societies that indoor fir trees equal Christmas and Christmas spirit. However, bringing a Christmas tree indoors during the Easter holidays is not considered appropriate, since expectations for the Easter holidays are associated with eggs, daffodils and Easter bunnies. Therefore, people's expectations are not to experience the Christmas spirit during the Easter holidays. Even if they have access to a fir tree and ornaments they will not value and use them to decorate for Easter, since this holiday entails completely different expectations. Furthermore, if an individual wishes to decorate their home with a Christmas tree for Easter they may face problems since the market does not supply many Christmas ornaments or trees at that time. Supply controls individuals' expectations and attitudes, as well as enticing different expectations and actions. Another aspect of this is that after the expectations of the Christmas spirit with a tree, ornaments and decorations have been fulfilled, it will not take long before the individual has gotten used to it and the expectations of the Christmas spirit subside. The expectations are saturated once they are fulfilled. Expectations seem to play a larger role than the experience itself since the striving subsides when the expectations are achieved and become ordinary. At this point, the individual finds something new to strive for. What does the neighbour, other people in one's surroundings, people in a TV-show or on Instagram have? What are my chances of achieving that? If the possibilities of achieving expectations are slim the individual experiences lower well-being. However, if the individual experiences their possibilities as equal to, or preferably better than others, they describe their happiness and experienced well-being as higher. Therefore, many things in life seem better before they have been accomplished. Before the Christmas or birthday presents have been opened, the expectations and presumed happiness still lie in the future. It is the journey which tickles many people's fancy when everything that is expected to satisfy us still lies ahead of us and our senses prepare for this satisfaction.

Society's, organisations' and the individual's attitudes and conditions related to the health effects of the work environment, financial incentives, social inclusion, support and sense of community, as well as the execution of work tasks, affect the possibility and decision of being able and wanting to work.

## WHAT DETERMINES A HEALTHY AND SUSTAINABLE WORKING LIFE FOR ALL AGES?

Four spheres of influence and determination, comprising nine determinant areas (Figure 3), crystallise in relation to different theories, empirical studies, and the four considerations previously described, which individuals more or less consciously make concerning their ability and willingness to work. Furthermore, the four spheres of determination are associated with employability, and also different definitions of age (read more in Part II *Age in Relation to a Sustainable Working Life*). The prerequisites for a good life are the same for younger and senior people, in other words, (A) the health effects of the surrounding environment, (B) financial incentives (C) relationships, social support and inclusion, (D) the possibility of executing meaningful tasks and activities.

In reality, the factors in the nine determinant areas that constitute the four spheres of determination overlap, affect and influence each other. However, it is important to specifically reflect and study each and every one of these factors with the purpose of making working life sustainable in all matters.

# Designing Sustainable Working Lives and Environments

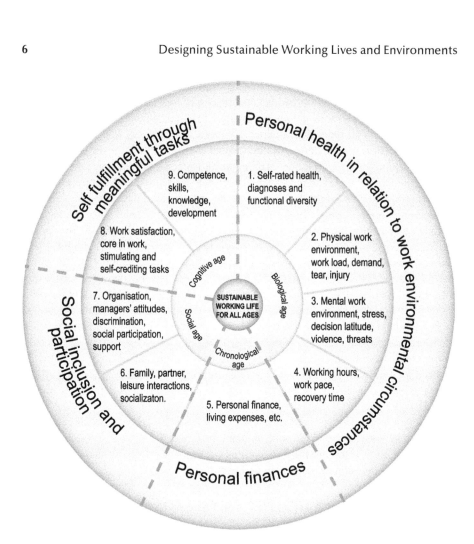

**FIGURE 3** The four spheres of determination related to the nine determinant areas that affect a sustainable working life.

**The four spheres for action comprise the nine determinant areas:**

Sphere for action **A: The health effects of the work environment** – health in relation to the physical work environment, the mental work environment, as well as working hours, work pace and time for recuperation. In other words:

1. Self-rated health, diagnoses, functional diversity, etc.
2. Physical work environment: physical workload, tear, risk of accidents, climate, risk related to toxic exposure, etc.
3. Mental work environment: stress, demands, control, risk of threats and violence, etc.
4. Working hours, work pace, time for recuperation, etc.

   Sphere for action **B: Financial incentives** – financial situation and security.

Determinant Areas for a Sustainable Working Life 7

5. *Personal finance*: providing for oneself, that is being able to afford living expenses, etc.

   Sphere for action **C: Social support, inclusion and sense of community** – social inclusion, sense of community, social support. In other words:

6. *Personal social environment*: partner, family and the social situation outside work, etc.
7. *Social work environment*: leadership, inclusion/exclusion, participation, social support, discrimination and disregard, etc.

   Sphere for action **D: Execution of work tasks** – the experience of stimulation in work tasks, work satisfaction, satisfaction with the execution of tasks, skills, competence and knowledge. In other words:

8. *Motivation, stimulation, work satisfaction*: the core of work, meaningful work tasks, etc.
9. *Knowledge, skills, competence*: the possibility of using one's skills, competence development, etc.

## CONSIDER AND REFLECT

Study the picture of the SwAge™ model when you consider and reflect on the following questions. Furthermore, you can use the SwAge™ survey regarding work ability as a support (you can find the survey in Part III *Practical Application Based on the SwAge™ Model*).

- Which factors primarily contribute to your ability of working in your workplace, based on the nine determinant areas in the SwAge™ model?
- Which factors primarily contribute to your willingness of working in your workplace, based on the nine determinant areas in the swAge™ model?
- Reflect on your health, functional diversity and well being in relation to your physical work environment, your mental work environment, and your working hours, work pace and time for recuperation.
- Reflect on your personal finances if you remain in your current workplace, or if you leave work.
- Reflect on in what way, how, when and where you best experience the safety of social support, participation and inclusion in your personal social environment and in your social work environment.
- Reflect on your possibilities to execute work tasks in which you experience motivation and stimulation, where you can use your skills, knowledge and competence and where you develop in your occupational role in relation to your leisure time.
- In order to improve your work situation and work conditions, what measures would be needed on the:
  - individual level?
  - organisational level?
  - societal level?

# Sphere A. Health Effects of the Work Environment

Sufficient health and well-being is a prerequisite for the ability to work and participate in working life. Individuals' health is affected by and must influence the physical and mental work environment, as well as the working hours, work pace and possibility of recuperation (Figure 4). The connection between health, working life and the health effects of working life relates to biological age since work requires certain biological prerequisites and maturity, as well as not being biologically worn out (read more in Part II *Age in Relation to a Sustainable Working Life*). The health effects of the work environment are significant to individuals' employability and to their ability and willingness to work [1,2,3,5–7,11].

Individuals more or less consciously consider their own health in relation to the physical and mental work environment, as well as to the possibility of recuperation, whether they are able to and can cope with working, or whether they should leave work, for example for a different occupation or for retirement (read more in the Section *To Stay or to Leave?*).

---

**Examples of Considerations that Matter to the Decision of Staying in the Workplace or in Working Life**

- If I feel good, or better, in my work situation and remain mentally and physically active through work.
- If my self-rated health and functional diversity are good or sufficient in relation to the physical and mental work environment, working hours, work pace and time for recuperation in working life.
- If my (potential) diagnoses do not complicate or risk deteriorating due to the physical or mental work environment, working hours, work pace or the possibility of recuperation at work. In short, I feel that my health is soundly based on conditions in the work environment.

*(Continued)*

**Examples of Considerations that Matter to the Decision of Leaving the Workplace or Working Life**

- If my physical work environment, mental work environment, working hours and work pace do not have excessively negative impacts on my health.
- If my health and well-being are too poor to remain working.
- If leaving working life may improve my health or decrease physical and mental strain.
- If leaving work probably will provide an improved possibility for the recuperation I need.

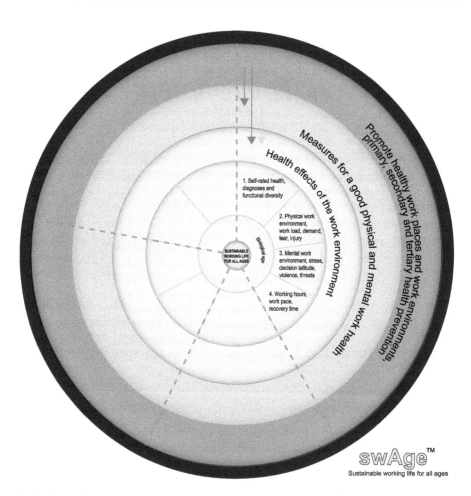

**FIGURE 4** The first sphere of reflection and action in the SwAge™ model, (A) *The health effects of the work environment*, which includes the determinant areas: (1) Health, diagnoses, functional diversity, (2) The physical work environment, (3) The mental work environment and (4) Working hours, work pace and time for recuperation.

# Area 1. Diagnoses, Self-rated Health and Functional Diversity

Health is a significant factor for the ability to participate in working life (please note Figures 1–4). Diagnoses, self-rated health and functional diversity matter to whether individuals state that they are able to and want to work [1–3,8,10,12–14]. There are age differences in individuals' anatomy and physiology, the biological ageing is connected to our health. Health affects employability, productivity and employees' ability of working until an older age. Illness, diagnoses and ill-health are common reasons to leave work. Society, organisations and companies strive to maintain good health in the population and among employees in order for them to be employable and productive as workers and citizens. Sociotechnical systems theory claims that ill-health provides a socially acceptable reason to withdraw from working life prematurely, both for the individual and for the organisation [15,16]. Furthermore, leaving working life can offer an opportunity for rest and recuperation to prevent the individual from becoming ill and worn out prematurely. The risk of being affected by ill-health in the present physical and mental work environment can result in individuals choosing to leave work, in order to not be affected by work injury and illness. A wish to experience better well-being and promote good health can be reasons to make an early exit from the workplace and working life early. People who enjoy their work seem to play down their health issues and work for longer, while those who are dissatisfied with their work may emphasise their health issues and retire earlier [6].

## DIFFERENT DIMENSIONS OF HEALTH

Health is not an easily defined concept. Health can be perceived as a tool, as a resource and as something we strive for in order to experience well-being in our life.

The concept of health can be divided into different dimensions, connected to different areas which can both threaten and promote well-being (Figure 5). For example, health can be divided into:

- The biomedical concept of *health classification*, where an individual's diagnoses, illness and functional diversity are assessed against a reference interval
- *Physical health* which refers to physiology, the mechanical functions of the body and functionality
- *Mental health* refers to the ability to cope with stress, tension, depression and anxiety

DOI: 10.1201/9781032616681-3
This chapter has been made available under a CC-BY license

**FIGURE 5** Areas to promote health and prevent ill-health, based on different dimensions of well-being and the nine determinant areas for a sustainable working life.

- *Health in relation to time and pace*, i.e. the ability to cope with the pace of one's lifestyle, for example to process the impulses that an individual is exposed to during a day in relation to the need for time to recuperate
- *Financial health* refers to the individual's financial circumstances, for example being able to afford and having access to a place to live and food
- *Emotional health* refers to the ability to experience emotions such as joy, fright, anger, sorrow and wonder and to express these emotions in an adequate way to oneself and to others
- *Social health* refers to the ability to create and maintain relationships with other people
- *Spiritual health* refers to life philosophy, self-actualisation, principles for behaviour as well as different ways to achieve peace of mind and serenity through different activities, rituals, etc.
- *Intellectual health* refers to the ability to think clearly and coherently, memory, creativity and developing knowledge

Furthermore, the individual's experience of health and well-being relates to health in society. Health is meaningful on different levels; on the societal level, on the organisational level, in working life and last but not the least on the individual level, in the lives of individuals. Because of this, we need to reflect on the ways we perceive and experience health. Additionally, health among citizens has different definition levels in society (Figure 6). This means that an individual's health is indissolubly linked to their surroundings, which means that it can be harder to experience health in a society which lacks the resources to provide for basic physical and emotional needs like food, clothes and a place to live. Furthermore, it can be difficult to experience health in a society which lacks basic human rights, basic health care, and possibilities for communication and recreation. Moreover, it can be difficult for an individual to experience sound health if they are unemployed and live in a society which highly values career development and occupational efforts.

FACTS

**Health Definition Levels**

Health is defined differently depending on who defines it. Health is often described as the opposite of ill-health. Overall, these different levels can be described to be based on different interests in health, that is the different interests in and needs of health on the societal, organisational and individual levels.

- Macro level: health on a societal (macro) level can be described as not deviating from the norm in society. The term *sickness* pertains to ill-health on the macro/societal level, which includes different societal measures to achieve health. The objective attitudes of a society towards the individual's health and lifestyle. The purpose of certain laws, rules, strategies, decrees, etc. is to promote health among citizens in order for them to contribute to society and avoid burdening the welfare system. Read more in Section *The Societal Macro Level and Health*.
- Meso level: health on the organisational (meso) level can be described as not having a diagnosed illness. The term *disease* pertains to ill-health on the meso/organisational level, which includes the practice of authorities, and potential organisational measures to achieve health. Terms such as diagnoses and functional diversity are used on this level. This level includes objective attitudes of health care, the social welfare system, employers, schools, etc. towards the individual's health. Read more in Section *The Organisational Meso Level and Health*.
- Micro level: health on the individual (micro) level can be described as experiencing well-being in one's life situation and the individual's subjective attitude towards their own health in their daily life. Many circumstances and factors contribute to what pertains to the philosophical term of holistic health. The term *illness* pertains to ill-health on the individual/micro, which includes potential measures for the individual to achieve health.

## HEALTH ON THE INDIVIDUAL LEVEL

Various studies conclude that the individual's subjective experience of their own health, in other words, their self-rated health, in many cases is a better predictor of their health development compared to objective diagnoses [14,17], as well as whether the individual is able to work until an older age [18]. The mortality risk for individuals who experience ill-health is several times greater than the mortality risk for individuals who experience sound health, even when the different individuals have the same rate of ill-health in terms of diagnoses, set by objective measurement methods in health [17].

In other words, the individual perspective, also known as the micro perspective, on health is subjective and includes the individual's own experiences. Health definitions on this level often emphasise well-being and focus on the individual's self-determination, aims, needs or wishes. Lennart Nordenfelt [19] describes individuals' health as the possibility of well-being through the balance and adaptation in life and life conditions. In this description, he refers to the *equilibrium model of health*, first launched by Caroline Whitbeck [20] and Ingemar Pörn [21,22]. The equilibrium model of health is based on the note that individuals' ability should be balanced with the individual's goals in order to be in good health. Furthermore, it describes that health depends on the equilibrium or balance between the individual's ability of action, depending on external and internal circumstances, and the individual's goals of action. The fundamental thought is that an individual who has the ability to realise their goals is healthy. Ingmar Pörn [21, p. 5] defines health in the following way:

> Health is the state of a person which obtains exactly when his repertoire is adequate relative to his profile of goal. A person who is healthy in this sense carries with him the intrapersonal resources that are sufficient for what his goal require of him.

The individual is healthy if, and only if, they have the second-order ability to, in standard circumstances or normal circumstances in the individual's life, realise their vital goals. The individual cannot be regarded as healthy if they only have partial ability to realise their vital goals. Nordenfelt connects the concepts of health and happiness: "to be healthy is tantamount to having the ability, given standard circumstances, to realise one's minimal happiness" [23, p. 86].

Conditions that tend to (but do not have to) impair the individual's ability to realise their vital goals, such as diseases (ongoing processes), defects (which have been present since birth) and injuries (which have been acquired), are referred to as *maladies*. Therefore, illness is not equal to ill-health in this theory but can be the cause of an individual's ill-health by impairing the individual's ability of action. In other words, an individual who has an illness but whose ability of action is not impaired by this is considered to be healthy. However, if an individual cannot fully realise their goals, they are considered to have some rate of ill-health according to this theory. Therefore, possibilities to adapt and the ability to make choices in favour of long-term health, rather than giving into short-term pleasures, may be needed in order to achieve health. For example, an alcoholic needs to avoid giving into the urge for another drink in order to be healthy in the long run; the young employee at a building site needs to avoid an excessively heavy workload and consider the long-term consequences it may have for their physical health in 20 years' time. The surrounding environment influences the individual's possibility to adapt, or lack thereof, in order to realise their vital goals. According to the equilibrium theory of health, the ability of *adaptedness* in the individual and society creates better opportunities for health among individuals and groups.

## BODILY FUNCTIONS THAT MATTER TO WORK

In order to understand how work and the work environment affect and influence individuals, we need to understand how the body works. First, we need to understand how the anatomy and organs in our body work, their structure and how they cooperate, then we need to understand physiology, the normal functions of the body and organs. Furthermore, we need to examine biological ageing during working life.

Functional and biological diversity is of great importance in the physical work environment. We all have different physical predispositions to execute work tasks based on our biological diversity. Therefore, it is important to take the individual and their functional diversity into consideration and consider how to best take measures, enable and make the individual employee want to work and deliver with maintained health. Based on chronological age, a human being generally peaks physically at approximately 20–25 years of age, after which muscle strength, oxygen uptake capacity, the skeleton, sight, hearing, etc. begin to deteriorate. Furthermore, increasing age generally entails an increased need of time to recuperate, while healing abilities deteriorate and the risk of developing chronic illness increases. Therefore, the risk of developing illness and the risk of long-term sickness absence increase for senior employees. In studies of individuals' work ability in different ages, it is apparent that experience and skill to a large part can compensate the potential deterioration of physical ability that generally occurs with increasing age [24]. Assuming a binary division of gender, women are generally shorter and have a different body shape (hands, feet, neck) than men. Furthermore, women generally have a lower oxygen uptake capacity and less muscle strength. Moreover, differences in muscle strength depend on functional diversity, genetic predisposition, illness, as well as whether we have suffered injuries, for example in our working life. However, all of these factors entail great individual diversity.

### THE CIRCULATORY SYSTEM

The circulatory system plays an important part for the body's ability to cope with work and workload. The circulatory system, or the cardiovascular system, consists of the heart, with different atriums, ventricles and valves, which have various functions when the blood is circulated in the body. Furthermore, the cardiovascular system contains the blood vessels, which are divided and connected in two systems; the pulmonary circulation and the systemic circulation. The heart circulates blood through these systems, carrying blood throughout the body. The blood is oxygenated in the pulmonary circulation. The circulating red blood cells have the task of oxygenating organs and muscles, as well as carrying oxygen and carbon dioxide in the blood.

Deoxygenated blood, which has oxygenated the body, is transported from the heart's right ventricle to the lungs. A gas exchange takes place in the pulmonary alveolus, where oxygen from inhaled air is passed to the blood, while carbon dioxide and other waste gases are passed over to the exhaled air. When blood is oxygenated, it has a lighter colour. The oxygenated blood is transported to the left atrium and ventricle of the heart, where the pulmonary system ends, and circulated in the body to oxygenate the organs. In the systemic circulation, oxygenated blood is transported from the left ventricle through the aorta to, for example, the brain, intestines, legs, joints and muscles, in order to oxygenate cells in all parts of the body. The blood vessels branch out and become smaller on the way in order to enclose and pass through the organs. In the smallest blood vessels, the capillaries, an oxygen exchange takes place between the blood vessels and the surrounding tissue of the organs. When oxygen has been passed to the organs, waste products and carbon dioxide are passed through from the tissue to the capillaries. Thereafter, the deoxygenated blood containing waste products, which has darkened in colour, returns through larger veins back to the right ventricle of the heart where the systemic circulation ends. The deoxygenated blood is then passed over into the pulmonary circulation, through the right ventricle of the heart, in order to be oxygenated yet again and return to the systemic circulation to oxygenate the organs of the body. Furthermore, in the systemic circulation, the blood passes through the kidneys and liver, where the blood is cleansed from toxins, the toxins and waste products pass on to and leave the body with the urine and faeces, respectively.

### THE RESPIRATORY ORGANS

The respiratory organs consist of the upper respiratory tract, including the pharynx, nose, sinuses and larynx, as well as the lower respiratory tract, including the trachea, bronchi and lungs. The respiratory organs provide oxygen to the body, which is needed not only in the cells of the organs but also to remove waste gases such as carbon dioxide, which emerge from cell activity, from the human body. Oxygen uptake capacity refers to the amount of oxygen which the heart, lungs and blood can deliver from the inhaled air to the muscles in the body. The more oxygen and the faster it passes through the body of an individual, the higher oxygen uptake capacity and therefore better cardiorespiratory fitness.

### BREATHING AND CIRCULATION DURING PHYSICAL ACTIVITY

Physical activity increases muscle and organ activity, resulting in an increased need and consumption of oxygen. At the same time, the amount of carbon dioxide and waste products from muscle activity increases and must be transported away at a faster rate. Increased cardiac volume affects the circulatory system, since the heart must beat at a faster pace and increase its frequency (pulse). Accordingly, cardiac volume varies significantly between approximately 6 litres per minute when resting and approximately 30 litres per minute during maximal physical work capacity.

The number of heart beats per minute, i.e. the pulse, varies depending on the pace of a physical activity or work task. An adult has a heart rate of about −60–80 beats per minute when resting. The heart rate during low-intensity physical work or activity amounts to 65–75 percent of the maximum heart rate, while the heart rate during high-intensity physical work or activity amounts to 80–90 percent of the maximum heart rate.

The maximum heart rate depends on genetic predisposition, however, it generally decreases with increasing age. The maximum heart rate does not really depend on how physically fit you are, it should rather be used as a reference for different intensity levels of physical activity. It is difficult to calculate the exact maximum heart rate for an individual, however, an estimate of maximum heart rate is 220 minus the individual's age.

Oxygen uptake capacity depends on the number of heart beats per minute, which is important since the organs in the body need more oxygen during physical activity and work. An adult needs about one-fourth litres of oxygen per minute when resting. The need for oxygen increases to about 1 litre per minute during a leisurely stroll. The body needs about 2.5–7 litres of oxygen per minute during maximum physical activity. The oxygen uptake capacity differs between individuals and changes during the life course. Furthermore, oxygen uptake capacity differs between men and women, this difference peaks at about 20 years of age, when men have a significantly better oxygen uptake capacity compared to women. The oxygen uptake capacity decreases with increasing age, which means that the capacity to oxygenate the organs in the body decreases with increasing age. This affects the ability to perform physically with regards to physical workload and activities.

The blood volume to inner organs like the liver, spleen, kidneys and the gastrointestinal tract decreases through vasoconstriction in these organs in order to increase blood volume in the muscles, in the circulatory and the pulmonary systems.

When the body is at rest, the blood volume to inner organs like the liver, spleen, kidneys and the gastrointestinal tract increases in order to take care of digestion and to cleanse the blood, while the blood volume to muscles decreases. Furthermore, blood contains white blood cells and lymphocytes which scan the body for injuries and foreign microbes in order to repair, remove toxins and renew cells in the various organs of the body. It is important for the body to relax to enable this process to run efficiently.

## Skeleton and Joints

We use our skeleton, joints, muscles and nerves during physical activity and movements. The skeleton is a changeable and constantly renewed body tissue. Physical load strengthens the skeleton, it becomes stronger and thicker to cope with the physical load. However, this is only applicable when the physical load is appropriate, an excessively heavy or sudden physical load can cause a fracture and weaken the skeletal tissue. If the skeleton is not subjected to physical load, it will weaken, since the body considers it to be an unused organ, which does not need to be strengthened. Furthermore, the skeleton weakens with increased biological ageing (read more in Part II *Age in Relation to a Sustainable Working Life*, Chapter *Biological Age in Working Life*).

The skeleton has several functions in the body. It works as a frame and support to keep the body upright. Furthermore, it contributes to mobility, since different parts of the skeleton work as leverage for the muscle tissue connected to the joints and bones. When the muscles tense and contract, this results in movement. Moreover, the bones of the skeleton provide mechanical support to and protect vital inner organs like the brain, heart and lungs. Additionally, the skeleton works as a storage for minerals, primarily calcium and phosphate. Blood cells are produced in the red bone marrow, for example in the pelvis and the sternum.

The spine can be described as the central part of the skeleton. The human spine consists of 29 vertebrae which can be divided into the cervical spine or neck, with 7 vertebrae, the thoracic spine with 12 vertebrae, the lumbar spine with 5 vertebrae, the sacrum with 5 fused vertebrae and the coccyx or tail bone with 4 vertebrae.

Between each vertebra we find a disc, which resembles a soft little pillow, to soften the pressure between the vertebrae. In other words, the discs work like shock absorbers for the back. The discs consist of a fibrous cartilage enclosing a gel-like core which changes depending on the pressure it is subjected to. When the spine is leaned forward, the core of the disc is pushed backwards and puts pressure on the tissue surrounding the disc, which can cause pain. When straightening the spine, the core of the disc is pushed back to its' centre and the pain diminishes. There is a constant pressure inside the disc, when sitting without supporting the back, the pressure increases when compared to walking or sitting with appropriate back support. The different vertebrae align in curves, the reason for this is that the spine must cope with different physical loads. In order to evenly distribute the physical load in a gentle way, the curvature of the spine must be maintained. Because of this, it is important that we avoid slouching or a lopsided posture when we sit or stand.

The extremities, in other words, the arms and legs, are often subjected to load and strain in the physical work environment. They are usually referred to as the upper and lower limbs. The upper limbs are constructed for light load and precision and consist of shoulders, upper arms, forearms, hands and fingers. The lower limbs are constructed for strength and stability and consist of hip bones, femurs (thigh bones), tibias and fibulas (shin bones and calf bones), feet and toes. In fact, all mammals actually have a similar skeletal construction. The skeletal construction of a giraffe may not look very similar to a human. However, despite their different appearances, a giraffe has seven cervical vertebrae just like a human being. Similarly, people's appearances can greatly differ in shape and form, and their skeletons can be of different sizes. It is important to consider their conditions for different types of physical labour and work tasks when individually adjusting the workplace to prevent occupational injuries.

#### Muscles

The function of the muscles in the body is to keep the body upright, protect the skeleton, distribute physical load and perform movements and mobility. A human being has more than 300 muscles in their body, the muscles constitute about 40–50 percent of their total body mass. One muscle never works alone, several muscles always work together. Muscle strength depends on age, gender, genetic predisposition and physical exercise. There are three kinds of muscles in the body, which differ in structure and construction. This is because they have different tasks in the body, even if they are all muscles. They have adapted according to their specific function in the body. The three types of muscles are as follows:

- *Striated skeletal muscle cells* are connected to the skeleton and are voluntary. They support, protect and enable our movements. We can, for example find these muscle cells in our arms, back, abdomen and legs. These are the kind of muscles we exercise through physical activity.
- *Smooth muscle cells* are found in our skin, blood vessels and inner organs. These muscles are involuntary and work without us consciously thinking about them.
- *Cardiac muscle cells* exist in the heart only. These muscles are striated like skeletal muscles but are involuntary and cannot be controlled by our will.

Muscles contract and relax to perform motions, these are divided into static and dynamic muscle contractions. Dynamic muscle contractions occur when the muscles or group of muscles both contract and relax during an activity.

Dynamic muscle contractions enable the muscles to maintain a sufficient circulation of oxygen and nutrients, as well as excretion of waste products which emerge from muscle contractions (for example carbon dioxide and lactic acid).

Examples of dynamic muscle contractions are, for example in the arm when cutting a board using a hand saw, in the legs when walking or running, in the back and legs when lifting a box, in the arms and fingers when writing on a computer or smart phone.

Static muscle contractions occur when a muscle or group of muscles contract continuously. There is no time or opportunity for the body to excrete sufficient amounts of the waste products resulting from the continuous contraction of the muscle. Furthermore, the body has no possibility of delivering enough oxygen and nutrients to the contracted muscles. The lack of oxygen and nutrients, as well as the surcharge of lactic acid and waste products, can cause the muscles to ache. Examples of static muscle work can be stationary work with static load of leg and back muscles, using a nail gun or a pressure washer when the muscles of hands and arms must contract continuously to keep the machine in the same position, when carrying heavy objects, the muscles in the hand and arm which hold the board still when cutting it with a hand saw, or the muscles in the arms, back, legs and abdomen when doing the plank exercise. The constant static muscle contractions result in a higher energy consumption and requires a longer time to recuperate.

The physical work environment and work tasks must be examined in working life and the workplace, static muscle contractions should be decreased to the greatest extent possible. A statically contracted muscle or muscle group tires faster than a dynamically contracted muscle or muscle group and results in numb muscle fibres. Blood circulation decreases, the delivery of oxygen and nutrients and the excretion of waste products are obstructed. If the sense of tiredness is ignored, pain emerges in the muscle and increases the risk of strain issues in the skeleton, joints, muscles and nerves because of the work situation.

The biological diversity and genetic predisposition of our muscles affect our physical conditions and possibilities to perform physically, just like other parts of the body. For example, muscle strength can be affected by functional diversity, genetic predisposition, illness as well as strain injuries caused by, for example, working life. Furthermore, muscle strength decreases with increasing age, people's muscle strength generally peaks between 20 and 30 years of age, after which it decreases. Muscle strength differs between men and women, also peaking between 20 and 30 years of age when men have approximately 40 percent more muscle strength than women on average. After this peak the curve flattens, between 60 and 70 years of age men have approximately 15–20 percent more muscle strength than women on average. In general, the tolerance for physical work demands decrease with increasing age. Consequently, the conditions of work, work tasks, the design and organisation of the work environment are of great importance. Therefore, the physical work environment must be considered and adjusted to individual biological ageing.

## The Sensory System

The sensory organs are important for our ability not only to execute our work tasks but also to alert us of dangers and risks. The task of the sensory organs is to convey different kinds of information for the brain to process. The sensory organs and the information they detect are:

- The eyes detect light and colour
- The ears detect sounds
- The nose detects smells
- The tongue detects taste
- The skin and ears detect temperature, touch, vibrations, pain and pressure
- The position sense or proprioception detects the positions of the body parts

The sensory system protects us and conveys information and ensures that cells receive the stimuli. Our sensory system has sensory cells or receptors, which respond to and register different types of stimuli in our body and in our surrounding environment. The stimuli can be said to contain information, which is converted into electric signals, which are conveyed as nerve impulses to the nervous system. The nervous system processes the information and conveys information through signals to the brain.

There are different kinds of receptors, for example:

- *Photoreceptors* respond to colour and light and are found in the cones and rods of the eyes' retinas
- *Mechanoreceptors* respond to pressure, touch and vibrations and are found in the balance and hearing organs
- *Chemoreceptors* respond to the chemical composition of gases and liquids and are found in the tongue's taste buds and the olfactory system of the nose and nasal cavity, as well as in certain blood vessels
- *Thermoreceptors* respond to changes in temperature and are found in the skin and various inner organs
- *Electroreceptors* respond to changes in electric potential of nerve impulses and are found in the nervous system

Furthermore, the sensory organs are important to store information as memories (read more in Chapter (D) *The Execution of Work Tasks*, [9] *Knowledge, Skills and Competence Development*). The sensory system and receptors can be affected by, for example, injuries and pharmaceuticals, they also deteriorate with increasing age. The deterioration of the sensory system results in decreased reactivity, which increases the risk of accidents. Statistics show that many senior employees in high-risk physical work environments have an increased tendency to suffer injuries. Therefore, the physical work environment must be examined for increased risks of injury, in relation to the deterioration of the sensory system, which should be a continuous part of the systematic work environment management in high-risk occupations.

## THE NERVOUS SYSTEM

The biochemical signals of the nervous system control and regulate the initiation of movements, communication to and between different organs in the body, as well as activities, reactions and emotions. The nervous system is divided into the central nervous system (CNS) which consists of the brain, spinal cord and retinas, and the peripheral nervous system, which consists of the other parts of the nervous system. Furthermore, the nervous system can be divided based on voluntary (somatic) functions of the motoric nerves in the skeletal muscles and involuntary (autonomic) functions of the nerves which control the basic vital processes in the organism, such as breathing, blood circulation, intestinal muscles, metabolism, as well as the secretion and excretion of the endocrine organs and glands. Processing stimuli includes chain of reactions in the CNS which mediates a series of neurological feedback loops to the hippocampus (which can be described as a switchboard for information) as well as synchronising cooperative somatic, cognitive and endocrine adaptive reactions in the body. When the individual faces insoluble problems, the chains of reactions are disturbed and become lasting loops of somatic, cognitive and endocrine adaptive reactions in the body, which risk resulting in illness and injury to the organism. Furthermore, there is a connection between CNS and the immune system.

### The Functions of the Nervous System

The autonomic nervous system triggers, affects and controls the reactions and actions in the organs and muscles of the body through neurotransmitters, proteins and hormones. The autonomic (involuntary) nervous system can in turn be divided into the sympathetic, the parasympathetic and the enteric nervous systems. These nervous systems counteract each other to some extent and strive to reach a physiologically stable equilibrium in the body called *homeostasis*. It is vital to our long-term survival that these systems are reasonably balanced during the course of a day. The enteric nervous system consists of a tissue of neurons, affected by the same neurotransmitters as the brain, which affects and controls the functions in the oesophagus, stomach and intestinal tract. The sympathetic nervous system is activated in situations where individuals must use force and are subjected to physical or mental demands and prepares the body for physical activity. The sympathetic nervous system activates the fight or flight response. The hormones and neurotransmitters of the parasympathetic nervous system counteract the sympathetic nervous system and are activated while resting. When the parasympathetic nervous system is active, the body digests food in the gastrointestinal system, repairs damage in organs, rests and recuperates and reproduces. This is also called rest and digest (read more in Section *Breathing and Circulation during Physical Activity*). Furthermore, in this state, the body prepares for the next activity and effort. Without sufficient time in this state, when the parasympathetic nervous system is active and dominates the bodily functions, we run the risk of becoming exhausted and ill.

Stress reactions are a vital function of the nervous system. The trigger causing a stress reaction is called a *stressor*. Different organs react to different stimuli like cold, heat, infections, trauma, nervousness and irritation. Even our positive and negative thoughts affect neurotransmitters and the communication between the brain and the nervous system.

When exposed to certain stimuli or a situation, a row of different physiological readjustments take place in the body. Innate and learned reflexes to different stressors initiate the secretion of biochemical neurotransmitters in the nervous system. In other words, what is described as stress is the organism's reaction to stressors (read more in Section *Stress*). The entire stress reaction is controlled by the CNS, where the hypothalamus can be described as the body's stress centre. Stressors automatically trigger physical reactions, even before we may have consciously registered and reacted to what has happened. When this occurs, the hypothalamus receives a great amount of information from external and internal sensors. Consequently, the hypothalamus transmits chemical and electric signals which, among other things, change the secretion of different neuropeptides and hormones from endocrine organs and glands (for example the pituitary gland, the thyroid gland, the pancreas, the pineal gland, as well as the testicles and the ovaries) to the blood.

There are common features of the so-called stress reaction in all organisms (read more in Chapter *Mental Work Environment*, Section Stress). If an individual is subjected to an unpleasant, exposed and demanding situation, the body prepares to fight, flight or freeze (avoiding being discovered or playing dead) in order to survive. The reaction to different stressors is followed by the experience of different emotions that the individual wants to remedy in order to regain homeostasis. Examples of feelings caused by stressors are not only pain, horror, hunger, fright and anger but can also be expressions of great joy, infatuation, desire, well-being and saturation, which also cause the systems in the body to react.

The need of achieving equilibrium between the sympathetic and parasympathetic nervous system is vital. Overload and imbalance between the systems can cause ill-health. Situations and conditions that cause constant and lasting reactions to stressors affect the different organs in the body through extensive and long-term demands of biological adjustment and homeostasis. If these situations are too intense or last for too long, they can result in lasting disturbances in the physiological functions of the body, which in turn can initiate the process of different types of illness.

There are above all two biological effects that can be caused by long-term stress. The first is that long-term stress can cause disturbances in the endocrine system's secretion of neurotransmitters like cortisol and thyroid hormones. This may cause a disturbed balance in the body's regulatory system and in the metabolism and energy mobilisation, in other words, between anabolism and catabolism. *Anabolism* is the constructive process of the metabolism where cells convert simple substances by combining several smaller molecules into more complex compounds to construct larger molecules like amino acids and proteins. *Catabolism* is the destructive process of the metabolism, where molecules are broken down to extract energy and material for other processes. This can affect the body's processes of construction, repair, nutrient uptake and the processes of scanning for and cleansing of toxins and waste products, bacteria, viruses and improper cell growth. Additionally, it may risk decreased fertility.

The second effect of long-term stress entails the risk of disturbances in the neurotransmitters of the endocrine regulatory system, which in turn can cause changes in the CNS, for example in the hippocampus, which controls impulses and memories. This may cause the individual to develop memory disorders, difficulties in focusing and thinking clearly, chronic fatigue, muscle pain in, for example, neck and shoulders, which are not allowed to relax due to constant alertness, and depression. These symptoms and states of illness can be perceived as diffuse and difficult to understand by the surroundings. Some individuals are doubly affected if the surroundings do not believe in the symptoms or that an illness is present, since being subjected to mistrust creates insecurity which in turn may risk causing new stress reactions for the individual.

Besides the serious risk of consequences in the neurotransmitters of the regulatory system and in the brain in cases of lasting and long-term stress reactions, there are other risks. These are risks of increased glucose, in other words, increased concentration of sugar in the blood, increased concentration of fatty acids in the blood, increased cardiac volume, increased blood flow to the skeletal muscles and decreased blood flow to inner organs and skin. The blood pressure increases and the increased amount of fatty acids and glucose increases the risk of plaque build-up in the walls of the blood vessels that may constrict and clot in the blood vessels and circulatory organs. Furthermore, the individual may find it difficult to fall asleep due to constant alertness and signals of increased wakefulness in order to cope with stressors. Consequently, the body's time for recuperation and the body's control to find abnormalities and repair damages decrease. This in turn can risk resulting in the constant errors that occur in cell divisions and mutations passing unnoticed. Therefore, a person exposed to increased stress may run a higher risk of developing cancer, cardiovascular diseases and other chronic illnesses.

## THE DUAL CONTINUUM MODEL OF HEALTH

As previously mentioned, there are different health definitions, just like age, which comprise different symptoms and dimensions. The concept of health has subjective as well as objective implications. In order to facilitate reflections and discussions regarding health, there is a tool called the *dual continuum model of health* [25] (Figure 6). The dual continuum model of health has four dimensions which intend to describe how the individual's subjective experience and rating of their own health relate to objectively diagnosed illness. In other words, the dimensions where:

- the individual subjectively experiences good health and does not have an objectively diagnosed illness
- The individual subjectively experiences good health, despite having an objectively diagnosed illness

**FIGURE 6** The dual continuum model of health [25].

- The individual subjectively experiences ill-health and has an objectively diagnosed illness
- The individual subjectively experiences ill-health, despite not having an objectively diagnosed illness

Individuals constantly move across the different dimensions of the model. Individuals who have a diagnosed illness, for example diabetes, can still experience well-being with appropriate medication. At the same time, an individual can experience ill-health and insufficient well-being despite not having a diagnosis. An individual can stay in one dimension for the majority of their life or move between the dimensions from one hour to another. For example, an individual can attend work feeling healthy but experience a stress reaction with palpitations and anxiety at work due to high demands, this will cause them to move into another dimension in the dual continuum model of health. Your health can decline rapidly if you are subjected to an occupational accident. You can experience well-being in spite of pain and disability, if you have exciting work tasks, feel included and happy in your work unit. However, this can change if an informal leader decides to disregard you, if they do not consider your efforts to be good enough.

## THE WORK ENVIRONMENT AS A PREDICTOR OF ILL-HEALTH

The work environment is an important predictor of ill-health which must be considered to enable a larger amount of people to work until an older age. In a study with 11,902 respondents, almost 1 in 4 respondents stated that they had diagnosed illness and injuries caused by their work situation and work environment [26]. Based on International Labour Organizations assessments, approximately 2.3 million workers die every year from work-related injuries and diseases. Furthermore, 160 million workers suffer from non-fatal work-related diseases and 313 million from non-fatal injuries per year. Additionally, it must be considered that for every person with diagnosed illness, a larger number of individuals have been exposed to risks, even if they did not develop illness or suffer an occupational injury. The metaphor of the ice berg is often used to describe this, the majority of the ice berg is hidden beneath the surface and is not immediately perceivable (Figure 7). For every work-related fatality, many occupational injuries and even more incidents occur, which risk becoming serious occupational injuries or fatalities. Working to decrease risks and the amount of incidents will decrease the other parts of the ice berg that remain hidden as well. Consequently, the risk of injuries, accidents, illnesses and fatalities will decrease. Work-related fatalities and occupational accidents should be reported to authorities and included in statistics or become perceivable above the surface to use the metaphor of the ice berg. However, sometimes occupational injuries and illnesses are not reported, which decreases the visibility of risks, incidents and occupational injuries in working life since they remain hidden under the surface. The number of incidents, in other words, what "almost happened", is staggering compared to the number of occupational injuries and fatalities that did occur.

However, it is important not to ignore the incidents, these must be made visible through workplace analyses and documentation in order to address risks in the work situation and to decrease the number of occupational injuries and work-related fatalities.

The proportion of diagnoses increases as people age. Moreover, chronic pain and pain conditions appear to increase with increased biological ageing. The health disparities between different age groups can be significant, however, there are health disparities between individuals in the same age group as well. A physically and mentally demanding occupation causes premature biological ageing and increases the risk of ill-health. Studies show that biological ageing and insufficient health is a significant reason for leaving working life in countries with a high retirement age [18,27], while health is not as significant a reason to leave working life in countries with an average retirement age, lower than 60 years of age that is. A probable reason for this is that a great part of the ill-health caused by the environment increases with the additional duration of exposure that a longer working life and an older age entail. A lower retirement age decreases the duration of exposure to issues or risks in the work and work environment which can cause ill-health for individuals.

Statistics show that many individuals in work-related sickness absence are on long-term sick leave due to mental issues, excessive demands at work, a poor physical work environment, insufficient social support as well as a poor connection to the workplace [28]. This applied regardless of the individual's gender, age, country of birth or marital status. Furthermore, exposure to excessively high work demands affects life expectancy. Since senior individuals usually have been exposed to work, environment and lifestyle for a longer time, they run an increased risk of developing ill-health. Physiological functions generally deteriorate with increasing age, regardless of whether the individual remains working or has left working life [29]. However, studies show that the health of seniors, who had been exposed to a demanding work environment, improved after leaving working life for retirement [30,31]. Another study did not discern that the risks of developing ill-health or passing away prematurely were affected by when individuals left working life, however, it was difficult

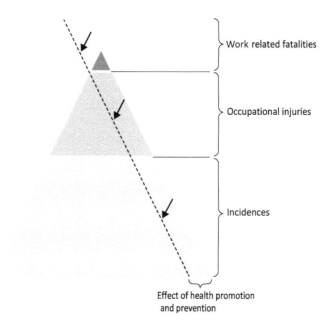

**FIGURE 7** The metaphor of the ice berg depicts how fatalities in working life only constitute the tip of the ice berg, the number of occupational injuries and incidents make up the rest. Furthermore, the figure also depicts how health promotion and prevention can contribute to decreased risks and improvements on all levels.

to prove whether the change of leaving working life affected mortality or if it simply was a health-related selection bias for early retirement [32]. This study featured the risk of a *healthy worker effect*, which occurs when statistics are affected by whether only the healthiest employees remain working until an older age, regardless of the risks and implications in the work environment, while employees with ill-health and injuries have already left working life prematurely and therefore were not part of the statistics. Physical and mental demands, wear and tear, for example caused by work, contribute to cell death and thereby to an increased need to repair injured muscles and organs through cell renewal. All cells have a limited life span, dead cells are replaced with new ones through the division of viable cells. Cell renewal occurs through cell division, in other words, that the cell replicates itself. Every time a parent cell divides, the chromosomes in the cell are copied into two exact daughter cells. However, when the cell is divided, the outermost end of the chromosome, the telomere, is not entirely copied. Therefore, the chromosomes are somewhat shortened for every cell division, because of this the chromosomes of senior individuals are shorter than the chromosomes of younger individuals (read more in Part II *Age in Relation to a Sustainable Working Life*, Chapter *Biological Age in Working Life*). This contributes to individuals' inability to live forever. Unhealthy demands in working life affect the individual's life expectancy [33]. Individuals who have been exposed to excessive mental and physical work demands and workload have more rapid cell death compared to individuals of the same chronological age who do not have equally high work demands and workload. Therefore, a mentally and physically demanding work environment with insufficient opportunities for recuperation increases biological ageing.

## MEASURING HEALTH AND ILL-HEALTH IN WORKING LIFE

Self-rated health and well-being, in other words, the individual's own experience of their health or ill-health, can indicate the individual's future health development and should be taken seriously. For example, follow-up studies of people with the same diagnosis and stage of illness show that people who experienced better well-being in the first study still experienced better well-being ten years later, despite having more severe medical conditions, compared to people who considered their health to be worse [17,34]. Furthermore, some studies show that self-rated health is a more accurate predictor for lasting disability and premature or disability pension compared to objectively diagnosed illness [35–37]. Additionally, several studies show that self-rated is a better predictor for mortality compared to objectively diagnosed illness [38–40]. Mortality has proven to be higher and more rapid for people who experienced lower rates of well-being and the same stage of severe chronic illness than for people who, despite their illness, experienced relatively high rates of well-being. Furthermore, subjective, self-rated health is, according to studies, a better predictor for the possibility of participating in working life compared to objectively diagnosed illness [14,36]. However, diagnoses are practical and useful tools when making decisions regarding treatments and rehabilitation in health care, as well as to assess insurance cases (read more in *Different Dimensions of the Concept of Health*).

High levels of well-being in working life contribute to higher self-rated health among employees. Furthermore, a good work environment improves health and delay biological ageing. A healthy lifestyle with physical, mental and social activities also prevents the deterioration of health, work ability and capacity. This applies to people with chronic illness in particular. People who thrive in their work situation and with their work tasks decrease their health issues and remain working despite their illness, while people who dislike their work situation and work tasks often exaggerate their health issues and state their ill-health as a valid reason to withdraw from work [6,41]. This also applies to the consideration of early retirement.

## CONSIDER AND REFLECT

- How do you define health?
- Consider whether illness and ill-health are the same.
- Consider whether being healthy and having good health are the same.
- When do you experience good health and ill-health, respectively?
- Consider whether health always has the same implication, regardless of situation, culture and historical era.
- Consider whether you have a diagnosis or ill-health caused by your work situation or work environment.
- In what way does your oxygen uptake capacity, breathing and circulation matter to your work?
- How do you function in your physical work environment, based on the physical ability and capacity of your skeleton, joints, muscles, nerves and sensory organs?
- What stressors do you experience as the most demanding in your work situation and work environment?

- What do you do to relax and recuperate after excessive or lasting stress reactions?
- In what dimension are you situated in the dual continuum model of health (Figure 6)?
- At what occasions have you been situated in the different dimensions of the dual continuum model of health?
- In what dimension do you think your surroundings consider you to be situated in the dual continuum model of health?
- When do you experience your health and well-being to be the best?
- When do you experience your health and well-being to be the worst?
- Consider how and what you can do in the future in order to experience well-being.
- In what supportive environment do you experience well-being, and what constitutes your supportive environment on these occasions?
- Consider whether you contribute to someone else's supportive environment for health.
- How and what additional supportive environments for health can be created in your surroundings and in your society?
- What specific measures can be taken in your workplace and work situation to increase your well-being and health?

# Area 2. The Physical Work Environment

The physical work environment matters to individuals' work ability and affects biological ageing [1,2,7,8,42–46] (Figure 4). Injuries and illness caused by the physical work environment are among the most common reasons for illness, sickness absence and sick pay. Heavy lifting, vibrations and repetitive work activities and movements cause strain and injury. Being exposed to physically demanding work activities and a physically demanding work environment on a regular basis can risk causing occupational injuries. Furthermore, these injuries often occur when an employee has reached an older age. Age-related changes in the body's anatomy and physiology already begin to show before 40 years of age, regardless of the individual's occupation. This is partly due to the biological ageing of the body, partly to the potential illnesses and injuries the individual has suffered in their life and working life.

However, senior employees usually have greater capacity than some stereotypical attitudes towards seniors assert. Productivity increases until individuals reach 50–55 years of age and generally only decreases negligibly when individuals reach 65 years of age, after 75 years of age productivity decreases more rapidly. Repeated measurements of work ability throughout life show that half of the labour force maintain their work ability until 63–65 years of age [47].

Working in a physically demanding work environment, which risks causing occupational injuries, work-related accidents and musculoskeletal issues, is a predictor of an early exit from working life. Health issues caused by a physically demanding work environment accumulate during an individual's lifetime. The age individuals exit working life in a certain country is partly explained by the differences between different countries, regarding how common particular occupations are in the different countries [48]. However, the differences are also explained by the differences in physical conditions for a particular occupation, regarding equipment, strain and the risk of developing occupational injuries, in the different countries. Consequently, these factors affect the differences between the average retirement ages in different countries. A poor physical work environment, or poor individual adjustment or a demanding work situation leaves people worn out, which causes them to leave working life prematurely. Studies using the work ability index show that decreased work ability with increasing age among individuals in physically demanding work situations occurs when they have low control of their work [47,49]. Due to this, employees with low work ability in their middle age showed a more rapid decline of health and functions than others, these health disparities were still apparent later in life. Examples of occupational groups where several employees describe it as difficult to be able to work until an older age due to physical demands are construction workers, industrial workers, police officers, fire fighters, electricians, painters, nursing assistants, secretaries and physiotherapists [10]. Because of this, it is important to examine and, when needed, take measures in the physical work environment (read more in Part III *Practical Application Based on the SwAge<sup>TM</sup> Model*).

## THE PHYSICAL DESIGN OF THE WORKPLACE

The physical and architectural design of the workplace affects the individuals' work, health and well-being. Therefore, the physical work environment should be adjusted and take the individuals' work tasks, work flow, health and well-being into account, rather than the other way around. Different regulations influence the design of the workplace, staff areas, lighting, climate, ventilation and the design of, for example, signs and signals, in order to contribute to a good work environment. It is important that the employer considers how and where the workplace is situated and the physical design of the workplace in their expectations of the employees' execution of work tasks. If the design of the workplace is perceived as beautiful, secure and possibly experienced as pleasurable, this will increase creativity, satisfaction with work tasks and activities and contribute to improved health and well-being [6,50]. Furthermore, people who work from home need to consider their workplace for it to be functional and not contribute to health issues. The employer is usually responsible for the work environment, even if the employee executes work tasks from their own home. Risks in the physical work environment and obstacles in the work flow can affect productivity and quality of the executed work tasks. An unsettling and noisy environment with loud people affects work ability. The design of the workplace can contribute to greater demands on the executive functions in the brain and result in cognitive limits for the employee. For example, a workplace in an open plan office or in a space with constant noise, bad lighting or where people constantly pass by complicates the possibility of quality-focused work, creativity and the swift and efficient execution of work tasks. All surrounding factors are potential stressors that the individual more or less consciously needs to cope with (read more about stressors in (3) *Mental Work Environment*, Section *Health Effects of the Work Environment* and in (1) *Diagnoses, Self-rated Health and Functional Diversity*, Section *The Nervous System*). The employee can usually not perform on their highest cognitive level of focus in an environment which demands parts of the cognitive capacity to cope with noise, people passing by, constantly being interrupted by questions, disturbances from co-workers occupied with discussions, etc. The employer must be aware that disturbances affect the brain's possibility to focus and adjust demands of the conditions, based on the design and environment of the workplace.

## PHYSICAL WORKLOAD, HEAVY LIFTING AND UNILATERAL MOVEMENTS

Problematic exposure in the physical work environment can, for example, be physically demanding workloads in lasting unilateral movements or static muscle work. This could be working above shoulder height in assembly or repair work, or working with raised shoulders in front of a computer screen due to insufficient ergonomics in the workplace [51–53]. This can affect sinews, nerves, blood flow and muscles in the shoulders, neck and wrists through increased pressure in sinews, in turn, causing insufficient circulation from sinews to joints. This can result in increased pressure, swelling, oedema, numbness, nerve entrapment and chronic pain (read more in (1) *Diagnoses, Self-rated Health and Functional Diversity*, Section *Bodily Functions that Matter to Work*).

Back pain is one of the most common reasons for the inability to work in the Western world. Approximately three out of five people experience pain in the lumbar region at some point in their life. Back pain causes suffering to the individual, as well as large costs to society, for example through sick pay, health care and rehabilitation. People with a physically demanding work situation generally develop back pain to a larger extent. There are no gender differences; men and women who are exposed to similar work conditions, in equal work situations develop back pain to an equal extent. People who work with manual labour, which, for example, can entail heavy lifting, work postures with a bent or twisted back, or with a physically demanding occupation, often experience back pain. Furthermore, these issues affect not only people who are exposed to full body vibrations in their work situation to a greater extent but also people who work outside ordinary office hours (for example shift work) [51]. Additionally, mental factors matter to whether an individual develops problems or pain in their back or not. A mentally demanding work situation and the experience of insignificant possibilities to influence the work content, combined with excessive demands and insignificant opportunities to develop in work, are connected to increased back pain or problems. However, people who state that they are able to influence their work content, experience support from managers and co-workers in their work situation and high work satisfaction more seldom develop back pain or problems. However, a purely physical risk factor to back pain is whether individuals hold on to and use outdated technology and equipment.

The bodily functions show signs of ageing before 40 years of age, and essentially all individuals show signs of biological ageing at 60 years of age, in all bodily functions and in the interplay between brain, sensory organs, circulation and inner organs. Muscle strength decreases with increasing age, which affects the ability to work. In general, it becomes more difficult to execute physically demanding and static muscle work with increasing age, since oxygen uptake capacity and the capacity of the heart and blood vessels deteriorate. The ability to transport oxygenated blood to the cells, tissues and organs of the body, and to transport waste products from the muscles deteriorates, which increases the risk of exhaustion. The employee's maximum oxygen uptake capacity should be approximately 30 percent higher on average during a work shift than the oxygen uptake capacity during the execution of work tasks, in order for the employee to cope and not be exhausted by the end of the work shift (read more in (1) *Diagnoses, Self-rated Health and Functional Diversity*, Section *Bodily Functions that Matter to Work*, and Part II *Age in Relation to a Sustainable Working Life*, Chapter *Biological Age in Working Life*). The required oxygen uptake for executing light assembly work while sitting down, driving a car and for office work is 0.2–0.6 litres per minute. The required oxygen uptake when alternately sitting down and walking, for example health care work, cooking and home care, is 0.6–1.0 litres per minute, and 1.0–2.0 litres per minute for activities like deep cleaning and heavy construction work. The required oxygen uptake for heavy manual labour, such as forestry work and smoke diving, is 2.0–3.0 litres per minute. The oxygen uptake capacity is estimated to decrease by an average of 30 percent, from an individual is 30 years old until they turn 60 years old, in other words, by approximately 1 percent per year. This entails an increased risk that the need of oxygen exceeds the oxygen uptake capacity with increasing age, especially in physically demanding work tasks and tasks involving static muscle work, as well as for women who have a lower oxygen uptake capacity in general. The employees' oxygen uptake capacity when executing work tasks must be considered in order to create a healthy work environment, to ensure that the need of oxygen when executing work tasks does not exceed 70 percent of the individual's maximum oxygen uptake capacity.

Young people with physically demanding work tasks show a higher level of physical ability, compared to young people who lack physically demanding work tasks. However, for senior people this is reversed. Senior people who lack physically demanding work tasks show a higher level of physical ability, compared to senior people with physically demanding work tasks. The reason for this may be that younger people with a high physical ability may be attracted to physically demanding occupations, compared to people in the same age group who lack this level of physical ability. Individual diversity regarding muscle strength depends on genetic predisposition to approximately 40 percent but can be improved and maintained through physical exercise and lifestyle choices. The physical work ability changes through the course of life, though these changes are individual. Furthermore, lifestyle choices that can cause cardiovascular disease and decreased oxygen uptake capacity can also contribute to decreased physical work ability. However, excessive strain of physically demanding work tasks contributes to the deterioration of physical ability. If conditions are good, working until an older age is not problematic in itself. However, almost every fourth individual of working age has experienced physical issues caused by their work at some point during the last year. A physically problematic work environment can cause several kinds of issues, or multiple issues, that primarily being to show after long exposure and with increasing age. The risk of leaving working life prematurely increases if the physically demanding work tasks are executed in a mentally demanding work environment, in a work situation with insufficient possibilities to take breaks and recuperate, in a work situation with poor leadership, if the individual has insufficient knowledge or if the individual feels excluded in the workplace [2,6,47]. It is important to manage and remediate issues in the physical work environment, for both younger and senior employees, to prevent employees from being prematurely worn out and affected by ill-health caused by working life. Improvements in the physical work environment can significantly facilitate and prolong working life, both for younger and senior employees.

## THE SENSORY SYSTEM AND THE WORK ENVIRONMENT

Our senses affect the work tasks we are able to execute, and a poor work environment can affect our senses negatively. Our senses are important, they protect us from and warn us of risks and dangers (read more in (1) *Diagnoses, Self-rated Health and Functional Diversity*, Section *The Sensory System*). Some individuals have innate or acquired injuries affecting their sensory organs that complicate the execution of certain work tasks or cause increased risks in the work situation. Senior employees run a higher risk of developing occupational injuries due to the deterioration of the sensory organs caused by biological ageing, for example decreased hearing and sight in general, compared to younger employees (read more in Part II *Age in Relation to a Sustainable Working Life*). Most people begin to experience symptoms of presbyopia when they reach 40 years of age and need to correct this by using glasses. Ageing affecting sight is a factor of significance to work ability with increasing age. Working with a computer, which particularly requires focused sight, can be complicated. Mucous membranes risk becoming frail with increasing age, tear production and blinking decreases [54]. If decreased tear production causes the eyes to become drier, this contributes to the individual experiencing eyestrain faster, especially in work tasks that require precision or when working in front of a screen. Furthermore, increasing age deteriorates the depth of field and depth perception and contributes to a large amount of seniors feeling tired, since declined eyesight requires more brain capacity to focus. Furthermore, ageing decreases and limits the field of view, which increases the risk of accidents since it contributes to decreased perception and makes it more difficult to perceive objects situated or moving from the side towards the individual. Furthermore, the risk of accidents increases with older age since it becomes more difficult to perceive different colours, for example of different warning lights, as the colour vision deteriorates. Consequently, professional drivers may find it more difficult to work in traffic as they age, since the risk of being blinded by lights increases when the pupil cannot contract as rapidly and night vision deteriorates.

A noisy work environment contributes to an increased risk of hearing loss and hearing impairment [49,55]. Hearing loss can in turn affect work ability and can increase the risk of work-related accidents, since it is more difficult for the individual to hear alerts from surrounding people, motors, warning signals, etc. Almost one in ten employees above 50 years of age, and one in five employees aged 70 years have impaired hearing. Furthermore, studies show that noise combined with a physically demanding work environment contributes to individuals having a shorter working life and leaving working life prematurely [49,55].

## CHEMICAL HEALTH RISKS IN THE WORK ENVIRONMENT

Toxic substances are all around us. Toxic substances can cause bodily injury, affect biological ageing and cause illness by contributing to chemical reactions in tissues and organs of the body, which results in cell injury. Furthermore, toxic substances can constrict oxygen uptake, or the uptake of other substances that are vital to the physiological and biochemical processes of the body. Toxic substances can enter the body through inhaled air, skin, the eyes and through the gastrointestinal tract. Individuals can be exposed to chemical health hazards in the work environment through unhealthy pollution, airborne particles, gas, radiation, liquids and solid substances. Furthermore, biological substances, microorganisms and contagions that can cause illness and injury constitute chemical hazards in the work environment. Occupational exposure limit values and threshold limit values are set to ensure that people are not exposed to hazardous doses of toxic substances in their work environment [56]. The limits are usually described as acceptable daily intake, tolerable daily intake or provisional tolerable weekly intake. Occupational exposure limit values and threshold limit values for countries around the globe can be found on the website of the International Labour Organisation. An exposure limit value can either be indicative or binding. An indicative occupational exposure limit value means that exposure in the work environment should not exceed the set limit value. A binding occupational exposure limit value means that exposure must not be exceeded at any time.

In the work environment and organisation of work, it is important to consider that different groups can have different levels of sensitivity to exposure of chemical and toxic substances. Particular attention should be paid to risk groups. For example children and adolescents, pregnant women and senior people are particularly sensitive to the exposure of toxic substances. Furthermore, individuals with certain diseases may have an increased sensitivity. The risk of ill-health and injuries can increase exponentially if the individual is simultaneously subjected to other toxic substances in the work environment, in their leisure time or through lifestyle factors like consumption of tobacco, drugs and alcohol. This increased sensitivity contributes lower levels of exposure to toxic substances causing injuries. Because of this, it is important to analyse individual employees' work environment and assess whether exposure is acceptable with regards to occupational exposure limit values and to reflect on whether the employee might belong to a risk group.

Approximately 2 million people in Europe are estimated to work in chemical industries. However, an even larger amount of people work with further processing of or handle, use or come into contact with chemical substances in their work environment in one way or another. Approximately one in five individuals of the entire European labour force is estimated to be exposed to hazardous substances in the workplace for at least one-fourth of their working hours, through their own or others' handling [57]. This can cause work-related illnesses, for example cancer, cardiovascular disease, breathing issues and eczema. This may risk causing the individual to suffer and their quality of life and health to deteriorate, cause costs to the organisation through sickness absence and loss of production and costs to society through health care, sick pay, occupational injuries and employees leaving working life. For example, deficiencies in the handling of hazardous substances and failure to use protective equipment can cause airborne particles from organic dust – like mould, pollen, wood and plant parts – and inorganic dust – like mineral fibres, glass wool, mineral wool, ceramic fibres, plastic, isocyanates, heavy metals, pesticides and stone dust – to cause allergies and issues in the paranasal sinuses and lungs. Inhaling dust particles, for example in occupations such as agriculture, stonemasonry and bakery, can result in silicosis, which causes the lungs to harden and decreased oxygen uptake capacity, which can increase the risk of developing lung cancer. The work environment for occupations in the beauty industry, for example hairdressers and nail technicians, entails exposure to hazardous substances and gases that contribute to allergies, occupational injuries and illness. Organic solvents in the work environment, for example in occupations such as painters and floor layers, can cause headaches, dizziness, nausea, affect mucous membranes, injure blood formation and affect nerves which can contribute to memory loss, affect sight, hearing and nerves in arms and legs. Aerosols, dust and airborne particles can also entail virus, bacteria and contagion, which together with other particles and hazardous substances can exponentially increase the risk of developing ill-health.

## CLIMATE AND WORK ABILITY

The climate in the workplace, for example temperature, radiation, humidity and draught, can affect individuals' thermoregulation, health and their possibility of executing work tasks. The temperature of hands and fingers affect dexterity, however, climate and temperature also affect concentration and learning [58]. The climatic conditions in which a working human being experiences thermal comfort depend on what kind of work is being executed and the required oxygen uptake. In physically demanding work tasks, the optimal comfort zone is often at a lower temperature than for sedentary work tasks. For example, the experience of an excessively cold work environment for physically demanding work tasks is often below an air temperature of 10°C, in comparison it is below 20°C for sedentary work tasks. Accordingly, the work environment for sedentary work tasks is experienced as excessively hot when the air temperature exceeds 25°C, compared to an air temperature exceeding 22°C for physically demanding work tasks. When assessing the air temperature comfort zone, the radiation temperature, humidity, clothing, activity, etc. must be taken into consideration.

## SAFETY RISKS AND OCCUPATIONAL INJURIES

Every year employees suffer occupational accidents and fatalities. In order to avoid individual suffering and financial consequences to organisations, companies and society, safety and risk evaluation is an important area of physical work environment management. To prevent ill-health and issues related to the physical work environment, the order of priority is as follows:

1. Eliminate harmful elements
2. Encapsulate and/or create barriers against harmful elements
3. Increase knowledge of risks and harmful elements
4. Organise to counteract harmful elements
5. Promote and implement personal safety, safety equipment, ergonomic and technological aids.

Even if occupational accidents generally have decreased since the beginning of the 1990s, the number of age-related occupational accidents has increased in recent years [56]. Occupational accidents often occur in occupations where different types of technology, equipment and/or other physical bodies are involved. When executing work tasks with different objects or when using different types of technology, safety risks often occur in two parts: (a) the machine, system or physical body the employee is working with, in other words, the physical object in itself with the functions and risks it entails, (b) the knowledge the employee must have of the technological procedure and technique to use it safely and functionally in a context.

However, safety risks and accidents often occur in the intersection between human beings and technology. Because of this, there is a third safety risk which constitutes (c) the human factor. Factors like the individual's functional diversity, physical and mental health matter to this safety risk. Occupational groups with high-risk and demanding work tasks often leave working life for retirement at a relatively young age [2,5,45,46,59–62]. If it is possible for an individual to leave a work environment where they do not feel good, many people choose this alternative above the higher pension benefits they would receive if they remained in working life. This probably contributes to the fact that the amount of occupational injuries appears to be smaller for the oldest age group, what is called a "healthy worker effect". People who feel worn out by working life and people who risk becoming or have been subjected to an occupational injury or occupational illness leave working life and disappear from the labour force statistics.

As previously mentioned in this book, senior employees generally run a higher risk of injuries in demanding work environments compared to their younger co-workers (read more in Part II *Age in Relation to a Sustainable Working Life*). This is because biological ageing deteriorates hearing and sight and causes a longer reaction time, difficulties with processing new information as well as preferences to use older equipment. Furthermore, senior employees need longer sickness absence after an occupational injury since biological ageing increases the time of healing processes and recuperation. Even if senior employees often suffer more severe and complicated injuries, studies show that younger employees more frequently suffer a larger amount of minor injuries [55].

It is important to pay attention to potential risks in the workplace and to introduce precautionary measures that take the employee's age in relation to their work tasks, surrounding environment and work situation, into account. For example, there may be an increased risk of occupational accidents for a senior operator and technology user, since biological ageing generally entails deterioration of sight, hearing and reactions. Or perhaps the operator is of an age when their family situation with small children causes increased stress or insufficient sleep, this also entails an increased risk of accidents due to insufficient ability to cope with and focus on the work task. Therefore, age and aspects related to the employees' age group should be considered in the risk assessment. Information, education and changes in work processes have proven to decrease occupational accidents and fatal injuries. Furthermore, regular health checks constitute a preventive measure for employees in high risk occupations.

Moreover, the employee can take measures to decrease the risk of physical overload in their work situation, for example:

- Use their legs and avoid straining their back when lifting
- Contract the abdominal muscles, bend knees and hip joints while keeping their back straight when lifting
- Work close to their body
- Avoid straining joints in extreme positions
- Distribute weight evenly when carrying
- Move feet instead of twisting the body
- Avoid contracting more muscles than the task requires

- Frequently take short breaks
- Stretch muscles
- Keep body in shape through physical exercise for at least 150 minutes per week with an intensity that causes perspiration, since a fit body is more resistant to strain injury

However, it is important not to put the entire responsibility of occupational injury risk prevention on the employee. It is the employer, workplace and design of work tasks that should have zero tolerance for occupational injuries and promote the possibility of good health at work.

## GENDER DIFFERENCES BETWEEN PHYSICAL DEMANDS IN WORKING LIFE

The gender segregation discernible in the labour market today was created more than a hundred years ago. The physical demands in male-dominated occupations have decreased in recent years. Therefore, the gender segregation that remains today significantly contributes to a higher risk for women to develop strain injuries [56]. Women in male-dominated occupations report sickness absence to a greater extent, which in part can be explained by these women working in conditions that primarily are adjusted to men's physical capacity, with aids and protective equipment supposed to fit the size of an average man. Furthermore, women experience the physical work demands as more difficult with increasing age to a greater extent than their male co-workers. This is probably due to the fact that women in general only have 40–80 percent of the muscle strength men in general have. Furthermore, muscle strength deteriorates with increasing age, for both men and women. Moreover, men and women who work in the same occupation often have different work tasks with different workloads. Women more often work with unilateral and repetitive work tasks and moving people. However, when women and men have the same work tasks in the same workplace, they still risk having different workloads if tools, protective equipment and workplace are not adjusted to both genders. Additionally, women in general still do the majority of unpaid household chores besides their professional work, which also affects their health.

## CONSIDER AND REFLECT

- Reflect on the design of your physical work environment.
- Reflect on whether your physical work environment and possible tools, ergonomic and technical aids, protective equipment, etc. are adjusted to the dimensions of your body.
- Reflect on possible risks of moving objects, falling objects, animals, vehicles, etc. in the workplace and in your work environment.
- Reflect on whether there are risks of contagion, environmental risks, chemical risks or risks related to climate or radiation in your work environment.
- Reflect on other possible risks of occupational injury in your work environment in the short- and long-term perspective.

- Reflect on whether you have been exposed to risks of occupational injury, occupational accident or occupational illness based on different situations that have occurred in your physical work environment.
- Reflect on whether your health has been affected by the physical work environment you work in or have worked in.
- Reflect on whether you risk being subjected to occupational accidents, occupational illness or occupational injury based on the design and organisation of the physical work environment in your workplace.
- What is the systematic workplace management and risk assessment of the physical work environment like in your workplace regarding hazardous risks related to ergonomics, chemicals and the sensory system?
- How can you contribute to preventing risks in the physical work environment and risks in the workplace?
- Reflect on how and what you can do in the future to experience well-being in your physical work environment.

# Area 3. Mental Work Environment

The mental work environment and work situation has been paid more attention in recent years. The mental work environment is significant to whether individuals consider that they will be able to work until an older age and affects biological ageing [1–4,8,10,63] (Figure 4). However, some people have focused on mental health in working life for a long time, for example the psychiatrist Marie Jahoda who discussed mental and psychosocial issues caused by work for girls working in factories in the 1970s [64]. The term *psychosocial work environment*, which is still used today, was coined by the psychoanalyst Erik H Eriksson as early as 1959 [65]. During the 1970s, competence development, influence and participation became important factors in the organisation of work, and therefore the psychosocial work environment as well. Today, the concept of psychosocial work environment is usually divided into the mental work environment and the social work environment. This divide also applies to the SwAge™ model. The SwAge™ model contains several determinant areas connected to the concept of psychosocial work environment, though they need to be assessed and rectified both separately and combined in order to create a healthy and sustainable working life (apart from *Mental work environment* in the SwAge™ model, you can, for example, consider the four determinant areas in the following spheres of determination: *Social support, inclusion and sense of community* and *Execution of work tasks*).

What are the arguments to create a healthy mental work environment? Humanitarian arguments aim to prevent occupational injuries and promote people's quality of life. However, business-related and organisational arguments aim to decrease costs of sickness absence, interim staff and rehabilitation. Furthermore, workplace health promotion facilitates recruitment, productivity, global competition, innovation and growth. Unhealthy stress is often mentioned in relation to deficiencies in the mental work environment.

The proportion of employees suffering from ill-health, who are absent from work due to work-related mental strain, has increased in post-industrial countries. Today, mental health conditions are described as the primary reason for work-related sickness absence [66–69]. There are significant connections between the work environment and mental health conditions related to stress [70,71]. Chronic stress in the work situation often results in suffering, social withdrawal and lower levels of performance than usual. Therefore, the experience of a healthy mental work environment is significant to well-being and long-term public health [72]. A work situation characterised by high demands, stress and an ill fit between the individual and the environment can cause depression, anxiety and fatigue. Furthermore, work that, for example, entails human suffering, exposure, powerlessness, exposure to noise, shift work, toxic substances or threats and violence makes individuals run the risk of developing work-related stress and ill-health.

Exposure to excessive mental demands can cause damage to cognitive abilities. In the worst case scenario, this can result in the individual loosing abilities to the point where they cannot function in the workplace again. However, some individuals can recover enough to return to the workplace, even if they do not return to their previous function and cognitive capacity. Others manage to fully recover and return to their former cognitive capacity. Furthermore, in some cases individuals are strengthened by the experience of stressors, trauma, injuries and illness to the point where they not only recover their previous function but proceed as stronger individuals with increased experience, ability, knowledge and capacity.

The mental work environment and work situation in younger years are important for the ability to cope with working life until an older age, since mental conditions, illness and long-term sickness absence in younger years can affect the individual's possibilities of coping with their continued working life. Furthermore, the mental work environment and work situation matter to senior employees since many seniors state an increased sensitivity to stress with increasing age. If a poor mental work environment causes fatigue syndrome, it often takes a long time to recover, despite medical and therapeutic treatment. Therefore, it is important to discover early symptoms of fatigue syndrome and to take appropriate measures as early as possible. A mentally sustainable working life is characterised by dialogue, communication, inclusion, a sense of coherence, clarity, trust and control of one's own work. People who have worked in healthy mental work environments and experienced control of their lives express that they feel younger than their actual age. Furthermore, a healthy mental working life often results in better mental health for senior individuals compared to retired individuals in the same age group, particularly compared to individuals who have felt forced to leave working life.

## STRESS THAT MATTER TO WORK

All living creatures experience what is known as stress (read more in the Section *Bodily Functions that Matter to Work – The Nervous System*). The Hungarian-Canadian physiologist Hans Selye is described to be the first to observe the bodily effects that occur as a response to different stimuli, he called this stress. His definition of stress was: "The non-specific response of the body to any demand for change". This definition is based on objective observations of the bodily and biochemical changes that different kinds of stimuli result in. In other words, stress is the term we use to explain the process that initiates when we process and adapt to the different strains we are exposed to.

Selye [73] studied how organs in the body react to different stimuli like cold, heat, infections, trauma and nervous irritation. He based his theories of biological stress, *general adaption syndrome*, on this. Other researchers have developed these theories further to include the relation between stress and what happens biologically in an individual who experiences excessive and lasting demands of adjustment. For example, the risk of developing cardiovascular disease and cancer increases with lasting and long-term stress. Too extensive and prolonged demands of biological adjustment and constant reactions to stressors cause chronic stress in the body with lasting disturbances of the physiological functions, which, in turn, initiate different kinds of illness. This has been described in Section *The Nervous System*.

The endocrine system in the body regulates and releases chemical signals, hormones, to different parts of the body. The most important stress hormones are adrenaline, cortisol and thyroid hormones. Adrenaline is produced in the adrenal medulla and in certain neurons in the brain. The release of adrenaline affects the body by:

- Increasing the concentration of glucose
- Increasing the concentration of fatty acids
- Increasing cardiac contractility and heart rate, resulting in increased cardiac output
- Increasing blood pressure
- Changing the blood flow, causing a larger amount of blood to pass through active skeletal muscles, heart and brain, and a lesser amount of blood to pass through inner organs and the digestive system
- Stimulating a part of the brain stem, the reticular formation, which increases focus and alertness

All kinds of stress increase the secretion of cortisol from the adrenal glands. Cortisol is one of the most important stress hormones in the body and affects cells of the entire body in different ways, for example by:

- Affecting the cardiovascular system, which, in turn, strengthens the effect of adrenaline by increasing blood pressure
- Stimulating the denaturation of protein and fat in many tissues to use as energy
- An anti-inflammatory effect
- Inhibiting DNA synthesis, decreasing production of antibodies and inhibiting the immune system to save energy

The thyroid gland increases the production of the thyroid hormones thyroxine and triiodothyronine. This affects all cells in the body by:

- Increasing the effect of the sympathetic nervous system
- Increasing nerve conduction velocity in the nervous system
- Increasing protein synthesis and cell division
- Increasing metabolism

Different reactions, emotional and physiological changes and changes in the surroundings activate the sympathetic nervous system in order to make the organism react and take necessary measures to survive [73]. Innate and learned reflexes to different stressors initiate the secretion of neurotransmitters (read more in Section *The Nervous System*, Section *Stress*). Stressors trigger automatic reactions in the body, before we even may consciously have registered and reacted to what has happened. When an individual experiences an unpleasant, exposed and exertive situation, the body prepares to flight, fight or freeze (i.e. avoid discovery or play dead) in order to survive. Different emotions follow the reactions to different stressors. Examples of feelings caused by stressors are not only pain, fright, hunger, fear and anger but also feelings of great joy, infatuation, desire, well-being and satiation.

Everyone has experienced what happens in the body in a stress reaction. For example, many of us have felt the reactions in the body when someone suddenly jumps out from behind a door and yells "BOO!". The reactions in the body are instant. The heart races and releases neurotransmitters to the organs, to increase the supply of new oxygen and energy to muscles and joints. Some people will start running, others will hit the person who scared them, while yet others will be petrified. Rapid reactions are vital when we face immediate danger. However, even minor events will activate and cause the nervous system to react. Stress surcharge and reactions are vital functions and protection not only for activities in the entire body but also to avoid being killed, eaten by wild animals or to avoid being excluded from the community and care of the pack, thereby running the risk of starving to death. Human beings are, physiologically speaking, still much the same as in the hunter-gatherer culture, when threats to our survival were not as subtle and diversified as they are today. Therefore, bodily reactions are essentially the same, for example in a case of potentially life threatening danger caused by two different mental stressors, like getting caught in rough brush woods and risk starving to death or like being tied up in a highly controlled work situation in which the individual has no control while simultaneously feeling threatened of losing their income due to an inability to cope with the excessively high demands.

Reactions to stress differ between different individuals, events, the severity of the reaction and different reaction patterns. There is a constant interplay between an individual's genes and the surrounding environment that sets the character and intensity of the stress reaction. Furthermore, the memories of stressors are stored in order to survive more easily and to be able to cope with future situations. Synapses in the nerve fibres become faster and, with time, turn into intrinsic reflexes after reoccurring stimuli of stressors. After burning your hand on a hot stove once, the body remembers the discomfort and you can relive and react in the same way without even getting your hand close to the hot stove. Furthermore, by observing how other people react to stressors, we can create intrinsic reflexes that can be triggered when the nervous system recognises a similar situation. Warning labels, visual information, films or seeing things we experience as unpleasant happen to others will trigger the sympathetic nervous system if we encounter a similar situation. The brain stores these memories of stressors, based on the experienced severity of a life- or health-threatening situation, and what the reaction should be in order to ensure survival. However, the brain's discernment can be rather blunt. When different stressors are lumped together in the memory storage, the intrinsic, automatic physiological reactions appear to be the same for all stressors of the same level of "life-threatening danger". The brain and body mix up different stressors. Consequently, the bodily effects and stress surcharge from reactions in the sympathetic nervous system can be equal for two completely different stressors.

To be in a situation with stress surcharge from reactions in the sympathetic nervous system for prolonged periods of time is hazardous to health, regardless of what stressors have caused it.

Stress can mainly cause two biological effects:

- Disturbances in the secretion of neurotransmitters, like cortisol and thyroid hormones. Consequently, this may cause a disturbed balance in the regulatory system, metabolism and energy mobilisation of the body
- Disturbances in the secretion of neurotransmitters can, in turn, cause changes in the central nervous system, for example in the hippocampus, controlling impulses and memories

This can risk increased biological ageing and cause illnesses and symptoms such as fatigue syndrome, anxiety, suicidal thoughts, cardiovascular diseases and even death. It is not only the bodily reactions that happen in stressful situations, or repeated reactions that affects health, but also the ability to unwind after acute stress reactions.

## STRESSORS RELATED TO WORKING LIFE

The work environment entails different kinds of stressors. Although stress is mainly associated with the mental work environment, there are stressors associated with all nine determinant areas in the SwAge™-model that all can trigger and activate a stress reaction (Figure 8). Stressors that directly affect the organism are genetic

FIGURE 8   Examples of potential stressors related to working life.

predisposition, injuries, as well as physical and mental fatigue. Stressors can also be artificial substances, toxic substances, drugs, radiation, etc. that have entered the body and impact health and the presence of illness by affecting the sympathetic nervous system and the bodily functions. Furthermore, there are stressors that affect us indirectly, since external stressors are registered and assessed through the cognitive process, resulting in a stress reaction. Physical stressors can, for example, be sudden sounds, light, temperature, sharp objects, dangerous animals and insects or objects or creatures that move towards us. Our senses, sight, hearing, feeling, etc. primarily trigger the sympathetic nervous system to react to physical stressors. Mental stressors can, for example, be exposure to or living in fear, being subjected to threats and violence and the risk of being subjected to an accident or illness.

A stress reaction can be activated indirectly through workplace conflicts, by excessive work demands the individual cannot cope with or has sufficient time to execute. Stress reactions are affected by the predictability of an event, if the individual has the possibility to control, launch or interrupt a task. Affirmations and visualising positive or negative events can also constitute mental stressors. Fatigue stressors are, for example, insufficient possibility to recuperate, fatigue, tiredness from shift work or lack of breaks. Financial stressors relate to an individual's ability to provide for themselves or their family, the ability to afford living expenses, to profits or losses and what their impacts on the individual's personal finance. Social stressors are changed circumstances in life that make us experience exclusion and the risk or experience of being left out of a community, of our pack, for example through the death of a close relative, divorce, leaving work or conflicts. This can also be experienced in new groups and with new co-workers. Other stressors relate to new work tasks, changed work circumstances and whether these are desirable or not. Furthermore, how an individual's competence is regarded in relation to what is expected, i.e. if an individual's knowledge and level of education is too high or too low in relation to the work task, this can constitute stressors.

## COPING WITH WORK STRESS

When coping with stress, the frequency and duration of stress must be regarded, as well as the ability to unwind after a stress reaction. If an individual finds it difficult to unwind after a day at work or experiences constant overload, the physical and mental resources deplete faster. An inability to unwind complicates the reconstruction of cells in the body and the recovery of different resources. How stressors affect an individual in certain situations differs depending on genetic predisposition, general health status, the experience of social support, previous experiences and coping strategies to deal with stress. Concepts traditionally related to coping with stress are not only *crisis and development, coping, locus of control* and *balance between demands and control in the work situation* but also *sense of coherence, self-efficacy, resilience, effort-reward balance, etcetera* (read more in (8) *Work Satisfaction, Motivation, Stimulation and the Core of Work*, Section *The Individual's Experience of Comprehensibility, Manageability and Meaningfulness in the Work Situation*).

## CRISIS AND DEVELOPMENT

Individuals react with anxiety to stressors such as unwillingness, meaninglessness, loneliness and overload. Much like pain, which is an individual's response to physical damage, the ability to experience anxiety is one of our most important protective functions and is vital to our survival. In 1844, Søren Kierkegaard described anxiety as an adventure that every human had to live through, learning to be anxious so as not to be ruined either by never having been in anxiety or by sinking into it. He wrote that whoever learned to be anxious in the right way had learned the ultimate [74].

It is a normal part of life that individuals at times, and occasionally for quite a long time, feel insecure, worried and question their own worth, abilities and the meaning of their existence. However, expectations of normality have to some extent contributed to anxiety being perceived as an illness. In order to be normal, the expectations from oneself and others is that a human being should have good work ability, good family relations, a functioning sex life, be generally optimistic and sufficiently committed to the outside world. The psychoanalyst Johan Cullberg [74] describes that people must live with polar opposites, for example good and bad, closeness and loneliness, creation and stagnation. Many people live harmoniously for long periods of time but sooner or later find themselves in a crisis, event or change that demands them to re-evaluate their lives and discover new sides of themselves that they were previously unaware of. This is often a painful process, but it is also a chance to understand oneself and the outside world better, and to mature as a human being. Development results in a more realistic perception of one's own capacity, in other words, increased self-knowledge, and an increased ability to discern between other people's needs and wishes and one's own needs and wishes. This sense of autonomy and independence results in an increased ability to evaluate one's life goals realistically, and to put them in relation to goals that commercial forces and other parties wish to impose on us.

The word *crisis* means a turning point and is also used in medicine to define an important change that results in recovery or decline in health. An individual faces a mental crisis in a situation where previous experiences, behaviour, learned reactions and reaction patterns are no longer sufficient to cope with the new situation. A crisis can occur in mentally demanding work situations, or in new, unclear events in the workplace. Furthermore, crisis reactions occur during large life changes in an individual's life, for example during the teenage years, when becoming an adult and entering working life, in re-organisations and changes in working life as well as when leaving working life and entering retirement.

According to Cullberg [74,75] crises can be divided into traumatic and developmental. A traumatic crisis is defined as severe events in the surrounding world that seriously threaten the individual's physical existence, social identity and security or other life goals. A developmental crisis is defined as events in the surrounding world that really are part of normal life but can be overpowering in certain cases and at certain times. The natural course of overcoming a crisis *always* consists of four phases. However, individuals can be in the different phases for varying periods of time, depending on the event, and on the individual's ability to cope with and overcoming the crisis. The first two stages constitute the reaction phase, while the latter two constitute the processing phase:

1. The sequence of events starts with the *acute crisis*, where the individual's entire energy is spent trying to rediscover their orientation in life. If the event has a great impact on the individual, they can react with nausea, tremors, dizziness and a rapid heart rate. Denying that the event actually occurred, feeling cut off or empty, experiencing that time stands still, feeling numb, experiencing difficulties thinking clearly and logically and experiencing derealisation are all common reactions. The individual can only execute daily routines, if anything at all. Some individuals display a calm exterior during this phase, while their inner world is a chaos where nothing is comprehensible or makes any sense. Other individuals react violently with anger or panic. Sooner or later the acute crisis turns into the second phase.
2. In the *reaction phase*, the individual suffers anxiety and uneasiness when the natural processing of the event begins. At this time, the individual experiences pain, which can be both mental and physical. Many people experience anger and attempts to find a scapegoat, alternated with feelings of guilt and self-blame. The extent of what has happened becomes apparent. In this phase, it is common to experience apathy, difficulties concentrating, insomnia, appetite disturbances, bodily reactions, mood swings, a need and wish to escape and run away. Increased sensitivity to sounds and experiencing heart palpitations at sudden sounds suggest that the body is stressed and alert. Memories and sequences are repeated and dwelled on in the individual's mind.

    Uneasiness and fears of what is associated with the event are natural reactions in the beginning. However, isolation may be the most dangerous reaction. The individual may easily cut off from the outside world when they experience ill mental health after a stress reaction in order to protect themselves. However, isolation postpones processing the event, the natural course and recovery. This entails a great risk that the individual gets stuck in the reaction phase and develops a depression due to prolonged stress. It is possible to develop a depression purely stemming from negative thoughts. Some individuals try to escape their despair and their situation using alcohol, drugs or overconsumption of different kinds. However, sooner or later the individual hopefully proceeds to the third phase.
3. In the *coping phase*, the individual begins to realise the need to structure the problem in order to understand what has happened. In this phase, the individual processes the event. The stress reaction gradually disappears which makes it possible to think forward. However, intrusive memories, sleep disturbances, high alertness and issues with situations that remind the individual of the event can still occur. Regardless, the individual can think more clearly of how they can and should best cope with the new situation.
4. When the individual reaches the fourth phase, *orientation towards the future*, the pain of the experience has decreased to a level where it is

manageable or even has disappeared. The individual no longer perceives everything in their situation or life as dark and threatening but can notice what is beautiful and positive too. At this time, the individual begins to function in their new situation, even if it is still difficult at times or in certain situations.

How the individual manages to process and cope with a crisis greatly depends on the support they find in their surroundings. A crisis leads to development. A well-processed crisis is an important chance for the individual to reach a higher level of strength and security. An individual who has experienced a crisis and has returned to ordinary life can have a deepened and increased understanding of life's terms and achieve wisdom (read more in (8) *Work Satisfaction, Motivation, Stimulation and the Core of Work*).

## COPING

An individual can perceive and experience stressors as both exertive and exhilarating, however, stressors entail a biological reaction that the individual must cope with. When an individual is subjected to a stressor or a crisis, in other words, a reaction to a stressful event or situation, the individual must be able to overcome the crisis in order to develop (read more in Section *Crisis and Development*). *Coping* can be simplified as an individual's continuous unconscious and conscious coping, processing and mastering of external and internal stressors, related to knowledge and emotions. In other words, coping means how we manage the stressors in the four different stages of processing a crisis.

Coping is sometimes perceived as a key concept in theories regarding adjustment and health [76–78]. However, coping strategies also control emotional and rational thinking when we make decisions on a daily basis. Coping can be divided into two different activities or systems to process the stimuli of a stressor.

*System 1* constitutes coping strategies that are automatic reactions initiated by the stimuli of a stressor. The reaction mainly depends on innate or early developed personality, and intuition developed by previous experiences that have been stored in the individual's neural pathways. *System 1* continuously scans and associates memories and constantly interprets events in the surroundings accordingly. System 1 constitutes rapid thinking. The individual's personality and intelligence, as well as their reaction patterns developed from previous experiences and knowledge, determine how the stressor and crisis are managed in the acute phase. The acute reaction response to a stressor made by system 1 is rapid, unconscious, affective and intuitive. The reaction comprises innate and learned reflexes. System 1 is often described as intuition and is what makes an experienced medical practitioner able to identify a correct diagnosis by just looking at a patient, or what makes an experienced chess player able to know how to achieve check mate just by looking at positions of way chess pieces. Sometimes individuals are categorised into different personality types based on their usual reactions when system 1 is triggered by stress. Some have an increased tendency to fight, others make sure to keep themselves safe, while yet others play dead. Sometimes different personality types, for example A- B- and

C-personalities, are described based on these reflexes. Type A represents people who usually react in an active, sometimes even aggressive way, type B represents people who usually take a step back and keep themselves safe, type C represents people who usually experience numbness and defencelessness in stressful or unexpected situations. However, an individual's system 1 often reacts differently to different stressors, and an individual has personality traits of all three personality types.

*System 2* refers to the coping strategy initiated by a stressor and develops into a long-term coping process specific to the situation in question. The system 2 coping process is not as rapid and unconscious as system 1. The coping process is not static but varies and adjusts to the character of the stressor and the context of the crisis. The coping process changes and becomes more problem-focused with time and the development of the experience. It focuses on problem solving and changing or sidestepping the stressor, the source of the problem that is. Furthermore, system 2 focuses on emotions and decreasing and coping with emotional anxiety through knowledge, practice and behaviour. Whether the individual unconsciously chooses to mainly use problem solving or emotions in their coping process depends on the character of stressor, context, how the individual assesses the situation, memories from previous experiences, social support and what the stressor means to the individual on a personal level. Furthermore, gender, age and personality affect the choice of a problem-focused or an emotion-focused coping strategy. System 2 represents slow thinking and is an exertive intellectual process. It requires attention and conscious thinking, where the individual weighs pros against cons, concentrates and makes conscious choices. Problem-focused coping strategies are common in work situations and situations that require additional information. However, at times the individual feels personally involved to a degree where it turns into an emotion-focused coping strategy. A mix of different coping strategies are usually initiated to cope with one or several stressors (read more in (8) *Work Satisfaction, Motivation, Stimulation and the Core of Work*).

### LOCUS OF CONTROL

The individual's experience of their possibilities to influence and cope with stressors affects their health and well-being. The general attitudes, beliefs and expectations of the causation between the individual's behaviour and the consequences affect the individual's choice of behaviour. In this, there is a relation between passivity and a belief in luck or destiny. The concept *locus of control* refers to whether an individual experiences that they themselves are in control of their situation, or whether other people or something else controls it [79]. When an individual experiences a high inner locus of control, they feel in control of what happens in their life and situation, and of their destiny. When an individual experiences a high outer locus of control, they feel like a victim of circumstances and of other people's wishes, and that they have little or no possibilities of affecting their own destiny.

Furthermore, *alienation* is a concept noted by several researchers in philosophy and sociology that plays an important role in the mental work environment. When an individual experiences alienation, they feel like an outsider, they do not participate in events in the environment and in the development of the entire organisation in the workplace and in society. This causes feelings of social isolation, powerlessness and abandonment. An individual who feels alienated in their work situation feels powerless to control their own situation. They experience themselves as an insignificant cog in the machinery of the organisation. The individual experiences that they are at the mercy of others and that they are controlled by powers too great or unclear for the individual to understand and to assert themselves. An individual believing that they lack control makes them experience alienation, while an individual believing that they are in control makes them experience a sense of control.

The choice of coping strategy an individual uses takes place after the primary acute assessment of the stressor, followed by a secondary assessment. Locus of control is an important factor in these assessments. The primary assessment controls whether a stressor or change in a situation is positive, irrelevant, negative or even threatening. A new or ambiguous situation provides an experience of minimal control through the lack of recognition and clarity to the individual. If an individual lacks previous experiences and information to use in a particular situation, the individual makes an estimation. In such cases, the individual draws conclusions based on their general previous experiences as well as their personal disposition, which includes their belief in themselves, inner control and the experience of manageability in the situation. In a particular situation, a person with more inner locus of control experiences less worry than a person with more outer locus of control. The more severe the individual experiences the threat and consequences, the more important it is to experience a sense of inner control. Even if the estimated probability is ambiguous regarding whether, when and how a certain threat will happen, and how common or uncommon the threat is, it matters to the processing of and experienced control of a stressor. In an ambiguous situation, the individual's estimation of their control of the situation most likely depends on the nature of the situation. The second processing of a stressor entails the estimation of coping resources.

The individual makes their choice of coping strategy in order to make a situational estimation of their actual locus of control. The nine determinant areas in the SwAge™ model (Figures 1 and 3) are part of the individual's assessment of a situation, since their health, functional, physical, mental, material, social and recuperative resources, and their knowledge and competence resources are estimated as coping resources in the situation. Even if individuals' locus of control mainly is important to the first assessment of stressors, it also affects their coping strategy. The more severe the individual experiences a threat or feelings related to threats, the more their coping strategy is controlled by their feelings and emotions. Emotional focus decreases the possibility of a problem-focused coping strategy, in other words, the efficiency and scope of a problem-focused coping strategy depends on the success of emotional focus when choosing a coping strategy.

Every individual has a frame of reference, established in childhood and adolescence, through which the individual perceives the outside world and the need of different coping strategies. The frame of reference contains knowledge the individual has developed of their own character, of others and of the surrounding world. The frame of reference is affected by genetic predisposition, all individuals have different innate prerequisites and levels of sensitivity. If an individual's early childhood has been characterised by a fundamental sense of security, their frame of reference is probably characterised by a sense of security. However, if an individual has had an insecure childhood, this will affect and influence their frame of reference to be unstable and characterised by negative expectations. Therefore, an individual who has developed an inner frame of reference characterised by negative expectations can react strongly to the surrounding world and experience threats where others do not. Furthermore, they can have a tendency to disparage and cast suspicion on their own personality in relation to the surrounding world and consequently experience themselves as less valuable. Therefore, an individual's locus of control, in other words, their belief in and experience of their own control, can contribute to a self-fulfilling prophecy regarding the manageability of stressors.

However, the frame of reference does not remain static throughout life, it can change in different environments and situations. The frame of reference is constantly re-evaluated, it develops and changes with increased experience. However, in order to choose solution-focused coping strategies, the individual must focus on what works and strengthens the positive aspects of a situation, in other words, they must have a salutogenic approach and focus on healthy aspects and reasons to experience well-being. Salutogenesis derives from the Latin word salus (health) and the Greek word genesis (origin). The salutogenic perspective focuses on aspects that cause and maintain health, rather than aspects that create ill-health and illness (pathogenesis). The individual must experience, develop and achieve sufficient self-esteem and autonomy in their inner locus of control to use solution-focused coping strategies. Individuals who primarily have outer locus of control run an increased risk of mainly using emotion-based coping strategies, with increased sensitivity that can make them feel bad, be self-destructive, burn out and ruin their self-esteem. They risk getting stuck in a loop of negative thoughts, concerning their lack of control, lack of capacity for action, and lack of authority in their situation. These individuals can simultaneously have a distorted interpretation of the expectations and perceptions the surrounding world has of their abilities. Therefore, they may need to increase their autonomy and strengthen their inner locus of control and frame of reference and focus on their own ability to cope with mental stressors (read more in (8) *Work Satisfaction, Motivation, Stimulation and the Core of Work*).

### THE BALANCE BETWEEN DEMANDS AND CONTROL IN THE WORK SITUATION

A healthy mental work environment is characterised by dialogue, communication, participation, clarity, trust and individuals being in control of their work situation. The individual's experience of control in their work situation, personal life situation and their ability managing demands on themselves and demands of others contribute to good health and to remaining in the workplace [14,80]. Furthermore, the experience of control in this sense has proved to be of great significance to the willingness to remain working, even if the physical work environment is dangerous and demanding [6]. For example, farmers and sole proprietors experience a great opportunity for autonomy in their work situation. As previously mentioned, agriculture and the construction industry are sectors with large amounts of work-related injuries and accidents resulting in fatalities, particularly for individuals aged 55 years and older [56,81]. Despite this, farmers and sole-proprietors are overrepresented among people who work until they are older than 70 years of age [82,83]. For example, the stress-reducing effects of nature and the physical activity of a farmer's work situation, compared to other occupations, have in some studies been described to contribute to lower rates of mental illness, cardiovascular disease and cancer. [84–86]. Furthermore, studies show that nature settings and farm animals are successful factors in rehabilitation when returning to work after long-term sickness absence, for different stress-related diagnoses and for cardiovascular disease [50,87,88].

The experience of control in the work situation in relation to work demands is significant to health promotion in the mental work environment. Karasek and Theorell [89] have developed a model where the dimension *work demands* is intersected by the dimension *control*, creating a chart of mental demands in the work situation and workplace (Figure 11).

The fundamental idea is that the individual strives to adjust and balance internal and external needs in their work situation. This is easier in a work situation characterised by routine, while adjustment is more difficult and demanding for individuals in an unstable and complicated work situation. This also increases the need of information. Moreover, the individual's need of feedback regarding their adjustment and work efforts matters to the experienced balance of external expectations in the work situation, for example expectations from managers, clients and co-workers (read more in (7) *Social Work Environment*).

|  | Low mental Demands | High mental Demands |
|---|---|---|
| Low Control | Active work situation | Tense work situation |
| High Control | Relaxed work situation | Passive work situation |

**FIGURE 9**   The Demand-Control Model [89].

The four fields in the Demand-Control Model (Figure 9) are the following:

1. *A tense work situation* refers to high mental demands from the organisation and surroundings, and little possibilities for the employee to make decisions, control and affect their work situation. This combination risks causing frustration due to the limited possibilities of using a problem-focused coping strategy in the work situation. In a long-term perspective, this can cause ill-health since the individual may feel tied up and unable to cope with stressors in a way that promotes the individual's health.
2. *A passive work situation* refers to low mental demands from the organisation and surroundings, and little possibilities for the employee to make decisions, control and affect their work situation. This forced passivity can result in learned helplessness and a lack of self-confidence and autonomy. This can contribute to an individual developing destructive behaviours, they risk getting stuck in an emotion-focused coping strategy due to the inability to

make decisions or of using a problem-focused coping process. Furthermore, this can contribute to decreased motivation and behaviours that promote the individual's health, and to increased feelings of abandonment.
3. *Relaxed work situation* refers to low mental demands from the organisation and surroundings, and a lot of possibilities for the employee to make decisions, control and affect their work situation. This may seem to be an ideal work situation. However, the lack of demands can contribute to the individual feeling demotivated and that no-one seems to acknowledge their efforts.

   Accordingly, an individual in this kind of work situation must have a strong internal motivation and will or ability to find inspiration, stimulation and motivation to keep going, in order to prevent the lack of demands in the work situation of causing stagnation, boredom and feelings of futility.
4. *An active work situation* refers to high demands from the organisation and surroundings, and high autonomy and possibilities for the employee to make decisions, control and affect their work. This combination enables the individual to be in control of and prioritise their work tasks. It increases the possibility of experiencing work satisfaction and the possibility of coping with stress and strain. High expectations combined with a high level of autonomy provide better conditions to grow and develop as a person. The employee's conditions to develop and learn new things benefit both the organisation and the individual. A sense of satisfaction can instil an urge to achieve new goals. However, there is a simultaneous risk that the individual, through this positive experience, remains working hard and fails to sense when they need to take a break and recuperate. In the long run, the satisfaction of their performance can become a sort of trap. The employee is triggered by and experiences satisfaction through their performance while simultaneously worrying about not performing well enough. In turn, this can cause performance anxiety, and that the individual continues to put excessively high demands on themselves until they burn out.

The willingness to work and participate in an extended working life decreases with insufficient autonomy, high demands, low control and dissatisfaction in the work situation. Feeling acknowledged and experiencing well-being in the work situation and profession are very important factors to an individual's experience of their own health and to their actual long-term health (read more in (8) *Work Satisfaction, Motivation, Stimulation and the Core of Work*, Section *The Individual's Experience of Comprehensibility, Manageability and Meaningfulness in the Work Situation*). It is important that work demands are balanced with the employees' capacity to enable their ability of remaining in working life. An individual ageing healthily and successfully is usually preceded by them having been in control of their work situation and of their leisure time [90,91]. People with low autonomy in their work situation retire earlier than people with greater flexibility to choose how their work tasks should be executed. Retirement is a socially acceptable way to leave mentally demanding working conditions.

Dissatisfaction with the mental work environment has been described to contribute to work-related issues and ill-health. Furthermore, a stressful work situation is an underlying cause for many accidents [56]. Work-related accidents are often associated with biological ageing, work peaks, stressful situations, high demands and low control, tiredness, employees' reluctance to take responsibility and initiatives and tedious work tasks [56]. Furthermore, a work situation characterised by stress and high mental demands can affect an individual's body and cause physical pain and ache. Moreover, physical health issues appear to be exaggerated in a mentally demanding work situations, since physical health issues typically have been a more legitimate reason to leave working life early compared to mental health conditions [3,6,44]. Particularly people with short-term education, poor physical and mental health, who have a bad relationship with their workplace, co-workers or managers, choose to retire from working life in order to avoid stress. Furthermore, having been employed in a mentally demanding work environment appears to partly contribute to continued mental health conditions after retirement.

## BULLYING IN THE WORKPLACE

In an increasing number of countries, stressful mental work situations are significant reasons for mental health conditions, sickness absence and exits from working life. A poor mental work environment with insufficient information, unclear goals and an insufficient sense of participation in the work group increases tendencies of bullying and scape goat mentality. Bullying in working life is a significant cause for developing mental health conditions, as well as leaving working life and premature retirement. An individual subjected to bullying often develops anxiety, depression and physical symptoms as a result, sometimes the individual has suicidal thoughts or even commits suicide. Being bullied is stigmatised, some people feel embarrassed to admit to being subjected to bullying. They rather express that issues in their work situation, that have been caused by the bullying, are the actual causes of their mental health conditions, such as high stress levels, an experienced lack of control in the work situation, that demands of work tasks exceed their ability to execute and that work tasks are poorly communicated [83]. It has been observed for a long time that individuals in a poor mental work environment, who experience exclusion, generally have a greater tendency to be on short- and long-term sickness absence. Furthermore, they tend to look for opportunities to leave work and exit working life in order to avoid social isolation and bullying in the workplace (read more in (7) *Social Work Environment*).

Work-related bullying has previously mainly been taking place in the workplace, leisure time has been a safe space. However, digital communication has contributed to a lack of boundaries between individuals' working life and personal life. Due to the technological development, bullying in the workplace can now occur through electronic resources, what is known as cyber bullying, which means bullying through digital channels, including aggressive or negative behaviours through e-mail, text messages, blogs and social media such as Facebook and Instagram. Digital channels can facilitate a rapid spread and distribution of smear tactics to a large amount of people. People an individual previously only met in the workplace can now be friends, followers or friends of friends on social media and thereby take part of posts the individual shares with people in their personal life and take part of the personal life of the individual's friends and relatives. Bullying appears to differ between different electronic resources. In working life, bullying through e-mail is mainly passive, through exclusion and withdrawing information, while bullying through social media mainly occurs through aggressive direct threats, offensive statements and smear tactics. In the past, it has been the managers and informal leaders who mainly have bullied their subordinates. However, cyberbullying in working life has reversed the roles and it is people in leading positions who are subjected to bullying instead. Furthermore, work-related cyberbullying subjects a larger amount of people in leading positions. People outside the organisations, like students, customers or citizens, can also become perpetrators in work-related cyberbullying, subjecting an individual they only "know" through their profession. Digital communication appears to affect power dynamics and enable people in lower positions to bully a manager [92].

## THREATS AND VIOLENCE

Threats and violence occur in different parts of society. Working life can potentially contain threats and violence that affect people's ability and willingness to work. In working life, threats and violence may occur between co-workers, however, employees can also be subjected to threats, violence and sexual harassment by clients, customers, patients, students, visitors, etc. In the long-term perspective, the risk or threat of violence, sexual misconduct or revenge can be more mentally demanding than an actual act of violence. A sudden, unexpected and dramatic event, for example a robbery or a patient lashing out unexpectedly, affects people more strongly than if the risk of being subjected to violence is expected or can be foreseen. Unexpected violent events risk undermining the individual's fundamental sense of security in the long run, since the individual's fantasy eventually can make them perceive potential risks of violence in most situations. The emotion-based coping strategy in the shock phase when reacting to a crisis or stressor, risk becoming dominant in a work environment characterised by threats and violence. Suspiciousness and irritation build up priming for conflicts. Some work situations are more prone to the risk of threats and violence than others, for example security guards and occupations in retail involving cash management, where managing money and valuable objects significantly increase the risk of being subjected to robberies. Moreover, the use of weapons has become more common in these situations in recent years. Furthermore, professions in health care, social services, education, police, transport as well as administrators of the state and municipality, like food and animal welfare inspectors, are often at risk of being subjected to threats, violence or actions that are experienced as aggressive. These actions can come from people who are arrested or taken into care, clients, patients or their relatives, parents, students or customers. The work situation can be experienced as particularly insecure, unpredictable and uncontrollable if recurring meetings at work involves people with, for example, brain damage, mental illness or people under the influence of drugs who can react in unexpected ways. In some professions, one or several people "take over" a public space, like a library, a school, an emergency room, a train or a bus. It can, for example, be groups of youths or different interest groups who want to use the space as their own. Threats and violence can be real and direct, though the threatening scenario can also be very subtle to employees in these work environments, if the threats are not explicit.

An employee's life experiences are usually an asset if they are subjected to threats and violence in their work situation. They may have experienced similar situations before or something that they can relate to. In this way, reflections and consequence analysis of an event can contribute to tacit knowledge regarding how to manage these situations. A longer time in and experience of working life can be an advantage in these situations. At the same time, biological ageing may have contributed to an increased sensitivity to stress. Furthermore, the anxiety created by constantly facing a potentially threatening situation, and the risk of being subjected to injuries and mental health conditions can be very demanding and tiring, particularly for seniors. Because of this, many individuals who have experienced a work situation characterised by a risk of threats and violence for years do not want to stay (read more in (7) *Social Work Environment*). They express the demanding mental work environment as a reason for their early exit from working life [5,8,9,93–95].

## GENDER DIFFERENCES BETWEEN PHYSICAL DEMANDS IN WORKING LIFE

Women aged 35–45 are primarily on sickness absence due to stress-related ill-health, mild depression, anxiety and fatigue syndrome [66]. Furthermore, it is primarily women who state their mental work environment as a reason for their inability of working in an extended working life. Women work in contact-based professions to a larger extent than men, professions that involve meeting and working with other people, and these people's need of help and support. Contact-based professions, or social service professions, generally entail mentally demanding work situations. Examples of these professions are health care professions, teachers, police, social insurance professions and insurance companies. People employed in these kinds of organisations make up the link between the student/customer/client, etc. and the organisation they work for and represent. In their position, as the face of the organisation, they are responsible for the treatment of the people they meet at work and may need to make unpleasant or unwelcome announcements that they may not personally agree with, as a part of their work tasks. The work environment is characterised by complex relations and contact with several different interest groups. This results in high expectations from people and groups on different levels, for example co-workers, patients, clients, students, parents, managers, authorities, laws and politicians. In their profession, the employee must manage other people's personal issues and grief, while they need to notify the patient or client of, for example, possibly life altering information, relating to political decisions, financial frameworks, regulations, laws and ordinations. At the same time, they need to cope with their own reactions to the client's or patient's dilemma and suffering [96]. Moreover, employees in contact-based professions are often subjected to criticism, threats or even acts of violence that in reality are aimed at their employers. Employees in contact-based professions generally run a higher risk of developing mental health conditions caused by their work situation. Furthermore, they run a higher risk of leaving their work prematurely due to high mental demands, almost every second employee experienced their work as too mentally demanding to remain working until the regular retirement age. However, stress factors are not only present in contact-based professions. Different professions entail different types of stress in their work situations, and since men and women traditionally work in different professions and sectors, they suffer from different types of stress. Men and women who are employed in the same work environment tend to develop similar kinds of illness [68]. Apathy, indifference, lethargy and a desire to stay at home in the morning primarily occur among employees who experience their work situation as stressful, as well as in reorganisations that are experienced as demotivating. An increased sensitivity to stress and increased stress symptoms relates to increased age and to changes in work. Among senior women in particular, mental symptoms increase the more work changes, though cardiovascular and respiratory symptoms increase as well.

However, both men and women exhibit better self-rated health after retirement if they express that they have been working in a poor mental work environment with high demands, low work satisfaction and insufficient possibilities to perform in their work tasks [6,30,90,97,98]. At the same time, senior employees with a healthy mental work environment often exhibit a better mental health than people in the same age group who have retired [6,90,98,99].

## CONSIDER AND REFLECT

- What is your mental work environment and work situation like?
- Reflect on whether your health is affected by the mental demands in your current or previous work environment and work situation.
- What are your coping processes in the mental work environment, related to the nine determinant areas of the SwAge™ model (Figures 1 and 3)?
- Reflect on whether you, through problem-focused coping strategies, can strengthen your inner positive expectations on your own ability.
- Reflect on whether and how the systematic risk assessment of the mental work environment and work situation is managed in your workplace.
- What risks of being subjected to occupational injuries, occupational illness or occupational accidents are present in your mental work environment?

# Mental Work Environment

- Reflect on whether you, in your mental work environment, are exposed to risks that, in a short- or long-term perspective, can cause occupational injury, occupational accidents or occupational illness.
- Reflect on whether your health has been negatively or positively affected by your mental work environment and work situation in other ways.
- Reflect on how and what you can do to experience well-being in your mental work environment in the future.

# Area 4. Working Hours, Work Pace and Time for Recuperation

Working hours, work pace and time for recuperation matter to whether individuals express an ability to work and affect biological ageing [1–3,95] (Figure 4). Different properties are attributed to time. Time can mean money in the production, time can heal all wounds or time can be a stick to keep up when many things should be executed in short amount of time. The time can be right, time can be out of joint and some people must kill time to make it pass. Time is equal for all – everyone has the same amount of time during the 24 hours of each day. The experienced lack or abundance of time depends on how we spend it, as well as what we prioritise during our lifetime. This also affects whether we experience that we are ahead of or behind time.

In working life, an individual's time for rest and recuperation depends on their working hours, work pace, and their access to breaks and pauses, it is closely related to the physical and mental work environment. Sufficient time for rest and recuperation has a positive effect on biological healing processes, cell division, physiological activity in the inner organs and cognitive functions like memory and learning. Insufficient time for rest and recuperation often results in tension, stress, anxiety and pain that, in turn, relate to general sleep disorders. Rest is vital in order to cope with and manage one's life and work tasks, and to heal. Lack of sleep and sleep disorders affect people's cognition and concentration and increase the risk of accidents – many occupational accidents and traffic accidents are caused by fatigue. Between 15 and 20 percent of all traffic accidents are estimated to be caused by tiredness or fatigue.

## WORKING HOURS AND WORK SCHEDULE

Sufficient time for rest recuperation during a day at work relates to a high level of self-rated health, regardless of the individuals' recuperation between work days [100]. If individuals are drained of energy during a day at work, they risk developing illness and perform and produce less in a long-term perspective. In order to make the work situation sustainable and promote health, it is important to enable employees to take breaks and recuperate during a day at work and to prevent too tight schedules.

A tight schedule in organisations and companies can be caused by financial and commercial circumstances and limits, for example goals to produce as much as possible with limited resources, increase production due to turnover requirements or profit requirements. A tight schedule can also be caused by industrial or organisational circumstances and limits, such as surges of activity in production, for example during certain times of the year when availability, demand or turnover of the manufactured product is higher. However, issues with work schedules can also be caused by increased absence from work due to illness, for example during flu season, or vacations during certain times of the year. Furthermore, it can also occur due to lack of sufficient and competent staff to employ.

There is an increased risk of labour shortages in certain sectors and in some countries due the demographic changes in the ageing society with a larger amount of elderly citizens and fewer people active in the labour force [101–103] (see Part II *Age in Relation to a Sustainable Working Life*, Section *The Demographic Situation*). For example, it is to be feared that labour shortages will affect public administration, education, care and health care. This risk may primarily affect depopulated regions, where the future proportion of pensioners risks exceeding the proportion of people active in the labour force [101–104]. Therefore, people who work risk having a tighter schedule in order to catch up with work tasks and maintaining the organisation.

## WORKING TIME REGULATIONS

Working hours are regulated by different types of laws, which differ between different countries. In the EU, working time and the right to breaks is regulated in Directive 2003/88/EC of the European Parliament and of the Council. Labour laws regulate the duration of working hours and night shift work. These laws regulate for how long it is acceptable to work each day, week and year in order to facilitate sufficient time for recuperation. These laws also concern on-call and stand-by time, the right to breaks and pauses, annual holidays and daily rest periods. Working time is partly regulated by other laws, such as laws concerning the work environment. Furthermore, in many countries certain rules apply to how people under 18 years of age are allowed to work.

Working night shifts affect the circadian rhythm that generally controls the human being's physiological functions between high metabolism during the day and low metabolism during the night, when the individual changes their activity to the night hours. This means that the employee working night shifts is exposed to a disability due to the change in their normal metabolism. Sleeping during the day hours that follow a night shift takes place outside the normal circadian rhythm and during a period of high metabolism, which disturbs the quality of sleep. Furthermore, shift work causes an increased amount of health issues, for instance the risk of chronic illness and occupational accidents increase due to irregular sleep, and so does the risk of exhaustion and sleepiness.

Employees and managers have expressed the possibility of adjusting the amount of working hours to enable a larger amount of people to work in an extended working life [3,11,105–108].

## WORK PACE AND BREAKS DURING WORKING HOURS

An appropriate work pace is important in order to experience well-being and avoid becoming too stressed. Even if there may be periods of a high or very high work pace, these must be followed by periods of a lower work pace in order for employees to recuperate and recover. Recuperation during a day at work is enabled through variation and change of work tasks, activities, work pace and workplace. Furthermore, recuperation during a day at work is enabled by taking breaks when needed. Support from co-workers, acknowledgement, helping one another, social conversations and laughter also contribute to recuperation during a day at work.

The individual can experience a lack of control in their work situation if they have no possibility of taking breaks when needed, and if they experience physical or mental tiredness. This results in what is described as a "tense work situation" [89]. The employee faces a work situation characterised by demands to perform and produce to high standards from the organisation and surroundings, even though they have little or no possibilities of taking a break when they need it. The possibility of using a problem-focused coping strategy and to listen to the signals of their body and their need for recuperation is limited in such a work situation, resulting in the individual feeling tied up and finding it difficult to cope with stressors in a manner that promotes health. This can cause ill-health in the long run. A work situation characterised by insufficient recuperation, where employees are unable to take breaks and pauses when needed, causes many occupational accidents, occupational illnesses, occupational issues and ill-health [6,56,61,109].

## EXHAUSTION AND SLEEP DISORDERS

Fatigue, sleep disturbances and insomnia are public health concerns in many countries. The proportion of people who suffer from sleep disturbances is continuously growing and is the largest in older age groups [110]. Feeling exhausted and needing time for and rest recuperation is completely normal. Physical work tasks and activities result in the body becoming physically exhausted, while mental work tasks and activities, for example communication, writing and cognitive work tasks, result in mental exhaustion. Furthermore, mentally demanding work environments and issues in the work situation can contribute to individuals finding it difficult to let go of work in their leisure time and cause sleep disorders and insomnia, which, in turn, contributes to tiredness and exhaustion. Primarily women state that they suffer from exhaustion and sleep disorders [66]. The risk of developing health issues is three times larger for women who experience lack of recuperation, fatigue and who rarely feel recuperated between work shifts.

If the time for recuperation is insufficient, constant exhaustion from too physically demanding work situations, mentally demanding work environments and poorly adjusted and demanding occupational circumstances can result in fatigue and excessive wear and tear. If the brain is too exhausted from high mental demands, lack of recuperation or overload of cognitive functions due to lasting stress, this can lead to mental fatigue which takes a long time to recover from, some people never fully recover. Furthermore, stroke, medication or head trauma for instance can risk causing mental fatigue and an inability to recuperate fully. Mental fatigue often requires measures in the work situation, for example adjustments of working hours and work tasks.

Exhaustion affects our judgement and ability to concentrate, many accidents are caused by exhaustion and sleep disorders. Exhaustion and sleepiness is frequently stated to be a cause for serious accidents [111]. About four out of five traffic accidents are connected to the driver's errors that are likely caused by inattention and misjudgements. Exhaustion and sleepiness, i.e. difficulties to stay awake and alert contribute to this (about 10–20 percent). A common cause for accidents related to fatigue is driving at night, in many cases night time driving also relates to irregular working hours. Pauses and breaks play a certain part, as well as whether the driver uses their breaks to eat, drink caffeine or sleep. The length of the drive can have significant indirect effects, since long distance driving and long driving shifts prevent sufficient sleep. Among other things, a wide, straight road with little traffic, possibly automatic transmission, reduced wind noise and cruise control may contribute to less disturbances, monotony and increased relaxation, which, in turn, can cause sleepiness. Furthermore, chronic diseases increase the risk of accidents related to fatigue about two to four times. Fragmented sleep due to chronic illnesses and mentally demanding circumstances where the individual finds it difficult to relax, the daily rest can in severe cases completely lack any recuperative value, which results in an increased risk of falling asleep in situations that can cause accidents. Furthermore, the duration of a period spent awake is of significant importance, which means that a late night and an early morning becomes twice as demanding, for employees working night shifts, for example. Starting a work shift early in the morning can imply that the individual executes their work tasks during the lowest level of the circadian rhythm, combined with a significantly shortened duration of sleep. The duration and quality of sleep directly affects the level of alertness and security.

The cause of fatigue can be related to biological ageing and issues with sight, hearing, mobility, chronic illnesses as well as mental stress and insecurity in hectic and demanding work situations. The risk of ill-health and chronic illnesses related to shift work increases with increasing age. Likewise, the risk of occupational accidents in shift work increases with increasing age. A study with 14,714 participants displayed strong evidence that mental fatigue, exhaustion and sleep disorders radically decreased after retirement [112]. The conclusion stated that the reason for this positive development was the elimination of occupational stress when individuals left working life. Other studies have brought to light that mental exhaustion and insomnia often decrease after retirement [31,113,114]. Another factor in accidents related to exhaustion, besides biological ageing, is cognitive age, in other words, the maturity of the brain, which generally varies between women and men. For example, there is a significant risk that a man aged 18–24 is involved in a traffic accident, which is described to be caused by poor alertness in this age group, though it probably is due to poor insight of one's own limits [111,115].

## TIME FOR RECUPERATION

How we spend our time partly relates to pace and recuperation. If an individual works in a fast pace they will get a larger number of tasks done in a particular amount of time. At least, most people believe so. However, what an individual does not prioritise and spend their time doing today, for example activities, work tasks, physical exercise, time for rest and recuperation and sufficient sleep, they have to spend their time doing another day. If the pace of different activities affects time for rest and recuperation, this must be compensated later on. In the best case scenario we rest during a pause or break, a day off, a weekend or vacation, in the worst case scenario we are forced to rest due to illness caused by the lack of recuperation. The 24 hours of the day must contain sufficient time for activities and sufficient time for rest and recuperation. In the age of social media and the fast pace of most Western societies today, one must reflect on whether the brain and body get sufficient time for activity, and sufficient time for rest and recuperation.

Furthermore, it is important to reflect on the variation or different types of recuperation, as well as pace and time for recuperation (Figure 10). After being exposed to a stress reaction in a sedentary or low-paced activity, the time for recuperation requires high activity, for example physical exercise, in order to let stress hormones, increased levels of glucose and fatty acids in the blood stream, etc. to be consumed and enable the body to regain equilibrium or homeostasis. In a passive and monotonous work

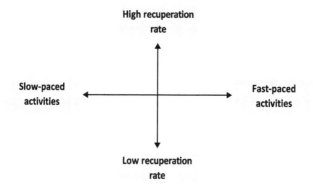

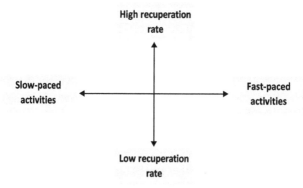

**FIGURE 10** Dual-axis model for reflecting on the need of time and type of recuperation after exertion through physical or mental/cognitive activities.

situation, time for recuperation requires a lot of creativity to stimulate our cognitive functions. However, if we spend our days doing cognitive work, with a lot of images, text and different impressions, a quiet nature experience, watching clouds in the sky, swaying trees, flowers in bloom and running water may be just what we need to recuperate. When our muscles are exerted due to physically demanding activities, we need to let our muscles and bodies rest with lower paced activities in order to recuperate, and to let the body repair damaged muscles, dispose of waste products and control the functions of the cells and inner organs. In other words, there is not one single way to recuperate, it entirely depends on the previous activity. Therefore, work pace and time for recuperation is based on the need for variation and for the body to achieve equilibrium or homeostasis between activities. However, equal time for physical activity and physical recuperation is important, just like for cognitive activity and cognitive recuperation. In other words, what we call recuperation is really the variation between activity and rest in strive to achieve homeostasis.

## WORKING HOURS, WORK PACE AND TIME FOR RECUPERATION RELATED TO AGE AND GENDER

Senior employees in general need more time to recuperate and to adjust their work pace in order to maintain health. Furthermore, individuals in general need longer time to recover and recuperate after an illness or injury, since their sleep tends to become more fragmented. Many studies show that senior employees appear to need more time to rest and recuperate in order to increase their possibilities of participating in working life, and that a slower work pace for ageing individuals can contribute to a longer working life [3,8,14,95,105,116,117]. Furthermore, studies show that the significance of working hours, in relation to when an individual leaves working life for retirement, is greater among women than men [5,8,9,94]. A larger amount of women express that inconvenient working hours have a negative effect on their family life.

It has been described as better for senior employees' well-being to gradually decrease the number of working hours, rather than to abruptly exit working life, since this enables them to make lifestyle changes with new routines and activities. Furthermore, enabling people to control the time and manner in which they retire facilitates them to plan their lifestyle changes and the new phase of retirement, which has positive effects on maintained mental and social well-being after retirement.

In an intervention program, employees had the possibility to decrease their working hours by 20 percent after 58 years of age, to provide more time for rest and recuperation between work shifts [105]. The evaluation of the measure showed that senior employees who participated in the program had a higher retirement age on average, a better experience of the work environment as well as decreased sickness absence. In the end, this resulted in higher productivity and profitability for the company.

However, studies have also showed that senior employees who can control their work pace and working hours do not need to decrease their number of working hours in order to experience good health and well-being, despite chronic health conditions [118].

## CONSIDER AND REFLECT

- Reflect on whether your time for recuperation is enough in relation to your time for activities.
- In what way do you best recuperate from your:
  - physical activities?
  - mental/cognitive activities?
- Reflect on whether the amount of work tasks and activities in your work are balanced with your number of working hours, in other words, if you have time to execute what is expected of you during a day at work when working in an appropriate pace.
- Reflect on the possibility to take breaks and pauses when needed in your work situation.
- Reflect on whether you have enough time to rest and recuperate between work shifts and work days.
- Reflect on whether your work pace and working hours can contribute to an increased risk of occupational injuries, occupational accidents or occupational illnesses.
- Reflect on whether your working hours and work pace affect your possibilities of participating in working life until an older age.
- Reflect on whether your work pace affects your recuperation and health.
- Reflect on whether your working hours and the duration of your work shifts affect your recuperation and health.
- Reflect on how and what you can do to experience well-being with your working hours, work pace and time for recuperation in the future.

# Organisational Perspective and Action Proposals that Matter to the Health Effects of the Work Environment

The employer and management are ultimately responsible and thereby very significant to the health effects of the work environment (Figure 4). Employers, management and managers who are knowledgeable and interested in the work environment, and who set time in the organisation aside for work environment issues provide better conditions for a healthy work environment. Furthermore, these conditions decrease the risk of occupational injuries for employees in the organisation.

## ORGANISATIONAL/MESO LEVEL AND HEALTH

Health and illness must be defined in order to enable the organisation of individuals' health in different administrative and bureaucratic systems on the meso level. Experts, authorities and organisations on the meso level have the power to define individuals' health and well-being. The individual's health is an important factor for the ability to participate in working life. Ill-health decreases work ability, however, a medical practitioner's diagnosis or assessment does not mean that the individual lacks work ability. Most individuals have health factors that can be strengthened and compensate for ill-health. At the same time, a diagnosis provides a legitimate possibility to decrease the number of working hours, work less, increase time for recuperation or leave working life.

Current assessments and efforts on the meso level are usually based on the biomedical or biostatistical perception of health. The biomedical or biostatistical health definition is objective. According to this definition, health is the lack of disease, this can be perceived to represent the classic biomedical perception of health. Christopher Boorse [119] has written one of the most well-known health definitions based on the biostatistical perception of health. Ill-health is perceived as the same thing as disease, a state that prevents normal functions in an individual's organ or system of organs. A disability or abnormality that decreases the ability to survive or reproduce is defined as *disease*. According to this definition of an individual's health and ill-health, a disease must be diagnosed by a medical practitioner or other authorised staff, by noting a deviation from a statistical standard value or normal distribution in the functions of the body and organs. An individual who displays values within the standard value or normal distribution is considered to be healthy. If the individual has been diagnosed with a disease by a medical practitioner, the individual can also suffer from *illness*, an unwanted internal state, experienced by the individual, which decreases physical or mental functions and motivates treatment. However, according to the biostatistical health definition, illness always presupposes an ascertained disease.

Critics express that the biomedical and biostatistical health definition causes individuals to be passive objects that are managed with minimal influence of their own, often without being asked or believed in regarding their symptoms or experiences, unless they have a diagnosis [23,120]. The biomedical and biostatistical health paradigm has received criticism based on physiological and anatomical fixation, where the discourse defines health as the individual's anatomical and physiological functions in relation to standard values. The individual as a patient risks being objectified as an organ or as their disease. If the individual has several ailments, the individual as a patient is sometimes divided into several anatomical objects, based on their symptoms when in contact with different experts or health care departments. For example, individuals in a care unit may be defined based on their illness, when health care staff refer to the individuals as "the appendix in room five" or as "the cervix cancer in the single room". Furthermore, the anatomical fixation is evident when an illness cannot be proven through reliable biomedical tests and contributes to scepticism towards symptom diagnoses. It is not uncommon that ill-health which cannot be proven as disease through physical examinations or deviations from the standard value in tests is rejected by health care professionals, authorities and insurance companies. This occurs despite the individual experiencing and expressing ill-health. The reason for this is that only disease, in other words, deviations from standard values, is regarded as ill-health in the biostatistical and biomedical perception of health. Therefore, efforts to cure, medical treatment, approving certificates of sickness absence and rehabilitation are mainly aimed towards ill-health that can prove disease through examinations and tests. The biomedical, anatomical and physiological fixation risk neglecting and overlooking the overall picture and causes of ill-health. If underlying causes are not taken into account, some treatments risk having short-term effects and will fail to eliminate the causes of illness and ill-health. When risks and causes of ill-health are not eliminated, for example in the work environment, the risk of recurring ill-health increases, just like the risk that others are exposed and risk developing the same ill-health. In recent years, an approach to health focusing on the individual has developed in the care, treatment and rehabilitation of health care, rather than disease-centred approaches to health.

## HEALTH PROMOTION AND HEALTH PREVENTION

According to WHO, health promotion is the process that enables people to increase the control of and improve their health. This requires individuals or groups to "identify and to realize aspirations, to satisfy needs, and to change or cope with the environment" [121]. Health promotion aims to maintain, contribute to or increase health. What can the individual do to promote their well-being and self-rated health, and to prevent ill-health and illness? WHO recommends physical activity to increase the individual's general physical and mental well-being, as well as to prevent many types of illness like cardiovascular disease, type 2 diabetes, cancer, anxiety, depression and arthritis [122]. Physical activity of an intensity that causes the individual to sweat for at least 150 minutes per week is recommended. All physical activity is better for health and well-being than none. Furthermore, the WHO recommends health-promoting food choices and a healthy diet, and healthy lifestyle choices to promote individuals' well-being and health.

Knowledge and insight are valuable resources for motivation and taking individual responsibility for one's own health. Furthermore, participation, empowerment and influence are significant to the process of health promotion. The individual who participates is valued in the process, is allowed to contribute with their own experiences and takes responsibility for their own actions in the changes that must take place.

The following policy instruments are used in health promotion and to promote individuals' health:

- *Enable healthy choices*, for example, by building bicycle paths to make more people ride bicycles rather than drive cars to work, thereby getting physical exercise and decreasing environmentally hazardous substances in the air
- *Complicate unhealthy choices*, for example, by introducing taxes on foods that contain high levels of fat and sugar, causing them to be more expensive
- *Coercive measures* through laws and regulations, for example, against harmful substances in food, tobacco control laws and seat belt legislation.

Health prevention concerns prevention, prohibition and protection against ill-health. Health promotion concerns enabling and encouraging public health and is separate from clinical health care which concerns prevention, treatment, care and rehabilitation of identified risk groups who run the risk of developing or who suffer from ill-health. Based on health promotion and the level of ill-health in question, prevention, in turn, is divided into *primary prevention* with measures to prevent individuals from developing ill-health or injuries, *secondary prevention* with measures to prevent individuals from becoming more ill or injured, and *tertiary prevention* to prevent individuals who have developed illness or injuries to suffer from sequelae caused by their illness or injury (Figure 11).

That individuals experience well-being, a sense of security and legitimacy in their context, contribute to positive experiences and create good conditions for health promotion. The discourse of health promotion points out the shortages of the biomedical and biostatistical perception of health. The discourse of health promotion emphasises the significance of strengthening health and healthy aspects rather than focusing on illness and deviations from the norm. It is focused on providing opportunities and *empowerment* to individuals to actively participate in the management, treatment and measures regarding their subjective experience of ill-health and functional diversity in their daily lives. Empowerment has different definitions in different contexts, however, the common thread is a focus on the individual's general autonomy regarding their own life, their health and the determinant areas that affect their health. A person-centred approach, which has a lot in common with empowerment, is used in health promotion and aims to identify the individual's unique situation and provide

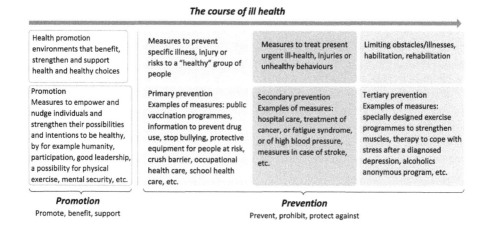

FIGURE 11  The course of ill-health related to the sequence of measure actions to promote health and different types of preventions to hinder ill-health and return to better health, i.e. health promotion, primary prevention, secondary prevention and tertiary prevention.

support for the individual's well-being, health, treatment and rehabilitation based on this.

Health promotion emphasises a holistic approach to health, much like Hippocrates' (460–370 B.C.) descriptions of how to create health and well-being among citizens. Hippocrates is considered to be the father of modern medicine, the fundamental idea in his writings is that medicine should be based on observations, experience and critical judgement. He connected the individual's lifestyle patterns with the influence of structures in society, in the workplace, in the environment, at home, etc. when he examined and remediated individuals' ill-health and lack of well-being.

The health promotion discourse is based on the same principles, however, it emphasises on strengthening healthy aspects and avoiding focusing only on individuals who already have developed illness. Instead, the aim is to prevent more people suffering from and being exposed to health hazards. The critique against the inclusion of people who are not ill in health promotion is based on the ambition to prevent ill-health, according to which it is paradoxical that all individuals, both ill and healthy, are considered potentially ill and at risk of developing ill-health [123]. This results in health becoming a temporary state that must be managed, through continuous preventive measures and self-discipline. Furthermore, critics describe a risk that the surroundings are ascribed to an excessive responsibility for individuals' health. For example, society, the workplace, environment and childhood conditions are responsible for and determine the individual's health, while the individual's own responsibility for their health and health choices is ignored. However, the discourse of health promotion states that it is the task of experts and society to provide knowledge and tools to strengthen individuals' responsibility and empowerment of their own health and to live a healthy life. Furthermore, experts and society must allow individuals to take responsibility for their own health. However, the fact that experts and society define what constitutes a health-promoting lifestyle poses an ethical dilemma [124].

Health promotion in general concerns the societal – macro – level, where many decisions regarding individuals' surroundings and opportunities in society are made. The spaces that largely constitute people's living conditions make up arenas for health promotion, for example child care, schools, workplaces and health care. Therefore, considering arenas for health promotion, a large part of health promotion and health prevention measures should be taken in these spaces and arenas. Work and working life are one of the most important contexts that affect individuals' opportunities for healthy and different life style choices for large parts of their lives, accordingly, the organisational – meso – level is one of the most important arenas of health promotion.

## HEALTH PROMOTION AND ORGANISATIONAL PRODUCTIVITY

The purpose of health promotion in working life is to promote and support employees' health. Furthermore, increased general well-being among employees results in increased competitiveness and organisational efficiency. The fundamental idea is that healthy aspects are strengthened by empowering individuals regarding their own life and health. Healthy individuals who experience well-being perform to higher standards in their work. Historically, bureaucratic organisations have existed for as long as we know. However, the development of bureaucratic organisational models has been stated to be an important condition for the demands of a fair division of resources and opportunities that constitutes democratisation in society and in workplaces. Bureaucratic organisations are authoritative hierarchies, where power is divided into different positions in the organisations, giving the power holder a particular position and rights. Furthermore, the division of labour states how labour and work tasks should be divided in the organisations, and coordination mechanisms with formal rules and approaches to the work flow [125]. Regarding employees as capable individuals, rather than passive and expendable units in the organisational bureaucracy, contributes to increased productivity and employees' well-being, this was first presented in the school of *scientific management* [126]. Unfortunately, this theory did not appear to work in reality, since many employees lost their scope for action in workplaces that introduced scientific management. Later on, researchers of working life and organisations in the school of *human relations* and the school of *human resource management* (HRM) proclaimed the need of considering employees as capable individuals and to focus on employees in recruitment, evaluation, rewards and development [127,128]. The development of HRM is based on experiments, perhaps mainly those executed by the Western Electric Company in Hawthorne, Chicago. The results of these experiments showed the significance of social and instrumental support, and that production increases when the management pays attention to their employees and makes them feel acknowledged and appreciated. This is called the Hawthorn effect. Later organisational theories of *post-bureaucratic organisations* state that individuals have greater possibilities of autonomy and self-monitoring through management principles like "freedom under responsibility" [129,130]. The post-bureaucratic organisation is, unlike the bureaucratic organisation, flexible, flat, non-hierarchical, dynamic and project-based, where employees to a greater extent monitor themselves. In these types of organisations, employees are described to have a more humane work environment owing to better opportunities to manage and adjust their work to their own ambitions and interests, and better flexibility with regard to their personal lives. Employers are expected to get more in return from their employees, since this approach is thought to increase work motivation, resulting in higher work performance. However, there are risks related to employee autonomy. The lack of boundaries in the work situation can make employees feel like their work is never done, and that work intrudes on their personal life. In a work situation with a lack of boundaries, the intensive needs of the organisation risk becoming predominant in the employees' lives. This results in difficulties for individuals to limit how much work they should take on, what is expected of them and when the work day is finished. This, in turn, causes feelings of insufficiency, stress and a constant bad conscience when working hours intrude on the employees' personal life, or if they feel they have not executed enough work tasks.

Enduring organisations build on the balance between individuals' demands and control. The employee, based on the holistic perspective of the post-bureaucratic organisation, is perceived as an individual with needs in and outside of the workplace. The manager, in their leadership and division of work tasks, responsibilities and resources, should not only regard the employee as a cog in the production machinery but also make sure that they have appropriate formal and tacit competence, physical and mental conditions and sufficient support to execute their work tasks. Furthermore, managers and management must regard how work and work-related decisions affect their employees' personal life, lifestyle, family and age-specific conditions, or work tasks and occupational challenges. Enduring organisations, with a holistic perspective of their employees, strive to create a family-like atmosphere when exercising power, rather than the approach of a formal management. They work to develop shared norms and values within the organisation, where employees are not only co-workers but also friends. Furthermore, the management actively contributes to the empowerment of employees and work units with sufficient possibilities and resources to manage their work tasks. An enduring organisation respects the need for work-life balance, and that employees' health and well-being are based on their possibility to set physical, mental and social boundaries between work and their personal life in order to recuperate.

The enduring, post-bureaucratic organisation appears to have a lot in common with the fundamental ideas of health promotion. However, applying a direct approach to health promotion in working life has been criticised, mainly since the employer risks crossing the boundaries of employees' personal integrity [120]. Furthermore, critics claim that measures of health promotion in the workplace and as a part of work can be a policy instrument and a way for the employer to control the employees' lives, lifestyles and habits. An employee who does not partake in the measures, or who develops illness or ill-health due to lifestyle choices like having a poor diet, smoking tobacco, drinking alcohol, using drugs or who lacks motivation, risks being blamed and called irresponsible. Furthermore, such a discourse includes discussions of whether the employer should pay for sickness absence or have a right to terminate the employment, in cases where an employee does not take care of their health, thereby causing costly sickness absence. According to this discussion, the employer appears to have a right to dispose of their employees' lifestyle, eliminating the individual employee's responsibility and decisions regarding their own body and health. Likewise, health promotion in working life is sometimes described as a way to make the individual employee responsible for their own health, and to avoid suffering occupational injuries. Through health promotion measures, for example wellness allowance, the employer shows a wish to promote employees' health and well-being, thereby morally acquitting themselves of responsibility for the risk of employees developing illness or ill-health, even if health hazards primarily are present in the work conditions and work environment.

Since the enduring post-bureaucratic organisation is based on the idea that the employees' interest and commitment should permeate work, that co-workers should be friends and that co-operation is expected to be informal, the boundaries between the employees' work and personal life dissolve. There is a risk that the boundaries between the employee's personal life and working life disappear, which presupposes the employee's self-discipline to make healthy choices and to set boundaries themselves. At the same time, in the enduring, post-bureaucratic organisation, the initiative of health promotion risks becoming a way to assert power. Increased individual empowerment can be a resource used for increased productivity. An employee who lives a better and happier life, with work-life balance, who makes healthy choices and who has a healthy lifestyle is beneficial to the employer. Bringing health promotion to its' head, the formal management of individuals' behaviour at work in the traditional bureaucratic organisation is replaced with exercising power over and controlling the individual's entire social identity, where they are pushed towards the expected responsibility of maintaining their health and development. The management's objectified perception of a good employee is spread through values that are expected to be shared, to become the employees' subjective perception of themselves. Accordingly, health promotion in working life entails an ethical dilemma, since it does not only regard the organisation of working life. Asserting power over and influencing employees, with the purpose of increased productivity, includes empowering the individual to manage and organise their own life in order to maintain their health and employability. However, health promotion and health prevention should primarily be perceived as a win win model for both employers and employees.

## THE HEALTH PREVENTION PROCESS: IDENTIFY – PRIORITISE – RECTIFY – REORIENT

The employee's physical work environment can pose risks to and have effects on bodily functions and organs, for example joints, muscles, hormone systems, respiratory organs, circulatory organs and sensory organs. Furthermore, the mental work environment can cause illness and ill-health. Diagnosed ill-health caused by issues in the mental work environment, an excessive work pace and lack of recuperation, has increased in later years. It is important to note that the deterioration of bodily functions, among individuals situated in a demanding work environment, has appeared to decrease when changes and measures are taken to improve the demanding work environment.

Since deterioration and ill-health can be prevented through changes in the work environment, the work environment must be managed, risks must be assessed and measures taken when needed, based on all nine determinant areas for a healthy and sustainable working life. Active and systematic work environment management does not only decrease the amount of occupational injuries, but it is also statistically significant to employees' working until an older age [10,80].

In health prevention management, it is important to consider the dissemination of information regarding risks and measures in the work environment, information must be clear and easily accessible for everyone, and perhaps even be available in different languages. Furthermore, it is not enough to implement an information campaign once, information and the dissemination of information must be repeated and continuously updated. An organisation with the reputation of a good work environment is a good way to facilitate competitiveness, innovation and development in order to recruit and keep staff and competence. Furthermore, it is morally and ethically correct to work with health promotion and health prevention in the work environment, as well as to follow legislation. It is important to troubleshoot and take measures when occupational injuries occur, however, suffering and financial losses can be avoided if risks are prevented and a healthy and sustainable work environment for all ages is created.

Taking measures to prevent negative effects of the work environment, in other words, the health prevention process, follows different steps. These steps are essentially the same as for other measures (for example read more in the Part III *Practical Application Based on the SwAge™ Model*):

- Identify and quantify issues
- Inform different parties
- Prioritise
- Propose measures
- Make a timetable and activity plan and decide who is responsible for a particular implementation
- Take measures
- Evaluate measures
- Start over, through reorientation and new examinations

In order to facilitate and make work more clear and understandable for everyone involved in the process, in other words, employees, managers, labour unions, occupational health care, etc., it is beneficial to base all processes in the work situation on the SwAge™ model. The model visualises, identifies and quantifies issues and their complexity and proposes appropriate measures in different areas and on different levels. Using the SwAge™ model as a tool and support in health prevention and health promotion management in the workplace increases the conditions to create good health and well-being and to prevent ill-health for everyone in the workplace. Please find more information on this in Part III *Practical Application Based on the SwAge™ Model*.

## MEASURES FOR PHYSICAL, MENTAL AND TEMPORAL WORK ENVIRONMENT ISSUES AND HEALTH

It is not bad luck that causes occupational injuries, occupational illnesses or occupational accidents. It is often caused by poor risk assessments in the work situation and work environment, safety routines and protective equipment. A healthy work environment comes with organisational and financial benefits since it decreases the costs of sickness absence, interim staff and rehabilitation, it also facilitates recruitment, productivity, etc. Using the different parts of the SwAge™ model as a checklist is a great tool for the work environment management and inventory, and to take measures to counteract issues related to the nine determinant areas for a sustainable working life, where employees are able and willing to work and remain employable regardless of age. Furthermore, the SwAge™ model is an excellent tool for active work environment management to comply with the EC Directive on the introduction of measures to encourage improvements in the safety and health of workers at work (89/391/EEC). Therefore, making a workplace analysis based on the SwAge™ model ensures both the inventory of the work environment and the active work environment management for organisations and companies. Take the following steps to conduct a workplace analysis based on the approach, tools and template of the SwAge™ model (read more in Part III *Practical Application Based on the SwAge™ Model*):

1. Examine the work environment, paying attention to the different determinant areas in the SwAge™ model, for example: the physical work environment with demanding work tasks, movements, climate, hazardous chemicals and substances, the mental work environment with stress, demand and control, threats and violence, working hours and work pace, the possibility of recuperation, etc. Make a list of factors with positive health effects and factors with negative health effects. Note whether these factors risk causing occupational illness or occupational accidents, and whether they risk affecting or being affected by biological ageing.
   - Reflect on, assess and make statements of the current situation with available data and studies, for example using surveys, observations, interviews, focus groups or other technology. The reflection can also be made in regular work unit or workplace meetings, or in an orientation of the work environment, when a manager studies a section and observes work processes in order to evaluate if there are obvious aspects of the work situation that can cause problems. For example, the execution of work tasks, the work pace or other conditions. This is most effective if combined with the employees' descriptions of their execution of work tasks. This approach entails reflection, both for employees and managers, regarding potential risk factors. Do not rely on *one* source of information to assess the risk in the workplace, regardless of the source. It is better to facilitate an overview of the situation by using data from several sources. List all sources of information and data used in the risk assessment.
2. Make a list of priorities regarding positive factors, acceptable and unacceptable risks in the workplace, as well as what measures should be taken and in what order. This list should be based on the examination and risk assessment of the work environment according to the different parts of the SwAge™ model.
   - When you have the results, ask employees and others if they agree that the results make up a plausible representation of the work situation.

Confirm your findings and examine what local implications they have. Reflect on and document possible underlying causes, even based on variations in the design of the workplace, work tasks, communication, age, functional diversity, etc. and depending on staffing levels at different times and in different situations.
- Make an assessment of the organisation's capacity for positive initiatives and management of the most significant causes of problems in the work environment and identify areas where the organisation appears to perform to lower standards. Acknowledge the fact that risks and problems may affect someone and make potential problem areas clear to employees and work units.
- Initiate and facilitate discussion and communication regarding results with and between employees and their representatives. Discuss and reflect on measures of health promotion and health prevention and try to find possible solutions with the help of a representative sample of staff. Facilitate and support proposals and agreements of what can be done to utilise positive conditions and to make improvements. Compare the organisation's current performance levels with standard performance levels. Identify positive routes to a healthy and sustainable workplace and appropriate measures to decrease and close the gap between the current situation and the goal of the organisation.

3. Make an action plan for measures to prevent risks and problems and to maintain good conditions. Make sure to inform everyone in the workplace of this action plan and make sure that the information is continuously clear, comprehensible, updated and easily accessible to everyone.
   - Set common goals for the mental work environment that the organisation can compare their progress and successful measures to. Pay attention to risks that can threaten this goal and the underlying causes of risks in order to address them. Find support in positive and healthy aspects of the organisation and strengthen these further by preventing excessive mental demands in the work environment. In other words, identify areas of improvement in the organisation and identify positive areas to strengthen and develop further.
   - Divide responsibility for implementing the action plan. Document who is responsible for what. Make a schedule for the implementation and for follow-up of the action plan.
   - Facilitate daily risk assessment of work-related demands in the organisation through monitoring the risk factors identified as the most important and frequent. However, do not forget to evaluate and reassess risk factors continuously.

4. Follow up and assess the results of the measures. Acknowledge and celebrate improvements! Make an adjusted action plan, schedule and division of responsibility for measures that have not achieved satisfying results.

5. Start over with step 1 above and examine risks in the work environment and the workplace once again, this is a continuous task. What is the work

# Organisational Perspective and Action Proposals

situation and work environment like now? Have initiatives of health promotion been maintained and strengthened? Have previous risks been rectified? Have new risk factors occurred?

In order for the effect of health promotion, health prevention and risk prevention to be good and lasting, employers and employees must cooperate. Reflecting on and paying attention to the different determinant areas of the SwAge™ model should be a natural part of the daily operations in the organisation, career development discussions, workplace meetings and systematic work environment management.

## THE DESIGN OF THE PHYSICAL WORK ENVIRONMENT

Physically demanding work tasks put a strain on individuals' bodies and physiques. Consequently, this increases the level of biological and functional ageing. Because of this, the design of the physical workplace is an important part of a healthy physical work environment. For example, the design of the physical workplace determines the postures in which work tasks can be executed. The design of the workplace should, to the greatest extent possible, be based on and adjusted to employees' unique body size and functional diversity. Work tasks which include operating vibrating tools and machinery pose a health hazard to the nervous system and should be limited based on occupational exposure limit values. Furthermore, a rule of thumb is to work with as small a physical workload as possible, carrying or lifting it as comfortably and close to the body as possible in order to avoid working with legs and arms strained and stretched from the body, or with the neck bent back, in order to decrease the risk of negative impacts on muscles, joints and skeleton. Moreover, technological tools and protective equipment should be adjusted to the employee's unique body size and functional diversity and designed in the best possible way according to this principle.

In order to come to terms with excessively physically demanding workloads and work postures, movements and lifting in working life that cause ill health, every work procedure and body part exposed to the workload must be examined regarding:

- How large the workload is (in kilos, degrees and work height)
- How long the duration of the workload is, in minutes at a time
- How often the workload is repeated, during a month, week, day, hour and minute

In order to execute work tasks in a manner that promotes the most possible health for the mobility of the body, take the following items into account:

- The execution of work tasks should be comfortable and not strain the body parts
- Try to execute work tasks as close to your body as possible and in neutral body postures, in other words, avoid stretching body parts in awkward positions

- Establish rotation and change body postures through varied work tasks and workloads, in other words, avoid locked positions with prolonged or unilateral, repetitive motions
- Take short breaks, stretch your body and change work postures at least every 20 minutes
- Pay attention to and avoid ergonomic improvements that contribute to new problems, for example if technical solutions to avoid heavy lifting cause the employee to become very physically inactive and sedentary, with nothing more than little repetitive motions such as manoeuvring a joystick to control lifting devices. Another example is when measures and solutions to decrease a workload affecting the knees in a work task result in an increased workload affecting the upper body instead, such as the introduction of robotic milking machines in cow barns.

*Climate* is an important part of the physical work environment. Cold, heat, air flow and cold or hot surfaces can be demanding and risk causing ill-health and impacting biological ageing. Therefore, the workplace, work tasks and work postures should be designed in to decrease the employees' risk of exposure. Appropriate work clothes and work shoes should be worn. Shelter from wind and sun should be available to decrease the risk of exposure.

As previously mentioned, dust, chemicals, biological and hazardous substances and radiation that the employee handles or is exposed to are important factors in the physical work environment. This became especially apparent during the COVID-19 pandemic, when, for example, hygiene routines and protective equipment were hot topics when discussing the risks of occupational injury and occupational illness among health care professionals, employees in elderly care, schools, restaurants and in other workplaces with work tasks that potentially could entail risks of contagion and spreading the infection [131].

Workplaces must display information, routines and protective equipment for secure handling and storage of hazardous substances. It is important to continuously examine the hazardous substances used in the organisation. Perhaps, there are new occupational exposure limit values for a substance, or perhaps new hazardous substances have entered the workplace. Perhaps, new routines or products from new manufacturers have contributed to different exposure levels. Perhaps, other, alternative materials and substances that are less hazardous but equally effective are available, and the organisation may be able to use these in order to decrease risks of hazardous exposure. Perhaps, there are even better routines, protective equipment and techniques for handling hazardous substances. Both employers and employees are responsible for safety and the protective equipment being used and updated continuously. The use of protective equipment should be a natural part of the daily routine. Furthermore, other measures to decrease the risk of being subjected to occupational injury and illness should be taken, for example routines before eating food. In order to decrease the exposure to hazardous substances, it is important to wash hands and change from contaminated work clothes before eating or drinking.

## THE DESIGN OF THE MENTAL WORK ENVIRONMENT

A mental work environment should support healthy mental well-being and not decrease the possibility of good mental health. Research supports the fact that people who are situated in and experience a pressured work situation develop symptoms of depression and fatigue to a greater extent [68]. An excessive mental workload and chronic stress often result in the experience of decreased well-being, dejection or anxiety and the individual limiting themselves socially and performing to lower standards than their full potential. Therefore, neither society, organisations nor individuals should accept the increase and normalisation of extensive mental health conditions caused by work environments. Preventive measures can decrease both the risk of developing illness and the risk of high sickness absence or early retirement, which are the individual's ways of leaving a problematic situation and avoiding a mentally demanding work environment. Reducing the stigma of mental health conditions among employees is an important measure to decrease mental demands. Information, mutual reflections and disseminated knowledge of mental health conditions and their causes enable individuals to avoid being stigmatised or bullied in the workplace. Furthermore, information and dissemination of knowledge related to causes and measures of mental health conditions are important to avoid employees feeling that they must hide their health issues in order not to risk being subjected or marginalised due to their functional diversity, health issues or need of support.

Acute and chronic stress caused by the work environment can be prevented and managed through secure, social alliances in the workplace and through social support. It has been suggested that it is more common for women than men to seek social support and social alliances in order to cope with an unhealthy mental work environment. Men and women traditionally, and to this day, mainly work in different professions and sections. Furthermore, a larger amount of women compared to men are diagnosed and on long-term sick leave with diagnoses caused by a mentally demanding work environment. However, women and men who share work environments tend to develop similar types of illness. Therefore, a possible interpretation is that the organisation of work in female-dominated occupations contributes to these differences, rather than gender. Individuals who, due to excessive demands in the mental work environment, suffer from fatigue, depression and anxiety often need a long time to recover, despite treatment and irrespective of gender. A good way for managers to examine the employees' or work unit's experience of mental demands is through informal discussions in the workplace. Do individuals appear to be unhappy for a lasting time period or do they not perform to their usual standards? Ask employees if they experience problems in an informal manner. In order to prevent, avoid and discover early stages of fatigue, it is important to take appropriate measures in time. This can be done by developing a strategy to assess risks in the mental work environment (read more in Part III *Practical Application Based on the SwAge™ Model*, Section *Measures for Physical, Mental and Temporal Work Environment Issues and Health* as well as the different areas of the SwAge™ model in general).

The following six areas are regarded as especially important when assessing risks related to the mental demands in the workplace:

- *Demands on the employee* – including questions of work demands, work pace and work content, other peoples' opinions of the employee's work, the ability to execute work tasks with pride and with products that the employee experiences as satisfactory.
- *Control of the work situation* – including the employee's experience of and possibility to manage their working hours, work pace, pauses, the execution and order of work tasks and coping with frustration and risk of exposure, since the employee relies on other people in the organisation to execute their work tasks correctly on time
- *Clear roles* – employees must understand their roles and places in the organisation and not have contradictory work tasks
- *Relationships* – which include promoting positive cooperation to avoid conflicts and counteract disregard and unacceptable behaviours
- *Changes* – organisations constantly face small and large changes, however, these must be managed and communicated in a way that makes employees experience a sense of security
- *Work support* – which includes encouragement, aids and sufficient finances to manage necessary costs and guarantee the survival of the organisation, which contributes to a sense of security for both managers and employees

## Coping with Stress

Learning how to cope with stressors, for example through consciously encouraging, strengthening and stabilising a desired behaviour or sequence of events that promote health, makes it easier for the individual to use this strategy in the future. This is described in *social learning theory* (SLT). SLT points out that the more clearly a situation is divided into what is caused by "luck" and what is caused by the individual's own efforts to achieve a goal, expectations of fate affect the individual's own behaviour and capacity to act to a lower degree. Furthermore, according to SLT, the belief in the need for control causes an experience of control and belief in oneself, in other words, inner locus of control, which contributes to a self-fulfilling prophecy of success or failure to achieve one's goals.

Feeling wound up by too many impressions of imagined or experienced stressors has negative effects on long-term health. It is possible to relax using conscious coping strategies and exercises, with the purpose of coping with stress reactions, in order to wind down after reactions and enable the individual to proceed through the four phases of overcoming a crisis, rather than getting stuck in a lasting shock or reaction phase (read more in the Section *Crisis and Development*). Different coping strategies can calm the sympathetic nervous system by affecting the cognitive processes when assessing a stressor. Furthermore, it is possible to consciously activate the parasympathetic nervous system in order for the individual to unwind and recuperate. Through consciously breathing slower, activating diaphragmatic breathing, turning the palms of the hands upwards, relaxing muscles, especially in the neck and shoulders, and repeating phrases like "I am calm, safe and relaxed", we can influence our brains to assess that there are no threats in the situation, but rather that we actually are calm and safe. Furthermore, a moderate level of physical activity promotes the

Organisational Perspective and Action Proposals 79

activities of the parasympathetic nervous system, for example spending time with family, friends and animals we like and do not feel threatened by, sex and going for walks. Exercising conscious presence and noticing how the surroundings affect and trigger different senses can also be a way to learn how to master the impressions and stimuli that trigger different reactions in the body. Furthermore, observing one's surroundings in peace and quiet with an open mind, while thinking of things one feels grateful and happy for, can have a calming effect. Moreover, individuals appear to cope better with acute and chronic stress through actively seeking social support and building social alliances.

### Measures to Prevent the Risk of Threats and Violence

In working life, an employee can experience that they are threatened, exposed to risky situations, being followed, sabotaged as well as violence with or without weapons. Furthermore, employees' close relatives can be subjected to violence. The reason is usually that the attacker experiences a conflict with the employee's employer, in other words, with an authority or organisation that the employee represents. Therefore, the threats and violence are not necessarily connected to the individual employee and can be very difficult to predict. Examples of professions and organisations that run a higher risk of being subjected to threats and violence are politicians, religious communities, the judiciary, managers, social services, the police force, schools, health care, customer services, receptions, authorities, airports, department stores, refugee camps, etc. Besides the occupation or organisation the individual represents, risks of being subjected to threats and violence can be caused by:

- Working nights or evenings
- Working alone
- Lacking knowledge and experience of one's responsibilities, profession and work tasks
- Lacking knowledge of treatment and conflict management
- Stress, lack of time and excessive work demands
- Work tasks and transports taking place in areas or places with high crime rates

The employer has a great responsibility for safety routines, education, preventive measures, action plans and taking care of individuals who have been subjected to threats and violence. Workplaces should have routines and plans that acknowledge risk scenarios in order to eliminate danger for employees.

## ORGANISATION OF WORK TO FACILITATE RECUPERATION

Recuperation is important in order to remain healthy, decrease the risk of occupational injuries and for healthy biological ageing. It is important to follow working time regulations to facilitate recuperation and decrease the risk of occupational accidents caused by fatigue and tiredness. Monotonous and repetitive work tasks, noise, bad lighting and ventilation as well as high temperatures affect tiredness and alertness.

A possibility is to take measures that contribute to variation in work tasks, work pace and the physical work environment in the workplace in order to prevent employees' tiredness at work. A sense of community in the workplace and relationships between co-workers, where employees can laugh together and discuss non work–related topics, contribute to wakefulness. Furthermore, control of and decision latitude in one's work situation with the ability to manage when a work task should be executed, and to prioritise and finish one work task before starting the next, contributes to the possibility of taking breaks to recuperate and decrease tiredness. Some longer breaks are needed during a day at work, in order to recuperate, re-energise and remain awake and alert. Furthermore, it is important to take micro breaks in the work situation. One should not work intensively for more than 20 minutes in a row, after 20 minutes a micro break is needed. For example, stretching the body or doing a small exercise every 20 minutes, or after finishing one work task and before beginning another. Take a couple of deep breaths every once in a while to increase the oxygen supply. Furthermore, stretching the neck, shoulders and back contributes to a small break in the work situation. Drinking water and exchanging a few words with co-workers regarding other topics than work tasks also contribute to a break in the work situation, and to coping with work.

Breathing fresh, cold air or listening to music or a podcast has a short effect, up to a couple of minutes, if the individual experiences tiredness. Antidotes to tiredness are short naps and scheduling one's sleep. For professional drivers, signs warning of dangerous roads and the use of rumble strips are important measures to prevent inadequate attention and warn of risks. Some vehicles and machinery have equipment that warns against tiredness. However, these pose a potential problem in the work environment if they are used to stretch boundaries, for example if the driver does not take a break until the equipment alerts them of their tiredness. An important, though sensitive, question is to identify individuals with chronically impaired capacity for wakefulness and regulations to decrease risks in the work environment related to fatigue and sleepiness [115].

## Working Time Models to Facilitate an Extended Working Life

Senior employees who have the possibility of managing and controlling their own working hours choose to remain working to a greater extent, despite chronic health issues and musculoskeletal tears. Measures like the possibility of taking breaks when needed, a lower work pace, shorter work shifts and an additional day off per week increase the possibility of working until an older age. Different working time models and adjustments of working hours and work pace have been introduced at governmental, municipal and private workplaces in order to adjust working time and work pace to senior employees' different needs. For example, a working time model called 80-90-100 has proved to increase labour force participation among senior employees [105]. The concept of 80-90-100 means that an employee works 80 percent of their previous working hours, receive 90 percent of their previous salary, with provision for occupational pension corresponding to 100 percent (full-time) employment. The concept was feared to increase salary costs for the employer and to mainly be beneficial to the employee. However, evaluations of the concept have

# Organisational Perspective and Action Proposals

described it to be beneficial for all parties. The evaluations show that the employers have profited through decreased sickness absence, better conditions for generational change and competence transfer and a strengthened brand for the employer likely to result in lower costs for recruitment. For example, the Swedish energy company Vattenfall has, since 2005, offered their employees to participate in the 80-90-100 concept when they turn 58 years of age. The first evaluation of the intervention showed that approximately 20 percent of employees had participated in the concept in the first 7 years and that the average retirement age in the company had increased, from 60 to 63.5 years of age [105]. Furthermore, evaluations of the measure showed that older workers who participated in the program had a higher average retirement age, better experiences in the work environment and decreased sickness absence. However, it has been difficult to calculate profits in concrete numbers, like for most preventive measures, since it is impossible to know what the situation would be like without the preventive measures. Nonetheless, the measure contributed to increased productivity and profitability for Vattenfall. The company had a secured competence supply and stated that, despite increased salary costs, the concept was profitable since it contributed to decreased sickness absence among senior employees and postponed exits from working life, which had facilitated generational change. The 80-90-100 concept contributed to more favourable employment for participating employees. Participants had more time for recuperation and leisure activities, which gave them more energy to work while in the workplace. The employees' well-being and satisfaction in their work situation increased. The increased time for recuperation, financial incentives, the social significance of work and the experience that the company valued their competence remaining in the organisation were meaningful for the employees' decision to stay in working life for a longer time. Employees had better conditions to cope and be willing to work until an older age, which generally increases the possibility of higher pension benefits when it is time to leave working life. However, working a lesser amount of hours simultaneously decreases other social welfare systems that are based on the actual salary accordingly, for example sick pay and public pension.

## OCCUPATIONAL HEALTH CARE

Occupational health care is a strategic party that can support and facilitate the employer's task to make employees experience well-being and health. For example, the task of occupational health care is to propose measures, both to prevent ill-health in the workplace and to facilitate a faster return to work when illness and injuries occur.

Occupational health care should use different approaches in their support of organisations, companies, employees and the processes of health promotion in the workplace:

- Take all nine determinant areas of the SwAge™ model into account to contribute to a healthy and sustainable working life, where employees are able to and want to work in all ages, in other words, consider different age perspectives in their support of employers and employees.

- Disseminate information regarding directives, occupational exposure limit values and recommendations that should be respected in the work environment and work situation.
- Be supportive of and help individuals to find the courage to express issues and risks in the work situation to the management and give their view of the cause of and possible solutions to the problems.
- Be confrontational and point out different interpretations and perceptions of work procedures, to make the individual better understand issues in the work environment and work situation. Thereby facilitating individuals' ability to find solutions to problems in the workplace themselves.
- Support individuals to deal with their situation and find better information regarding work-related issues and risks in their work situation. Through this, employees can better understand the causality between problems and solutions and become more focused on measures and solutions regarding their work environment.
- Educate and disseminate knowledge to individuals regarding risks of ill-health and injuries and measures for health promotion based on theories and empirical knowledge. This approach will make the individual more conscious of risks in their work environment and work situation, able to analyse their current and new situation and have better possibilities to cope with their problems.
- Regular physical examinations can contribute to discovering work-related ill-health in the early stages and prevent ill-health among employees. Regular physical examinations can increase the possibility of working until an older age, especially for individuals in professions with high safety risks, among ageing employees and for employees in work situations characterised by high physical and mental demands.
- Physical activity and physical exercise often contribute to better abilities to cope with work, both physically and mentally.
- There are differences, for example whether individuals are genetically predisposed to be able to cope with demands or not. Regardless of the reason, it is important not to blame employees for poor or decreasing work ability.

## CONSIDER AND REFLECT

- What significance does health have for the ability to work?
- What significance does an individual's health have to the organisation where they are employed?
- Reflect on whether health has an equal or different significance to the organisation depending on the profession, position or age of the employee.
- Is there a well-functioning systematic work environment management in your workplace? Consider using the SwAge™ model as a checklist to cover all nine determinant areas for a healthy and sustainable working life (read more in the *Systematic workplace analysis and action plan for a sustainable working life for all ages – the SwAge™ model*, in Part III *Practical Application Based on the SwAge™ Model*, pp. 346–350).

# Organisational Perspective and Action Proposals

- Are work tasks and the work environment well-adjusted and take the employee's functional diversity, biological and physiological impacts of age, diagnoses and general health into account?
- Are there well-functioning career development discussions in your workplace? Consider using the SwAge™ model as a checklist to cover all nine determinant areas for a healthy and sustainable working life (read more in the SwAge™ model dialogue tool in Part III *Practical Application Based on the SwAge™ Model*, pp. 346–350).
- Does the organisational culture contain and promote the use of ergonomic aids in order to prevent occupational injuries in your workplace?
- How carefully are occupational exposure limit values followed in the work environment, to ensure that employees are not overexposed and risk suffering from occupational injuries?
- Use and comply with work environment laws in your country. For example, member countries of the EU should comply with the EC Directive on the introduction of measures to encourage improvements in the safety and health of workers at work (89/391/EEC).
- Are there rotation, variation and change of the employees' work tasks in order to decrease risks of unilateral movements and static strain in your workplace?
- Are there measures to prevent negative stress reactions caused by an excessively high work pace in your workplace?
- Are there measures to prevent negative stress reactions caused by too many work tasks and fragmented work efforts during a work shift in your workplace?
- Do employees have sufficient possibilities, resources and time to execute their work tasks in a way that is satisfactory and of sufficient quality to them at the end of the work day, in order to prevent ethical stress in the work situation (in other words, when the employee, due to organisational obstacles, experience an inability to perform to a standard they feel satisfied with and proud of)?
- Are there measures and support to balance the demands on the employee with the control of their own work situation?
- Are there measures to minimise the risk of threats and violence in the work environment?
- Are the working hours and work schedule well-functioning and do they facilitate time for recuperation?
- Is the work pace reasonable in relation to the physical demands, as well as the conditions and ages of individual employees?
- Is the work pace reasonable in relation to the mental demands, as well as the conditions and ages of individual employees?
- Are there possibilities to take breaks (to, for example, eat) during working hours, adjusted to the demands of the work situation, individual needs (for example health and age) and are they proportional in relation to the duration of the work shift?
- Does the workplace promote employees' opportunities for physical activity and exercise?

- Does the workplace promote employees' possibilities of a healthy diet and nutrient intake during a work shift?
- Do employees have access to occupational health care, as a support to prevent ill-health, to manage age-related risks in the work environment and for measures of health promotion and health prevention in the workplace?
- Reflect on whether the measures in the organisation contribute to employees' employability based on their health, physical work environment, mental work environment, working hours, work pace and possibility of recuperation in their work situation.

# Societal Perspective and Action Proposals that Matter to Health Impacts of the Work Environment

The societal level constitutes the regulatory framework for the health effects of the work environment, for a sustainable working life, as well as what is allowed or not allowed and what measures should be taken on the organisational- and individual levels (Figure 4).

## SOCIETAL/MACRO LEVEL AND HEALTH

On the societal level, also known as the macro level, the definition of health is based on citizens not deviating from the norm in society. The definition is based on society's objective attitudes towards individual health, functional diversity and lifestyle. The purpose is for all citizens to experience well-being, and, based on their own ability, contribute to society and minimise the strain on welfare systems. Laws, rules, strategies and regulations are used to exercise power and strategy in order to promote health and prevent risks of ill-health among citizens.

Internationally, the most well-known definition of health, based on a societal perspective, is the WHO constitution of 1948: "Health is a state of complete physical, mental and social well-being and not merely the absence of disease or infirmity" [132]. The definition is connected to the UN universal declaration of human rights. The way the WHO defines health has developed in subsequent years. Since 1948, the WHO recommends that health should be perceived in relation to the rate at which individuals and groups in society are capable of realising their goals and satisfying their needs, as well as coping with or changing their surroundings. In this developed definition, health, based on a societal perspective, is perceived as a resource in daily life and emphasises social and personal resources, as well as the capacity of the individual.

How is health supposed to be supported and promoted in society? The *Ottawa Charter for Health Promotion* by the WHO defines health promotion as "the process of enabling people to increase control over, and to improve, their health" [122]. Health and well-being is perceived as a dynamic concept that relates to the situation and conditions of the individual's life. The Ottawa charter is a milestone for the macro perspective on health and for the dominating paradigm of health and disease. The charter contains some criticism of the biomedical and biostatistical perception of health, and the medical discourse of health, that have had a great influence on

Western societies from the end of the 19th century up until today. The discourse of health promotion began as a critique towards only using diagnoses as definitions of what constitutes health and well-being. Furthermore, the WHO promotes environments that support health, in other words, that society should create environments that support the occurrence and maintenance of health and well-being in the population.

In all eras, society has been dependent on citizens' health and health development, since these factors are fundamental to the development and growth of society, for example through professions like workers and soldiers. During the 20th century, many countries were worried about their citizens' quality and the development of health on a societal level. Problems with overcrowding, low housing standards and alcohol consumption were emphasised, and these factors contributed to bad environments for children and risks of physical and mental injuries. These ideas were, to a large extent, the foundation of social engineering with the purpose of increasing health and welfare in the population. The question of nature versus nurture was debated. It was advocated that children and elderly people should be placed in some sort of institution with educated staff, while men and women of working age should participate in the official labour force and be active in working life. The idea was that this would provide economic advantages to the society and citizens, and educational advantages for children.

The sphere for action *health effects of the work environment* in the SwAge™ model includes the determinant areas of health, physical work environment, mental work environment and working hours, work pace and time for recuperation. These are some parts of the work situation that contribute to different conditions for the individual to make choices that promote health. Therefore, WHO has included lifestyle-related illness related to occupational illness since 1985. Important functions in society are to create supportive environments to facilitate healthy choices, and to take supportive measures for particularly vulnerable groups, who risk being subjected to illness and injury in working life. Furthermore, society should support groups who need adjusted measures to participate in working life based on their functional diversity or who are to return to working life after an injury or illness. Therefore, society should promote supportive measures that enable organisations to create work environments that promote health. The nine determinant areas of the SwAge™ model can be used as a tool for reflection and the development of these preventive and adjusted measures on a societal level, in order to promote organisations' and individuals' healthy choices (read more in Part III *Practical Application Based on the SwAge™ Model*).

## THE SIGNIFICANCE OF SOCIETY FOR HEALTHY CHOICES

In order to strengthen the individual's possibility of good physical and mental health, it is important to promote attitudes that facilitate individuals to make choices that promote health in society and in work places. Healthy choices are facilitated through *nudging*, a mental push in a particular direction that makes a person choose one option over another [133]. Employee dining rooms with subsidised lunch, wellness allowance or the possibility of physical exercise in the workplace, and prohibitions against smoking in or in the vicinity of the workplace are some activities that increase the conditions for and promote a healthy diet, physical exercise and healthy choices for employees. Society has an important role to play in the design of a sustainable life and working life for employees. Supportive environments are important in the creation of a sustainable working life, where employees have the opportunity and are motivated to maintain their physical and mental capacity and health. For example, research has shown that people who work in a physically and mentally demanding work situation but do not get any physical exercise outside work suffer from lifestyle diseases to a larger extent and generally have a more unhealthy lifestyle compared to those who participate in some sort of physical activity [134]. However, appropriate conditions and at times support for the ability and willingness to prioritise physical activity are required. Even if most of us know that physical exercise and a lifestyle that promote health are good for us, it can feel challenging to do a physical exercise rather than sit down on the couch after a day at work. Even if many people understand dietary advice that contributes to the maintenance and energy of the body, we do not always follow them. Furthermore, even if one knows that smoking tobacco, drug use, etc. are unhealthy, some people still choose to do this at certain times or more often. Conditions to make healthy choices are created and supported by attitudes in society and by the cultural surroundings the individual is situated in, both in their leisure time and in the workplace.

## THE SIGNIFICANCE OF SOCIETY FOR THE KNOWLEDGE OF AND COMPLIANCE WITH LAWS AND DIRECTIVES

How a work environment should be designed to avoid employees being exposed to risks of occupational injuries, occupational accidents and occupational illnesses is regulated through different types of laws which differ between different countries. The Luxembourg Declaration on Workplace Health Promotion in the European Union (November 1997) describes workplaces characterised by health promotion as a joint venture of society, employers and employees in order to improve health and well-being for people in their working life. The Luxembourg Declaration is based on EU legislation and Directive 89/39/EEC by the Council of the European Union. Furthermore, work environment laws in member countries of the EU are based on this directive.

Legislation, regulations, general advice and disseminating knowledge of these are some of the most important tools in society to contribute to health prevention in work environments. The employer should draw the employees' attention to several regulations. This knowledge should be easily accessible and continuously disseminated to all employees. The information may need not only to be simplified and communicated in different ways to reach different groups and to reach a constant flow of new employees but also to remind all employees of applicable conditions, alert employees if new information has been added or changed or if the information is no longer applicable. Furthermore, the dissemination of knowledge regarding regulations should also contain easily accessible information about the fact that different groups may face different risks. For example, information to draw attention to generally increased risks of occupational injuries, occupational accidents or occupational illness with increasing age, since senses like sight, hearing and reaction ability deteriorate. No one should be subjected to threats and violence in the work situation, however, senior employees may experience themselves as particularly exposed due to their decreased physical ability and muscle strength. Furthermore, senior or very young individuals, and people with different types of chronic illness, can be more sensitive to different kinds of unhealthy exposure, for example to chemicals in the work environment, which should be noted when making decisions regarding occupational exposure limit values in the work environment. The design of the physical work environment is significant to biological ageing. A physically demanding work environment, with strain and demanding or poorly designed work conditions, causes tears, accidents and occupational injuries.

It is important that society contributes to long-term health prevention to prevent or decrease hazardous exposure if working life is to be sustainable for all age groups. Informational support and inspections from authorities and society are required in order to achieve this. Continuous updates and information campaigns are needed in order not to forget exposure prevention in the daily operations in the work place, and to update and provide employees with new information. Information must be conveyed in a language that is comprehensible to the recipient, and on an accessible level in order for both new and experienced employers, managers and employees to remain updated. For example, society and authorities must disseminate information regarding occupational exposure limit values and working time directives in a comprehensible way. Examples of information from society and authorities, that should be accessible and comprehensible, are occupational exposure limit values and working time regulations. Furthermore, compliance with laws and directives should be supervised continuously by the society and authorities and managed to decrease non-compliances. In principle, all employers want the best for their employees. However, employers, managers and employees might find it difficult to know where and how to begin their risk assessment of the work environment and work situation. It's easy to become blind to flaws at home and not realise the risks and issues present in the workplace and work situations. Once again, the nine determinant areas of the SwAge™ model can be a helpful tool and used as a checklist.

## SOCIETAL PERSPECTIVES ON FRAGMENTATION, DEMANDS AND CONTROL IN THE WORK SITUATION

Stress and high mental demands in the strive for perfection, combined with expectations to perform to the highest standards in all aspects of life and in all ages, are general problems in society that must be taken into account in working life, but also in personal life. Research shows that people employed in professions with work situations characterised by high mental demands and stress leave working life early or want to decrease their work engagement. Furthermore, studies show that a stressful and mentally demanding work environment is a frequent reason for individuals leaving working life early [3,5,6,8,10,135]. Sickness absence and retirement become socially acceptable ways to leave work conditions characterised by excessive mental demands, or because people simply cannot cope any longer after many years of high activity and lack of recuperation. For example, individuals employed in contact-based professions state that they neither can nor want to work until an older age to an extent equivalent to individuals employed in professions who are not exposed to as much contact with people in need [136]. Furthermore, almost every third senior manager with a stressful and exposed position in working life tends to retire early to decrease their work engagement and to have time for recuperation. Many people who have long-term sickness absence in their 30s due to stress reactions affecting their body with symptoms like fatigue and memory issues worry about how they should be able to cope with working until their 70s. Because of this, society must contribute to changed attitudes, with lower demands on individuals, to make it socially acceptable not to perform to the highest standards in all aspects of life all the time. Many people must slow down for some years and in certain aspects of life, perhaps not working full time in order to avoid suffering from mental health conditions and stress reactions. This is only natural considering the entire life course of an individual and should not be perceived as deviant or less desired in society. Because of this, it is important to create opportunities for people who need to decrease or increase their work engagement based on their life phase, without risking their continued labour force participation.

## SOCIETAL SUPPORT IN CHOICES AND CHANGES OF PROFESSION AND WORK ENVIRONMENT

Some individuals may want and be attracted to a certain profession or work situation but do not have the required biological, physical or mental prerequisites, based on their health, functional diversity or innate conditions. In these cases, guidance counsellors have an important role in mapping not only interest but also prerequisites for sustainable occupational choices. Tools and support can be helpful to create adjusted environments and to improve conditions for everyone, but all aspects may not be possible to adjust. Because of this, guidance counsellors also have an important task in supporting people who are about to choose an occupation or career to consider this, based on their individual mental and physical prerequisites. For example, individuals with bronchial diseases may not be suited to choose a profession where they are exposed to air-borne exposure that may risk increasing their health problems and cause occupational disease. Individuals who experience themselves as sensitive to stress should perhaps be aware of professions that contain increased risks of negative stress in the daily work situation, which can worsen their problems and cause occupational injury. Furthermore, problems can occur later in life and changed prerequisites can result in a need to change professions and work environments. Because of this, increased opportunities for the individual to invest in their career and profession later in life are important. Perhaps, after the demanding years of caring for small children, or because experience and realisations of one's own mental and physical capacity and the "right" or new career choice has become apparent after 50 years of age. In these cases, guidance counsellors have an important task to support competence development and career change, even for people aged 45 years and above, as well as to find alternative routes back to continued sustainable working life for people who do not participate in the labour force.

A healthy mental work environment is characterised by the ability to cope with demands and experience control of one's work, the ability to discuss the execution of work tasks and experience a sense of coherence in the work situation. This is currently affected by working life partly appearing to move towards the increased division of labour, where work becomes more differentiated and divided into different institutions and specialist departments that are responsible for different parts and functions, such as financial departments, HR departments, administration and production. However, the trend moves towards flatter organisations of work, where every employee in a department is responsible for supportive functions to a greater extent than previously. Consequently, every employee is responsible for supportive functions to a greater extent than before, in order to execute their work tasks. Examples of supporting functions can be making sure the printer works, installing updates and new programs on computers, ordering snacks for meetings, emptying the dishwasher and making coffee. In such a division of work, it is important to consider the time it takes to switch from one thought and work task to another, what is known as set-shifting. Because of this, a work situation containing many fragmented work tasks requires more time for set-shifting to be considered and included in the working hours. One option is for society to promote an organisation of work where a work shift is not as fragmented so that the employee can focus on the core tasks of their work for a longer time.

While the work situation has developed to an increased shared division of work and flatter organisations, it has become more difficult to rehabilitate people who cannot return to their previous work tasks, to find entry-level jobs on the labour market and bridge employment for people who want to decrease their work engagement as they approach the age of retirement. Because of this, the number of work tasks related to supportive functions must be included in the entire planning of employment in order to show the correct number of work tasks to be executed in the division of work. When an employee's working hours are fully packed, it is not possible to add a new work task unless another one is removed. To use a machine as a metaphor, excessively high demands, with a lack of recuperation due to excessive work procedures, can risk wearing out the individuals' resources to cope with stress being and cause the "machinery" to break down sooner or later. Furthermore, the demands on an individual include the amount and range of activities in their leisure time. Individuals in their entirety include both their personal life and their working life. Thoughts, communication, reflection, relationships and development are also parts of work and take time. Therefore, an individual should have sufficient working time for these tasks, in order to feel that they execute their work tasks to satisfactory standards. Otherwise, the individual runs the risk of being exposed to what is called *ethical stress*, meaning that organisational obstacles cause individuals an experienced inability to do their work tasks in a way and of a quality that is satisfactory and make them feel proud of their efforts.

Attitude changes in society may be needed in order to slow down the trend that everyone should do everything and increase acceptance that some people need more limited work tasks rather than that all work tasks and work procedures should contain supportive functions. This can, in turn, contribute to a larger amount of people being able to participate in working life through entry-level jobs, vocational rehabilitation and bridge employment in the workplaces, since someone must execute the supportive functions.

## CONTACT-BASED PROFESSIONS AND MENTAL DEMANDS

People employed in contact-based professions often express that they experience high demands and low control in their work situations. Furthermore, statistics show high rates of sickness absence and a wish to retire early in these professions. Contact-based professions are usually female-dominated, however, the high rate of stress-related ill-health in these occupations applies to both men and women [5,8,9,26,135,137]. These professions face particularly large recruitment problems in the current demographic development.

In recent years, industrial working time models have been introduced to contact-based professions in order to minimise production waste and decrease costs in public organisations. However, studies including employees in contact-based professions show that strict, objective calculations, regulations and planning of every work task in these professions do not work in reality, since work tasks entail contact with, support of, and helping people on an individual and subjective level [95]. Furthermore, both parents of young children and seniors employed in these professions may need a lower work pace in order to cope and have sufficient time for recuperation. Because of this, working time should not only be calculated in hours but also be based on work content in relation to working hours and the life phase of the employee.

For all professions, biological age, number of work procedures, work content and time for set-shifting should be considered and included in the assessment of what is reasonable for the employee to be able to do and cope with during their working hours and during a day at work. Additionally, the mental fatigue which may be caused by constant meetings and contact with people in need should be taken into account in contact-based professions. In recent years, there have been proposals that full-time employment in contact-based professions should be shorter than 40 hours per week. Reducing the number of working hours not only increases the possibility of recuperation with the purpose of facilitating an extended working life but also makes contact-based professions more attractive when recruiting new employees. As a suggestion, the reduced number of working hours could contain mandatory activities that research has shown to significantly prevent mental health conditions and stress reactions, such as physical activity and an appropriate level of exercise. Research shows that this measure can increase the possibility of employees in contact-based professions to work until an older age [8,11,138,139]. For example, physical exercise has since long been mandatory for fire fighters and police to facilitate coping with physical and mental work demands. Mandatory physical exercise should perhaps be introduced in other physically and mentally demanding contact-based professions in order to prevent ill-health and to decrease the costs of sickness absence in society.

## WORKING TIME MODELS

Society is responsible for regulations and laws concerning working time, breaks, pauses and the possibility of parental leave and working fewer hours while caring for small children. Employers use different working time models in workplaces since employers are interested in offering solutions like decreased working hours for people who need it. For example, a question for society is the possibility of decreasing the number of working hours in the years leading up to retirement, which can contribute to a larger amount of people being able to, wanting to and feeling motivated to work until an older age, while they simultaneously have more time for recuperation and leisure. Working hours related to a sustainable working life have been discussed previously in this book (read more in the Chapter *Working Hours, Work Pace and Time for Recuperation*). Working hours, with breaks and pauses, and an appropriate work pace, are factors of significance to maintain good health in working life. Furthermore, working hours and work pace have proved to be factors of significance to the employee's self-rated health [18]. The ability to adjust and manage one's own work pace appears to contribute to the ability to maintain productivity for people with chronic health conditions [118]. As previously mentioned, studies have shown that some people feel better when decreasing the amount of working hours step by step leading up to retirement, rather than ending their working life abruptly. This is beneficial to health since it allows the individual to make gradual lifestyle changes. The individual's possibility to manage their decreased working hours and time for retirement facilitates the planning of lifestyle changes and the new life phase as a pensioner. Furthermore, this has positive effects on mental and social health that last for a couple of years after retirement, which increases the possibility of healthy and active ageing. Working time models for senior employees are needed in order to increase the possibility of sufficient recuperation for seniors and to decrease the risk of becoming physically and mentally worn out. An example would be the 80-90-100 working time model, which, according to evaluations, results in decreased sickness absence and increased productivity among senior employees (read more in the Section *Working Time Models to Facilitate an Extended Working Life*).

## THE SIGNIFICANCE OF OCCUPATIONAL HEALTH CARE TO DECREASE THE NEGATIVE HEALTH EFFECTS OF THE WORK ENVIRONMENT

From a societal perspective, it is important to acknowledge the complexity of making working life sustainable for all ages, as described in the SwAge™ model. This complexity must be acknowledged in occupational health care to support employers, organisations, companies and employees.

### CONSIDER AND REFLECT
- What significance does citizens' health have for society?
- What health promotion measures and environments that support health are present in the different areas of society in your life?
- Reflect on what measures can be taken on the societal level, related to the sphere for action *health effects of the work environment*, in order to increase possibilities of employability:
  - on the organisational level.
  - on the individual level.

# Societal Perspective and Action Proposals

- Reflect on measures taken in society to support health promotion in the work environment.
- Reflect on what measures can be taken on the societal level in order to increase the possibilities of making healthy choices:
  - on the organisational level.
  - on the individual level.
- In what ways can decisions on the societal level contribute to employers, management and managers, who are ultimately responsible for the work environment to:
  - have sufficient knowledge of and interest in questions regarding the work environment.
  - set aside time in the organisation for questions regarding the work environment, in order to provide better conditions for a healthy work environment.
- Reflect on the significance of society for the knowledge of and compliance with laws and directives concerning the sphere for action *health effects of the work environment*.
- What significance does society, the societal culture and societal development have for the employee's experience of:
  - their health, diagnoses, illness and functional diversity in relation to their work environment and work situation.
  - their physical work environment.
  - their mental work environment.
  - their working hours, work pace and possibility of recuperation.
- Reflect on primary, secondary and tertiary preventive measures in society with the purpose of improving conditions for labour force participation.

# Sphere B. The Financial Situation

All individuals need food, a place to live and access to health care, communication, leisure activities, etc. Financial compensation for executed work tasks is crucial for the possibility and ability to afford these factors. Employability, the ability and willingness to work are prerequisites for participating in working life and strongly relate to individuals' personal finances (Figure 12) [1–3,5–8]. Furthermore, financial compensation systems relate to chronological age.

Individuals more or less consciously consider their financial situation in relation to whether they are able to or want to remain working, or whether they should leave the workplace, for example to find new employment or to retire (read more in the Section *To Stay or to Leave?*).

**Examples of Considerations that Matter to the Decision of Remaining in the Workplace**
- Whether I cannot afford to leave my workplace and work situation.
- Whether my employability is sufficient to execute my work tasks.
- Whether I experience better personal finance, through salary from work, which provides additional possibilities and a higher living standard than if I leave the workplace.

**Examples of Considerations that Matter to Decision of Leaving the Workplace**
- Whether my personal financial situation is sufficient in order to maintain my living standard even if I leave the workplace.
- Whether another employment or activity contributes to better personal finance.
- Whether my employability is insufficient.
- Whether I want to adjust to a lower living and financial standard rather than remaining in the workplace.

DOI: 10.1201/9781032616681-9
This chapter has been made available under a CC-BY license

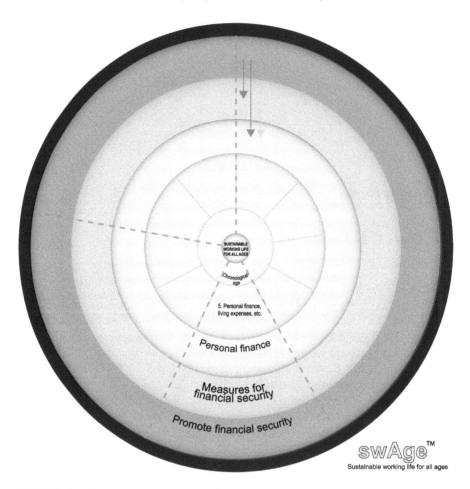

**FIGURE 12** The second sphere for reflection and action of the SwAgeTM model, (B) *Financial incentives*, which include the fifth influence and determinant area: Personal finances.

# Area 5. Personal Finance and Economics

This determinant area mainly concerns financial incentives. An individual's personal finance is significant to whether they are able to and want to work and relates to an extended working life [1–3,6,8]. The word economics derives from the Greek oikos 'house', and nemein 'manage'. The term "economics" refers to the study of managing limited resources in circumstances of scarcity, in other words, managing the resources and assets that are available.

Economics concerns different systems for resource management, for example what is usually called *economics*, *business administration* and *personal finance*.

*Economics* comprises theories of supply and demand, interest rates, etc. Economics is divided into two main parts: *microeconomics* and *macroeconomics*. In microeconomics, individual choices are studied, from the perspective of both the production and the consumer, as well as theories on how individuals reflect on and make decisions regarding production and consumption. In macroeconomics, entire economies are studied, for example the economy of a municipality, region or country. The focus is on flows and effects in different economic systems, for example how political decisions affect production, unemployment, inflation, etc. Another area that sometimes counts as a third part is called *international economics*, which concerns the study of trade and capital flow between countries.

*Business administration* includes studies of the financial management of an organisation or company, for example accounting, acquisition, production, organisation and marketing. These studies include external analyses of the market the company is active in, and how the company or brand should position themselves in order to be successful.

*Personal finance* regards an individual's income, expenses, assets, debt, etc. Large factors in an individual's expenses are usually living expenses like rent or interest on mortgage, food expenses, transport expenses as well as expenses for clothing and consumables. Income primarily refers to salary from work, return on invested capital or different kinds of benefits like pension, sick pay, sickness compensation, child benefits, and housing benefits.

Economics and economic systems relate to individuals' chronological age in several ways, through the classification and grouping of individuals in social welfare systems. For example, access to child benefits and pensions is controlled through the individual's chronological age. On the individual level, financial incentives are significant to the choice of profession and the possibility of leaving working life. For example, financial incentives can be used to keep individuals in the labour force through the threat of poverty. However, financial incentives can also contribute to individuals leaving working life, for example, in countries with state pension that enables individuals to retire with a sufficient living and financial standard. Financial security and pension benefits are usually utilised as a tool to regulate working life in society. Furthermore, socioeconomic status is directly connected to the individual's chances of healthy and successful ageing, while financial hardships increase the risk of premature death among adults.

Insurance companies and authorities often try to predict the best time, in financial terms, for an individual to leave working life for retirement. However, critics express that most people cannot predict different aspects that determine these calculations, it is difficult to predict one's own demise for example.

Among the most successful and competitive measures an organisation or company can take is to protect employees' health and promote a healthy and sustainable working life for all ages. In order to understand the long-term significance of this, employers must consider and reflect on the effects and benefits of the following:

- Costs to prevent risks of ill-health in the employees' work situation versus costs that arise as a consequence of accidents
- Financial consequences, fines, loss of production and other costs for the organisation or company when laws concerning health risks, security and occupational regulations are violated
- Employees' health and well-being as important assets and marketing tools for the company or organisation

It is beneficial for organisations and companies to promote employee health and minimise the risk of occupational injuries, occupational illness and medical costs, as well as costs related to high turnover of staff and premature retirement. Activities that promote health, risk assessments, measures and education to prevent ill-health increase long-term productivity and quality of goods and services. Global movements and ethically inclined entrepreneurs have introduced commercial labels, for example fair-trade, on products that have been produced in a manner that does not injure health and the environment, with the purpose of stimulating organisations and companies to better work environments, among other things.

## FINANCE AND EMPLOYABILITY

The concept *employability* refers to individuals' ability to gain employment, maintain employment, be relocated in the organisation when needed, and possibly gain new employment when needed [139]. Employability is significant for employees' ability to make a living through their work. Employees' employability is connected to the four spheres for action for a sustainable working life in the SwAge™ model, that is: *health effects of the work environment*; *financial incentives*; *social inclusion, support and sense of community* and *execution of work tasks* [1,2]. However, it is mainly the health effects of the work environment, social inclusion, support and sense of community and the execution of work tasks that affect individuals' employability, which affect their personal finances as a result. However, financial incentives and personal finance can, in turn, affect what employment an individual can gain and thereby the health effects of the work environment; affect employees' social environment with relationships, support and sense of community; affect employees' possibilities of executing work tasks and activities since the economy affects their possibility of education and access to knowledge. Both material resources (financial incentives) and immaterial resources (health effects of the work environment, social inclusion, support and sense of community and execution of work tasks) are significant to a long, productive and satisfactory working life. To ensure sound employability, it is important to maintain:

- *Vitality resources* – through good mental and physical health and well-being
- *Material resources* – through a stable ability to make a living
- *Transforming resources* – through self-awareness, support and help from managers, management and co-workers, one's own ability to reach out to different networks, including positive relationships with friends and family, and openness to new experiences and contacts
- *Productivity resources* – referring to the utilisation and acknowledgement of individuals' skills, knowledge, development and cultures that promote individuals' productivity, success and maintained employability

## HEALTH EFFECTS OF THE WORK ENVIRONMENT AND INDIVIDUAL EMPLOYABILITY

Sufficient health is a prerequisite for participating in working life. For that reason, health, functional diversity and diagnoses strongly relate to employability. Unfortunately, individuals do suffer injuries in working life. Since approximately one in four employees expresses that they have diagnosed illness or injury caused by their work, work can be perceived as a significant risk to public health in society [26]. At the same time, work is an important activity and the main source of income for most individuals. Organisations and companies depend on individuals' work ability and compensate for their effort, energy and time spent working accordingly. The entire society is built on the efforts and potential of the labour force, and the effect this has on the nation's economy, taxes and welfare. The labour force is crucial to social welfare and social welfare systems. There is a certain level of financial compensation available through social insurance systems and insurance for individuals who, due to ill-health, cannot make a living through working. At the same time, financial incentives are built into most of the systems that aim to rehabilitate individuals to return to working life. Individuals who are forced to leave working life due to ill-health, thereby being unable to contribute to their own or their family's personal finances in the same way, also risk losing status in their own perception and being excluded in society. Therefore, it is important that the physical and mental work environment, working hours, work pace and time for recuperation do not contribute to individuals' risk of developing ill-health or cause premature biological ageing. It is in everyone's (individuals, companies and organisations as well as societies) interest that the work environment supports individuals' employability.

## SOCIAL SUPPORT AND SENSE OF COMMUNITY IN RELATION TO EMPLOYABILITY

Social support, sense of community and the surroundings' attitudes towards the individual affect their possibility of having employment. The possibility of having and maintaining employability is affected both by attitudes in society and in the organisation or company where the individual is employed. If an employee receives support from their manager and co-workers, gets feedback on their executed work task, participates in decisions made in the organisation or company, participates in the goals of the organisation, has the possibility of using their competence and skills, access to new knowledge and continuous competence development, this increases their possibility of good employability. However, if an employee is disregarded or even discriminated against in the workplace, this will risk causing decreased possibilities of maintained employability. The workplace must have a context that facilitates and enables employability. Furthermore, support through age-conscious and situational leadership is required, since it increases the connection between individuals' work ability, work content and areas of responsibility. However, the employee must want to contribute to their own employability and feel motivated in order to utilise their possibilities.

The individual's own network is an important asset for the ability to develop employability and change, transform or switch work tasks and workplace. An individual broadening their reference group, in other words, the people that the individual compares themselves with, increases their possibility of self-awareness, identity, perception of their own abilities and better understanding of who they are, and even more importantly, who they can become. In this aspect, networking in and outside working life is significant to personal development which increases the possibility of employability.

### EXECUTION OF WORK TASKS IN RELATION TO EMPLOYABILITY

It is necessary to be aware of one's own employability throughout working life, not only when it is needed in times of change, in order to maintain sustainable employability. Furthermore, an individual must be aware of potential future demands and conditions in order to maintain lifelong employability. An individual with good employability can be described as having the skills, knowledge and capacity requested by the labour market, and who easily can find new employment in another organisation, but who remains attractive in their own workplace where they are motivated and want to remain. The merit list of an extremely employable person has been defined as having high human capital, good possibilities in the labour market, sought-after education, good support for career and competence development, a high level of professional skills and being willing to change work tasks and workplace, willing to develop new competencies, open to new possibilities, aware of opportunities, driven, has a high level of self-awareness, has the future ahead of themselves and has emotional intelligence [140]. It can be experienced as stressful to continuously remain employable in a changing labour market. However, employability does not simply lie in the interest of the employee, it is also significant for the organisation and for society. In order to increase senior employees' employability in an extended working life, the senior employee needs support in order to develop their own capacity for initiatives, motivation to work and perception of the future. Education and competence development must be adjusted to the capacity to learn, cognitive ability and age.

## ECONOMICS AND FINANCIAL INCENTIVES

A foundational principle of microeconomics is that individuals are economically maximised and make a cost-benefit analysis in every possible situation (i.e. what is in it for me?), weighing pros against cons before making a decision in order to maximise their own profit. According to this principle, all humans always make the decision that results in the largest possible profit at the lowest possible cost, what is known as the *maximum expected utility*. According to this concept, humans are assumed to weigh the worth of something happening, against the likelihood that it actually will happen in every given decision [141]. However, all people are probably not this rational at all times. Individuals' behaviour is not only regulated by their intellectual considerations but also by their feelings, and by avoiding making a choice at all and following the stream instead, where chance or other people make decisions for them. Maximum expected utility is first and foremost a theory and does not describe how all people act in reality [142–144]. The microeconomic maximum expected utility model is a normative model for how people, according to economic theory, should act in order to gain maximum profit, rather than a descriptive model of how people actually act.

Economic analyses describe pull factors, i.e. attractive factors, and push factors that affect when individuals leave their workplace. The *push* and *pull* theories describe mechanisms that counteract employability. The pull theory describes that if financial benefits of the social insurance system or other financial compensation systems, for example sick pay, unemployment benefit or pension, are equal to or more beneficial than work, this will contribute to a larger amount of individuals being attracted to leave working life of their own accord [145,146]. Furthermore, previous studies have noted that salary is an important factor as to why people work, however, salary does not increase individuals' internal motivation to work [18,59,63,108,147–149]. However, good personal finance does increase self-rated health and the possibility of healthy ageing, for both men and women. However, financial hardships affect the risk of illness and mortality, for example among unemployed individuals and elderly people with insufficient means. Registry studies of the entire Swedish population aged 55–65 show that above all, women with short-term education, but also men, in occupations with high physical and mental demands, are particularly sensitive to changes in social security systems [18]. Decreased possibilities of sickness benefits and sick pay appear to increase premature retirement. Furthermore, changes in social security systems for people who cannot cope with their work appear to increase the transfer of these individuals to other financial security systems. The push theory describes how less productive individuals can, against their will, be pushed out of working life due to health issues, bad work environments, unemployment, financial difficulties in the organisation or company, economic downturns in society, austerity policies or other economic conditions [18,150,151].

It is much better for employees' health if they work because they want to, and not because they cannot afford to leave the workplace, particularly if their work situation has negative effects on their health. The proportion of people who retire prematurely is most common among men with high income and long-term education. Many people receive pension benefits while simultaneously remaining working, in order to increase their financial possibilities. Furthermore, how much income and wealth affect the decision to work or retire seems to depend on civil status. This appears to mainly affect men and unmarried women, married women do not appear to be affected to the same extent. The influence of personal finance and salary is particularly important for women living alone with a low income who have not worked full time. Women who live with someone or are married and share the financial responsibility for their household and living expenses more seldom state their personal financial situation as significant to the decision to retire, despite them having a low income or not working full time. However, financial incentives appear to be more important to men than women in general, when comparing factors of importance to the decision of leaving working life [9,26,94]. The personal financial situation can be a significant factor to stay in working life despite ill-health, poor well-being, lack of stimulation, motivation and development in work, for individuals who have difficulties providing for themselves. Individuals who dislike their work situation are more optimistic in their expectations of their future retirement and make larger financial sacrifices to retire prematurely. Individuals who however experience good health and well-being in their social work environment and find their work tasks stimulating and developing rather perceive the financial incentives related to the possibility of working until an older age, thereby improving their personal finance, as a bonus.

## CONSIDER AND REFLECT

- What are your possibilities of good employability?
- Reflect on your employability in relation to your physical and mental work environment, and to your working hours and work pace.
  Does it:
    - facilitate sufficient time for recuperation during and between work shifts?
    - prevent occupational injuries, occupational accidents and occupational illness?
    - decrease the risk of you being prematurely mentally and physically working out?
    - eliminate the risk of you being pushed out of working life through sickness absence, sick pay, unemployment, etc.?
- Reflect on your employability in relation to:
    - the social support, inclusion and security in your personal social environment.
    - the social support, inclusion and security in your work social environment.
    - the leadership in your work situation, access to help in your work tasks and your autonomy and take your own initiatives.
    - are you not being subjected to disregard, exclusion and discrimination?
- Reflect on your employability in relation to ability, knowledge and updated competence:
    - that increases the possibility of new employment.
    - that tolerates replacement and rotation.
    - that decreases the risk of unemployment.

- Reflect on whether your personal finance includes:
  - reasonable salary increase, regardless of age.
  - the ability to afford fewer working hours when needed.
  - national insurance like social insurance, pension insurance, occupational injury insurance, etc.
  - personal insurance, for example personal occupational injury insurance.

# Organisational Perspective and Action Proposals that Matter to the Financial Situation

Most organisations and companies depend on sound financial management. In order to achieve this, the purpose of the employee's work is to contribute to the organisation's production and budget goal. The manager's task, in turn, is to work for the achievement of the organisation's goals and ensure that the budget is kept in order. Contributing to employees' employability is an important step in achieving these goals.

### OCCUPATIONAL INJURIES AFFECT ON FINANCE

If an employee suffers occupational injury, the risks affecting the organisation's finances through loss of production, loss of competence, sickness absence, loss of capacity, loss of productivity, interim staff, new recruitment or due to costs of fines and penalties related to work environment deficiencies, etc. Furthermore, the brand of the organisation or company risks being shown in bad light, which, in turn, can contribute to the surrounding society having a negative perception of the organisation, a decreased interest from customers and that potential individuals to recruit are not attracted to the organisation.

Severe individual injuries, several people suffering injury at once, fatalities and severe incidents that entail severe risks to health or fatality count as severe occupational injuries. When an injury or illness occurs at work, or when an incident that entails a risk of injury occurs, the injured individual, or a safety representative, should report the incident to their closest manager or employer immediately. Certain infections caused by work can also be perceived as occupational injuries.

A school is perceived as the students' workplace, and a student suffering an injury should be reported by the school. Self-employed individuals should report their own injuries the same way an employer does. It is important to remember that some occupational injuries may not appear as serious at first but can cause consequences in the long-term perspective. Therefore, it is important to document and report all occupational injuries to avoid the employee risk of being left without compensation, and in case the occupational injury causes problems in the future.

It is better for everyone to prevent risks of occupational injuries, mainly to prevent individuals' suffering, but also for organisations and companies not to risk financial losses and consequences.

## THE SIGNIFICANCE OF LEADERSHIP FOR GOOD HEALTH AND EMPLOYABILITY

The manager's leadership should promote good health and employability among employees, in order to avoid financial consequences for the individual employee. However, the purpose of the manager's leadership is also to avoid financial consequences for the organisation, for example due to new recruitment if employees become physically and mentally worn out, ill and injured, or if they simply do not want to stay in the workplace due to discrimination, disregard, lack of support, lack of motivation and stimulation, if their competence is not utilised or if they lack competence and do not have access to appropriate competence development. The employer and manager should be supportive and resourceful and help employees to better employability and to be prepared to face a labour market in constant change, characterised by unpredictability and insecurity. Immaterial areas that have been identified to measure an organisation's readiness for an ageing labour force and to cope with demographic changes are commitment to a work environment that promotes health and the health effects of the work environment, social inclusion, support and sense of community, information management, networking, life-long learning and career planning, knowledge and motivation to execute work tasks, age diversity and work units with mixed ages where different ways of defining age are significant. Work units with mixed ages report higher productivity, which is positive for the financial conditions of the organisation or company.

Employees in general appear to return to the workplace to a lesser extent when they have been on long-term sick leave or unemployed if they cannot be relocated when needed [152]. However, if their health is restored, their competencies desired and they have good social networks and support to return to work or find new employment, it is usually not a problem to continue working, regardless of their older age. However, seniors who are on long-term sick leave or unemployed usually have fewer opportunities to return to the labour market and to an extended working life. This appears to mainly be associated with employability (read more in Section *Employability*). A senior employee risks being less prioritised in decisions regarding supportive measures in the workplace and in working life since they are assessed to have fewer active years left on the labour market. Accordingly, expensive measures are taken less frequently, since financial benefits are weighed against invested capital. Furthermore, senior employees have chronic illness and worse health to a greater extent than younger employees. This increases the risk of senior employees' sickness absence, premature retirement and leaving working life due to deteriorated health. Furthermore, the senior employee may have shorter and outdated education that may no longer be desired in workplaces, resulting in the individual being unable to relocate to different work tasks. Because of this, continuous measures to prevent and manage the risk of occupational injuries and ill-health, and to promote updated and developed knowledge in order for all employees to remain employable are important, regardless of the employee's age.

## ATTITUDES IN ORGANISATIONS AND EMPLOYABILITY

Norms and attitudes in the workplace are behaviours that can limit or increase employability. Even if there are opportunities to maintain or increase employability in the organisation, attitudes and workplace culture can contribute to these opportunities not being utilised. The prevailing social norms in the workplace have a great influence on whether the employee takes measures to manage their employability, for example using ergonomic or technological aids or not, which can affect their long-term health and their biological ageing. Furthermore, attitudes and norms in the workplace affect the employee's employability through support of or mistrust in the employee's learning abilities and capacity to participate in competence development. Moreover, employability is affected by whether disregard and discrimination are present in the workplace, by whether the social support in the workplace enables employees to try new things, dare to fail, network and expand their network and by work-life balance, time constraints and expectations from others – both professionally and personally.

Norms are institutionalised when the three following conditions are met [153]:

- A large amount of members in a system or organisation accept the norm
- Several members who accept the norm take it seriously and internalise it
- The norm becomes sanctioned, meaning that some members are expected to follow the norm in certain situations

Individuals naturally adapt to institutionalised norms if they want to adapt. However, even if they do not want to, many individuals still adapt in order to avoid feelings of guilt or to avoid the surroundings expressing their discontent. This would result in a loss of status, both for the individual themselves and for others. Every individual takes on different roles in different situations, for example, one person can simultaneously have an occupational role, a leader role, a co-worker role, a mother role, a daughter role and a role as a political spokesperson. Role conflict occurs when norms do not match an individual's different roles in life and everyday life, which contributes to deviant behaviour in relation to one or several roles. At the same time, it is beneficial when one role motivates another role in an appropriate and useful way. Some norms and regulations in relation to different roles are taken more seriously than others, and some norms are generally more accepted than others.

Organisations and companies appreciate employee loyalty [6,11,107,108,136,149]. At the same time, a strong commitment to, for example, activities outside the organisation risks competing with the employees' productivity and can therefore be perceived as disloyal. Employees wanting to develop skills, take courses and engage in competence development in a field not strictly related to the current occupational role can be perceived as a waste of working hours and disloyalty. A loyal employee

Organisational Perspective and Action Proposals 107

may not be expected to deny working overtime or working on days off, regardless of whether the employee needs a day off in order to recuperate before their next work shift. An individual prioritising their health above their loyalty to the organisation and work unit can, according to the norms of some organisational cultures, result in an employee being perceived as weak, for example in a work unit where risk awareness is perceived as silly. Employees who remain in the workplace for a long time, who engage in and are willing to prioritise work above other interests and commitments are perceived as loyal. The acknowledgement of being loyal can give the employee a false sense of security. Furthermore, a culture that highly values this kind of loyalty runs the risk of impeding the employee's ability to prioritise themselves, at the expense of their health, social network and competence development, which can risk contributing to lower employability and premature exits from working life in the long run. Moreover, there are ethical and moral aspects to this, since some organisations contribute to employees' work health and allow their employees to spend time and energy to develop their employability. Even if the employee's loyalty has served the organisation well in the past, praising the loyalty culture in an organisation excessively can be diametrically opposed to an employability culture that encourages employees to develop their individual employability. The workplace needs an environment that supports health in order for the individual not to take unnecessary risks in the work environment, to use ergonomic and technological aids and to ask each other for help and support to cope with demands and stress, as well as to receive sufficient recuperation through breaks and pauses during and between work shifts. Furthermore, a supportive environment is needed to ensure that employees are allowed to use their skills and knowledge and that they have access to competence development, regardless of age and whether the employee works full or part-time. Furthermore, an employability culture in the workplace can contribute to better conditions of resilience for the organisations in times of crisis and change, through the current employees' experience, through conditions for better health, networks and knowledge to cope with adjustments and change. Empowering individuals through conscious actions, to enable individuals to strengthen positive and healthy factors contributes to greater possibilities for individuals to have a healthy and longer working life. Furthermore, this increases employees' possibilities to overcome obstacles and to adjust their own behaviour and feel in control of their own situation, which, as previously mentioned, also benefits the organisation through lower costs of sickness absence, occupational injuries, dissatisfaction, employees leaving the organisation, new recruitments, training new employees, etc. Increased employee employability also contributes to an environment where the boundaries between working life and retirement are less significant when it comes to living life to the fullest throughout the entire lifespan.

## ORGANISATIONAL COSTS OF SICKNESS ABSENCE

The economic costs to companies and economies are significant because of work-related injuries and diseases. Sickness absence, due to illnesses and injuries, is expensive to organisations [29]. Regardless if the work environment or the employee's lifestyle causes sickness absence, the employer is responsible for rehabilitation.

In this regard, large financial differences are perceived between employers who work with health prevention and those who do not. Costs for sickness absence are significantly lower for employers who have a properly developed rehabilitation policy and plan for measures with well-functioning routines, compared to employers who lack these. The conclusion is that even if measures to prevent sickness absence can be expensive, decreased short and long-term absence is financially profitable for the organisation. This is because preventive measures counteract costs for rehabilitation, interim staff, new recruitment and the risk of excessive demands and tear for remaining employees when their co-workers are absent due to illness or injury. At the same time, work efficiency increases when employees are not absent due to ill-health, or if they are present with ill-health resulting in lower capacity.

## CONSIDER AND REFLECT

- Calculate the organisational costs for sickness absence:
  - What is the hourly cost for presence and sickness absence in your workplace?
  - What are the total costs for sickness absence in your workplace?
- Could the manager take any measures in order to increase the possibility of employees':
  - maintained employability through changes, relocations and re-organisations?
  - ability to execute their work tasks in a better way?
  - performance, based on the needs of the organisation?
  - appropriate and sufficient competence?
  - the sense of security and inclusion in the organisation?
  - motivation and stimulation to contribute to the execution and development of the organisation and productivity?
  - avoidance to suffer ill-health caused by work?
- Execute preventive risk assessments in order to increase employability and decrease health issues, sickness absence, sick pay and unemployment, for example through workplace analyses based on the nine determinant areas of the SwAge™ model (read more in Part III *Practical Application Based on the SwAge™ Model* and Table 2).
- Reflect on employability with regards to the organisation providing adequate and sufficient occupational health and safety routines, as well as technical and ergonomic aids to prevent ill-health, illness and injuries.
- Reflect on whether activities and measures are taken to ensure that employees continue to be employable and are not forced to leave working life if they suffer from ill-health which makes them unable to execute their current work tasks.
- Reflect on whether the organisation provides opportunities to learn additional work tasks, in order to increase the possibility of career development and stimulate continued development.

# Organisational Perspective and Action Proposals

- Reflect on the opportunities for continuous knowledge and competence development, enabling employees' maintained employability in the workplace, in re-organisations and changes in working life, and in new types of work tasks, work situations and professions throughout their entire working life.
- Are the salary and financial benefits reasonable in relation to the work efforts in the workplace?

# Societal Perspective and Action Proposals that Matter to the Financial Situation

## SOCIETAL PERSPECTIVE AND ACTION PROPOSALS THAT MATTER TO PERSONAL FINANCE

The sphere for action *financial incentives* in the SwAge™ model comprises factors that influence individuals' financial security. On the societal level, a fundamental prerequisite to the individual's possibility and willingness to participate in an extended working life is based on norms in society and the financial incentives of social security systems, insurance and paid work [154–157]. If the financial compensation of the pension system and social security system are almost as high as the income from working, the pull factors of working for a living disappear. What is considered normal in society influences organisations' and individuals' different choices, for example if there is a culture, attitude and societal order to enter the labour force at 15 years of age and work until 70 years of age. Or if it is considered normal and acceptable to have a longer education where one enters the labour force at approximately 27 years of age and stops contributing to the social welfare by working already at 55 years of age. However, the diversity between what ages citizens work in different countries is affected by the most common kinds of occupations in the country and the health risks these occupations entail.

## FINANCIAL PERSPECTIVE ON SOCIETAL SUPPORT FOR EMPLOYABILITY

Health, safety and well-being in working life are important aspects of economies across the globe. The WHO has developed strategies to improve employee health, including self-employed business owners, and to decrease the financial consequences of occupational ill-health [71]. Hundreds of millions of people work across the globe and approximately 2 million people die as a consequence of occupational accidents and occupational illness and injuries each year. An additional 268 million non-fatal occupational accidents result in 3 lost work days per accident on average, and 160 million new cases of occupational illness occur every year [71].

If an individual suffers ill-health and injuries as a result of their work, it causes suffering not only to the individual but to their family as well. However, ill-health caused by working life also causes issues in the productivity, competition and sustainability of the organisation, and in society and the national and regional economies as well. For society, this results in increased costs for health care, the social welfare system, loss of occupation taxes, etc. Occupational health issues result in an economic loss corresponding to an average of 4–6 percent of GDP in most countries [71]. The International Labour Organization estimates that more than 4 percent of the world's annual GDP is lost because of work-related injuries and diseases.

A senior individual is often assessed based on their having fewer active years left in the labour force, and expensive measures are often focused on groups where the financial profit can compensate for the invested capital. Because of this, a senior individual risks not being prioritised for supportive measures in order to return to work. However, if individuals should be able to work until an older age, the measures in society must not decline or stop working as a supportive environment for the senior individual's employability. Society must strive for economic safety and security, by decreasing the risk of ill-health and contributing to employees' ability and willingness to remain employable until an older age. This can be achieved by supporting and increasing the possibility of appropriate and updated knowledge and competence, in order to prepare individuals for re-organisations and changes. Furthermore, it is important to encourage activities and measures of physical and mental health promotion and prevention in the workplace. Moreover, society needs to ensure that the physical and mental work environment prevents occupational injuries or employees being prematurely worn out, for example by encouraging environments that support health and preventive measures, updating directives and making follow-ups on compliance with laws and directives.

## SOCIAL SECURITY SYSTEMS

In many societies, different financial social security systems function as a safety net and contribute to a sense of security for employees, to avoid the risk of poverty, starvation, etc. Examples of social security systems are the social insurance system, unemployment insurance and the pension system. As previously mentioned, these systems also affect individuals' ability and willingness to work.

High compensation levels in social security systems, that come close to the level of income from work, appear to be a pull factor, in other words, it attracts people to leave working life. Social security systems are sometimes used as a tool in society in order to manage and increase the amount of individuals active in the labour force, and the number of worked hours in the economy.

However, changing the compensation levels in social security systems alone does not increase individuals' employability. If welfare cuts are too excessive, this will risk contributing to a transfer from one social security system to another, unless measures are taken to promote employability among individuals in order to facilitate a larger amount of people's ability and willingness to work. If an individual is not able to cope in working life, but who, for example, lacks the possibility of receiving sick pay due to stricter requirements, this can result in a permanent exit from working life. For example, changes in the Swedish social insurance were launched in 2008, these measures made it more difficult to receive sickness benefits and sick pay and contributed to a larger amount of people leaving working life for premature retirement [18]. This mainly concerned women with short-term educations, which, in turn, risk contributing to increased financial gaps in society, since the benefit levels in the pension system are set based on the chronological retirement age.

## EMPLOYMENT DEVELOPMENT IN SOCIETY IN RELATION TO SOCIAL SECURITY AND PENSION SYSTEMS

The social security and pension system in society are based on the individual's chronological age, however, the individuals' own experience of their ability and willingness to work in relation to their age are mainly based on their functional age, in other words, their biological, social and cognitive age combined. The retirement age regulated by law says little about people's health, their own experiences and decisions regarding their ability and willingness to remain in working life. Furthermore, the individual's thoughts and decisions are based on cultural and social perceptions and values. Moreover, politics in society as well as national and international economic fluctuations are significant to the employment development in different age groups.

However, reflection is needed from a national economic perspective regarding the fact that people who retire from working life with maintained health and work ability for an additional couple of years often contribute to the national economy through volunteer work, work in associations, care work, etc. [154–157]. Therefore, pensioners continue to contribute to society financially and socially through work and efforts that otherwise would have had to be executed by a paid employee.

## CONSIDER AND REFLECT

- Reflect on what measures can be taken on the societal level regarding the sphere for action *financial incentives* in order to increase conditions for employability:
    - on the organisational level
    - on the individual level
- What significance do society, societal culture and societal development have on individuals' experiences of their personal finance, financial safety and security?
- In what way can the societal level make employers, management and managers, who are ultimately responsible for the organisation, contribute to their employees' employability?
- Reflect on the significance of society for the knowledge of and compliance with laws, directives and support that contribute to individuals' employability.
- Reflect on the influence social security systems in society have on:
    - employment and employment development in society
    - organisations' and companies' possibility to recruit and dismiss employees
    - individuals' possibilities and motivation to work

# Sphere C. Social Inclusion, Support and Sense of Community

Social inclusion, support and a sense of community are factors of significance to individuals' employability, ability and willingness to work and relate to social age [2,5–8,80] (Figure 13). Humans are social beings, inclusion in a group and not feeling excluded is very important. Being excluded from the community of a group was once equal to mortal danger since it meant that the individual did not have the joint protection against wild animals and enemies, it also increased the risk of starvation and hardships. We still carry these instincts, and the experience of exclusion from a community can still trigger a stress reaction since it is fundamentally unnatural to us. Therefore, to this day, it is still important to include a group with social support and a sense of community in the personal social environment, and in the social work environment, in order for individuals to be able to and want to work (Figure 13). Relationships, support and a sense of community affect individuals' social capital, in other words, the return of resources available to support an individual or a group. The aggregated social capital available to an individual or a group can be described as the number of contacts and networks that the individual or group has, multiplied with the degree of mutual trust and acknowledgement in these contacts [158]. Social inclusion, support and a sense of community relate to social age in several ways, since individuals are categorised into different, sometimes stereotypical, roles based on their social age. Furthermore, individuals are often expected to provide and receive different types of social support at different social ages. Individuals more or less consciously consider their social inclusion, support and sense of community, and their ability and willingness to work or whether they should leave the workplace, for example in favour of another employment or retirement (read more in the Section *To Stay or to Go?*).

**Examples of Considerations of Significance to the Decision of Staying in the Workplace**
- If I am included in a social group that matters to me at work.
- If I experience good cooperation with co-workers, customers, clients, patients and managers.
- If I feel acknowledged, appreciated and included in my workplace.
- If I experience the manager and management in my work as safe and trusting.
- If I experience well-being in my social work environment.

**Examples of Considerations of Significance to the Decision of Leaving the Workplace**
- If I want to leave the social group, managers, management, co-workers, customers, clients, patients, etc. that I am exposed to in my workplace.
- If I am disregarded or discriminated against in my workplace.
- If I experience more appreciation, social inclusion and a sense of community in my personal social environment compared to the social environment in the workplace, for example with my partner, family, friends, and leisure activities or through volunteer work.
- If I want better possibilities of social participation outside working life and the workplace.

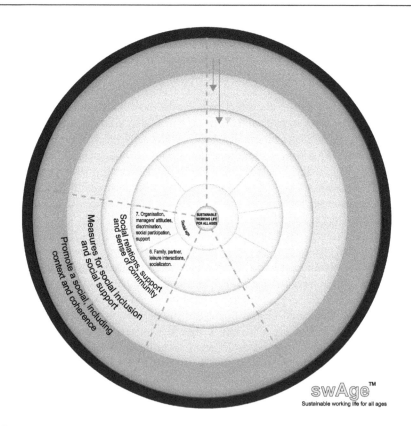

**FIGURE 13** The third sphere of reflection and action in the SwAge™ model, (C) *Social inclusion, support and sense of community*, which includes the influence and determinant areas: (6) Personal social environment with partner, family, leisure activities, etc. and (7) Work social environment with leadership, participation, social support, exclusion and discrimination.

# Area 6. The Personal Social Environment

The personal social environment is significant to whether individuals experience an ability and willingness to work and relates to their social age [1–3,8,26] (Figure 13). Individuals who participate in working life have a personal life separate from their work. The significance of personal life and leisure time for the ability and willingness to work should not be disregarded in discussions concerning employability, employment and participation in working life. Individuals can get a new job or quit working, but they still need to experience a sense of community. Accordingly, we need healthy social communities and social interactions outside the workplace and leisure time that provides rest, relaxation and recuperation to restore our physical and mental energy.

Entering working life through an occupation and workplace is a significant step in life; the individual leaves the social age of adolescence and enters the social age of adulthood. Similarly, leaving working life to become a pensioner is another significant step; the individual leaves the social age of active participation in the labour force and enters the social age of retirement. Accordingly, retirement is an important event and turning point in life. Factors such as the prevailing attitudes in society, civil status, whether the relationship with one's life partner is good or not, and whether the senior employee wants to spend more time with relatives and in leisure activities affect if and when individuals leave working life. The individual must prepare and plan for their retirement, with regards to their personal social environment, in order to ensure that they have a role as a pensioner in their personal social life. Some people experience that they lose their identity when leaving their profession and role as a teacher, medical practitioner, carpenter, nurse assistant, prosecutor, researcher, chimney sweep, seamstress, baker or NHL professional. In retirement, all individuals are pensioners, regardless of their former occupation. A study shows that men with larger social networks were more inclined to work after 70 years of age compared to men with smaller social networks [159]. Many people want to retire prematurely if work has a negative effect on leisure activities and family life. However, working life can also have positive effects on leisure activities and family life, which often contributes to individuals wanting to continue working regardless of their older age.

## SOCIAL INCLUSION OUTSIDE WORKING LIFE

The personal social environment is significant to the individual's ability and willingness to work. Some people in the social age between adolescence and adulthood face the decision of where and in which workplace they will begin to work, which has a great influence on where they will reside and perhaps even meet a partner and start a family. Others may choose their occupation and workplace based on the proximity to their home, what their personal life is like and where their friends live. Some employees are in the social age where many people have small children or adolescents to care for. These factors influence how and in what circumstances they are able to and want to engage in their work, and what working hours they are able to and want to work. They may have a child with time and energy-consuming leisure activities and interests that the parent wants to support. They may have a child who has problems with school work, bullying, drugs, alcohol or smoking that affect the parent and can contribute to their bad conscience, feeling insufficient and torn between work and personal life. Other employees may have a relative or close friend with functional diversity, or a parent with dementia, who needs support and caretaking. Furthermore, the employee may be of the social age when they approach the retirement age themselves. They may reflect on what life in their personal social environment will be like when they no longer have a workplace and social work environment to attend and participate in. What will my partner do and when will they retire? What are the attitudes towards an extended working life in my personal social surroundings and society? Is it acceptable to retire prematurely, or will I be perceived as lazy? Is it acceptable to work until an older age or will I be perceived as odd and that I lack a personal life? If I live alone, will my social life be limited without my social work environment? Are my possibilities of an inclusive social environment through leisure activities more stimulating than my social work environment? Will my personal social environment be more stimulating if I still have a social work environment?

## BOUNDARIES BETWEEN WORK AND PERSONAL LIFE

Many driven people, both in sports and in working life, know that it takes hard work to achieve success. However, what is sometimes forgotten, with devastating consequences, is that time for rest and leisure is vital if one aims to be successful [160,161]. This is applicable regardless of what goals one wishes to achieve and is significant to a sustainable working life for all ages.

There are often no definite boundaries between working life and personal life. We often experience that something from our personal life crosses our minds during working hours. For example, a discussion with the partner or teenager, trying not to forget to get a new raincoat for the four-year-old child, trying not to forget to pack lunch boxes for the children's day trip, worrying about a relative's illness, trying not to forget to book an appointment at the medical practitioner or dentist, trying not to forget to copy documents for the housing cooperative's board meeting, etc. Similarly, we often think about work-related issues during our leisure time. For example, worrying about a train being late, causing us to be late for an important meeting. Perhaps, an individual wakes up in the middle of the night thinking about work or finding it difficult falling asleep because they forgot to send an important e-mail as promised the day before. An individual might suffer from impatience or anxiety due to stress related to a work task that can affect their partner, children or friends. An individual can spend the evenings worrying about a patient, client or student. Furthermore, working from home can contribute to a lack of boundaries between work and personal life, between working time and leisure time. The boundaries between personal life and working life can also blur if some co-workers become close friends, while others remain to be only co-workers. The attempt of combining and balancing all activities and aspects of an individual's life metaphorically resembles a puzzle. This puzzle involves several people – not only oneself and one's family but also co-workers and other people in working life. In a society with blurred boundaries, individuals must try to separate their working and personal lives. Treating each other like human beings at work is not the same as being a family. In a family, an individual can rightly expect that others should be interested or at least listen to the personal things they want to share, as opposed to work. Privileges and duties simply differ between working life and personal life. It is possible that someone at work may find you interesting – but it cannot be expected. At the same time, a great part of people's waking hours are spent in working life, and there are no absolute boundaries between thoughts regarding personal life and working life. This may risk resulting in confusion and insecurity. Feelings of insufficiency to have time for, to cope with and to be able to perform one's best standards both in working life and personal life can occur, together with unreasonable demands put on ourselves and our surroundings. This is a great cause of stress and stress-related ill-health.

We all experience the influence that work has on our personal lives differently. For some people, working life can be experienced as burdensome and a necessary evil that takes time from their personal life where they may want to spend all their time, for example with children, partners or friends. For others, working life and work tasks, as well as the social context with co-workers, managers, clients, etc. are experienced as stimulating and motivating, while personal life can be experienced as burdensome and taking time they rather would spend at work.

## HOW PARTNER AND FAMILY AFFECT LABOUR FORCE PARTICIPATION

Caring for relatives, regardless if they are children, elderly parents or an ill partner, affects the ability and willingness to work, and at what age an individual retires, especially for women. Furthermore, people who express that their work has negative effects on their family life and leisure activities usually want to leave working life prematurely. To this day, some women still have a greater responsibility to care for children, grandchildren and ageing parents compared to some men. This has been described to affect women's working hours, and there are some differences between men and women regarding part-time work and the age of retirement [106,138,162]. The time spent doing household chores increases with part-time work. Furthermore, the time spent doing household chores increases after retirement, for both men and women. Retired men whose partner still works, in particular, spend more time doing household chores after retirement compared to their partner, and compared to when they were active in the workforce.

Partner, family, social context, volunteer work and leisure activities have significant effects on labour force participation and choice of workplace, especially if these factors are difficult to combine. For example, at what age friends and people retire in the individual's social context affect their retirement age. This also applies to age gaps between partners, since many couples plan their retirement together. Furthermore, the time of a partner's retirement affects retirement planning for both men and women

[9,163,164]. Satisfaction in a romantic relationship increases expectations of spending more time together. Both men and women are less inclined to continue working if their partner has retired. If the partner remains in the labour force, this has effects not only on the return to the labour market after retirement but also in cases of unemployment. Heterosexual couples who are most individually and mutually satisfied with retirement are when the female partner has not been influenced to retire by the male partner. Not having a partner can contribute to both early and late retirement. Being single causes people to be less interested in the financial incentives an extended working life entails, especially for men with reasonably good personal finance [162–164]. Women who live alone are more inclined to be willing to work until an older age, often due to financial reasons, compared to women who live with someone [162–164]. If individuals without a life partner experience their leisure time as boring and lonely, this can contribute to them remaining in working life until an older age, if the possibility of social interaction and inclusion is better at work.

## HOW THE PERSONAL SOCIAL ENVIRONMENT AFFECTS WORK

An active personal social life and leisure activities have positive effects on work ability and work satisfaction [13,165,166]. Examples of leisure activities that have shown to have positive influences are some kind of artistic hobby or physical activity. Combining a paid occupation with volunteer work has been shown to reduce the risk of mental health deterioration [167]. Furthermore, volunteer work and commitment to associations can also be a substitute for former employment, with the purpose of maintaining a sense of meaningfulness, usefulness and productivity, as well as maintaining routines in one's life. Individuals' state of health, education and religious commitment often affect whether they participate in and commit to unpaid volunteer work [168].

However, some people describe that working contributes to them not having time or energy to spend on their leisure activities and volunteer work, since they need to recuperate between work shifts with increasing age [6,11,162,166,169–171]. In contrast, a study with 28,780 participants in 18 different occupations from 27 different European countries showed that younger employees found it more difficult to adjust their working hours to family and social commitments, compared to employees aged 50 years and above. People above 50 years of age usually no longer have small children to pick up, drop off and care for, which results in them having more time to control their own lives and leisure time. Many people return to hobbies they did not enjoy when their children were young, and many want to put more effort into their careers and own lives.

Retirement does not appear to affect the interest in volunteer work among people who have not previously been committed to this, even if they have more time for such commitments when they have left working life. However, people who have previously been involved in volunteer work often increase the time, resources and commitment to this after retirement. Individuals who have been forced to retire often experience worse mental health than people who have retired voluntarily, therefore, it is not as likely that these people try to increase their personal social life and commit to volunteer work and leisure activities as a pensioner [168].

## WORK-LIFE BALANCE

Work occupies a great part of people's waking hours and affects their family life and leisure activities, at the same time, family life and leisure activities affect an individual's possibility to work. The need for work-life balance is often discussed, like an equilibrium or homeostasis between work and personal life would occur during the years we are active in the labour force. However, it is not quite so simple, since no such formula exists. Work-life balance refers to the experience of having sufficient time to do what one wants and wishes in life, paired with the opportunity for varied tasks, stimulation and sufficient time for recuperation. During certain times in life, an individual's personal life can offer possibilities for recuperation, however, there are other times or situations when the routines in working life can be experienced as recuperation from an individual's personal life. Furthermore, some individuals, during certain periods in life and during the life course, need to spend more time in working life, whilst their personal life demands more focus during other times.

*Activity balance* is a concept that refers to an appropriate mix and balance between different activities in life. There are mainly three indicators of an appropriate mix between activities in order to achieve activity balance [171]. The first indicator concerns *abilities and resources*, in other words, the activities an individual is committed to should not exceed their personal resources. These resources are, for example, the individual's physical and mental health, and contextual resources, for example if an activity is supported by working hours, the organisation of work and the attitudes of managers and co-workers. The second indicator concerns a *harmonious mix of activities*, in other words, variation, a reasonable amount of activities, activities that are challenging versus relaxing, activities executed alone versus for/with others and a lack of activities in work or leisure time that draws energy from all other activities. The third indicator of activity balance concerns *consistency with the individuals' values*, in other words, whether the individual experiences an activity as meaningful and challenging.

The activity balance can be poorer for individuals in working life if they have an excessive workload, contributing to their inability or lack of energy for activities in their leisure time. However, an individual's activity balance in working life can be better if their work tasks are stimulating and they have good work conditions, with social support from the manager and possibilities to influence work tasks and working hours [172]. Individuals who leave working life may experience activity imbalance, however, by developing new meaningful routines that challenge or develop their intellectual abilities, positive aspects of working life can be replaced with new activities. Experiencing activity balance between working life and leisure time has proven to be significant in order to experience sound self-rated health [172–174].

## CONSIDER AND REFLECT

- Reflect on whether your efforts and rewards in your personal social environment are balanced.
- How do you currently experience your possibilities of personal social life in relation to your work?

- How many hours do you spend thinking about work each day?
- How many hours do you spend thinking about your personal life each day?
- How do you achieve clear and structured boundaries between your work and personal life? Do you put your work phone away and close your work e-mail at the end of the work day to focus on your personal life? Do you put your personal phone away and close your personal e-mail during a work shift to focus on work?
- How available for work are you expected to be outside your ordinary working hours? If you are expected to be available at all times, do you have the right to compensation in time or salary? Is it clear or unclear what the conditions are?
- Reflect on whether you, during different social ages, have had a need to spend more time at work than expected based on your working hours, with negative effects on your personal life.
- Reflect on whether you, during different social ages, have had a need to spend more time in your personal life than expected based on your working hours, with negative effects on your working life.
- What consequences does constant work readiness have on your personal life, your health and your long-term productivity?
- Reflect on how and what you can do in the future to achieve or maintain good social support and a sense of community in your personal social environment.

# Area 7. The Social Work Environment

The work social environment and inclusion of a workplace community are factors of significance to individuals' ability and willingness to work and relate to their social age [1–3,8] (Figure 13). An individual's identity in the work unit, workplace and organisation is affected by the attitudes in the social context and by the individual's social age. How employees experience their occupational role, occupational culture, hierarchical position and possible discrimination is determined both by people in their own organisation, for example managers and co-workers, and by people the employee meets in their specific occupational role, for example people from other organisations, customers, clients and patients. The manager holds the power to influence the culture of inclusion and a sense of community, however, so do co-workers and the individual themselves. Furthermore, the social work environment is affected by the current culture, the individual's potential exclusion from society and by-laws and supportive measures to promote inclusion in society.

## IDENTITY AND SOCIAL CONTEXTS

Human beings are social creatures, and as previously mentioned, experiencing exclusion in a social context is threatening from an evolutionary perspective. Experiencing affinity and inclusion in a social context strengthens mental health since it is necessary for individuals to feel needed and included.

Individuals take on different roles in social contexts. An individual may have one role in the workplace, for example manager, salesperson, farmer, surgeon, carpenter or teacher, while they take on different roles in their personal life, for example daughter, husband, live action role-player or football coach. Individuals create a personal facade, an identity, based on the social context they are in. Identity is created through constant interaction and dialogue between the developing person and the social environment in their surroundings, history and culture [175,176]. Through interactions with others, and through verbal and non-spoken communication in social interactions, individuals display themselves and each other in different discourses. These discourses contribute to individuals' ability to take on different roles.

Since social contexts contribute to different roles, this is described as creating or "doing" different roles. For example, "doing gender" is described as the expectations of different genders to act and behave in a certain way within a discourse and in a social context, otherwise, individuals may be perceived as unfeminine or unmanly [177,178]. Social age is done through individuals in a certain age group, for example children or pensioners, being expected to behave in a certain way in different situations [1,2,7,80,106]. How individuals act in a certain situation is affected by their sense of self-identity. Self-identity is primarily a mentally and socially constructed story that integrates the reconstructed past, the perceived present and the expected future into a discourse of the own self [176]. During different life phases and social ages in life, the sense of self-identity and different roles are developed in a world where adult life is divided into three parts: education, work and retirement. The individual's self-identity is based on subjective perceptions. At the same time, other people in the individual's surroundings create their own perceptions of the individual, based on how they objectively experience the individual. However, both the subjective self-identity and the objective identity an individual is attributed to are characterised by stereotypical perceptions, created in the discourse and context of the surroundings, so-called norms. If an individual does not belong to the norm, it contributes to them being stigmatised [179]. When using the same routines for a certain social context, these tend to become institutionalised and create stereotypical expectations [175]. Everyone has, in some context, experienced stigmatisation through breaking the majority's norm of dress code, occupation, taste in music, table manners, age, gender, socioeconomic group, ethnicity, sexual orientation, nationality, etc. For an adult individual, the norm is to participate in the workforce and in working life. Therefore, not participating in working life can be stigmatising, for example if the individual is in sickness absence, unemployed or retired. This stigmatisation can cause exclusion, an inferiority complex, poor self-esteem and poor self-confidence, which, in turn, can cause mental health issues. Therefore, the social contexts the individual finds themselves in must be allowing safe, socially supportive and affirmative in order for the individual not to feel stigmatised as a norm breaker. Identity is shaped through repeated exposure to similar social situations, and preparing to take the same actions in a generally similar manner [180]. This creates socially constructed patterns of our self-identity that are difficult to break, which complicates acting in different ways in the future. The culture of the education we have or do not have, the culture of the sector and the workplace we are employed in and the culture of the society we live, in strengthen our self-identity. The expectations we have on our identity can complicate stepping outside of these expectations, for example, a man can find it difficult to go on parental leave if the expectations in the workplace are to prioritise working life above one's personal life, this threatens the individual's social work role and identity. For instance, business executives, development engineers, financial executives or high-ranking researchers may find it difficult to decrease their number of working hours as they approach retirement since their self-identity can be affected if they experience themselves as insufficient and weak, that they lose face and damage their reputation as strong and invincible. Decreasing the number of working hours, changing positions or work tasks can entail other social work identities and patterns, with different social frameworks, contexts and discourses for the individual in the future. People who remain in working life until an older age often do so because they experience satisfaction in their social work environment, and with managers, co-workers, students, suppliers, customers, clients, patients, etc. in the workplace [6,63,80,116,166,172]. However, individuals who remain in working life with increasing age are usually part of a larger social context; they are participating socially both in working life and in their personal life. Furthermore, healthy social contexts are often expressed as a reason for the better mental health people who remain in working life until an older age often have, compared to people who retire at a younger age, and that they maintain a high level of mental health for longer in life. However, we must acknowledge the healthy worker effect, i.e. that it probably is easier for people with good mental and physical health to remain in working life until an older age.

Since a larger amount of people live until an older age, we will probably have to re-evaluate the thoughts and attitudes of identity in relation to the predetermined stages of education, work and retirement. Furthermore, we will probably have to re-evaluate the perception of these three stages as divided life phases and stigmatising social ages throughout life. The division of adulthood in these three phases, social frameworks and contexts has been institutionalised and caused abstract stereotypical expectations of individuals' self-esteem and identity and is no longer applicable when working life must be extended and changed. In an extended working life, it is probable that the division of the social age phases of education, work and retirement is no longer sustainable. It will probably be more sustainable if these phases are integrated and overlap. After finishing education and after working for a while, the individual may need to return to education in order to remain employable in an extended and more sustainable working life. If the health effects of the work environment have made the individual unable to continue their current career, they may need to create a new identity in a new profession, sector and workplace. Perhaps, the individual needs to work less during certain life phases, even if they earn less financially in a short-term perspective, for example during the stressful life phase of caring for small children in the family, or in other cases of personal social commitments. In these instances, the individual can invest their time in relationships, their future health, new skills and competence development. In the long run, this can contribute to better personal finance, improved health, happier relationships in the personal social environment and in the social work environment, and better employability and possibilities of working until an older age in the future.

Furthermore, at certain times in working life, the individual's capacity and ability may not be up to date, or perhaps reorganisations and technology development have rationalised away former work tasks that have previously been meaningful for the individual's self-identity in their career. In these instances, a new education and a new working life can create the context and discourse of a new identity. Furthermore, when the individual is approaching the age of retirement, a new education can contribute to a new self-identity through becoming a student again, or through the possibility of creating a new career besides being a pensioner, perhaps as a self-employed expert in a particular field or as a wise generalist. Being allowed to educate oneself continuously, based on the social work situation, is fundamental to a sustainable working life for all ages. An extended working life presumes that individuals invest in their future selves and their identity in one way or another, in order to secure their employability in a social work community for a longer time.

## GROUP DYNAMICS

A group is a collection of individuals who experience themselves as a community apart from others. They are psychologically aware of themselves and have a common goal. At the same time, they are mutually affected by and dependent on each other to reach their goals. The individuals in the group must be able to tolerate and handle each other's differences and value these as a resource. In order to participate in a group, the individual must have personal characteristics like empathic ability, a willingness to consider other people's perspectives and to participate and contribute. Furthermore, it is expected that the group member should have a stable identity so that the others know who they are and what they stand for and be able to take on a role in the group. In order for the group to be lasting, group members must tolerate tensions and conflicts and have the ability to cope with them constructively and accommodate their own and the group's mutual interests, since conflicts and disagreements more or less explicitly always occur in a group.

A group consists of its participants and their egos, feelings and reactions. Because of this, the group is shaped not only by the individuals' different characteristics like age and gender but also by the individuals' abilities, for example their social ability, tact, cultural and ethical behavioural patterns, motivation, creativity, rationality, knowledge, as well as their ability to assess and predict risks and future events. Furthermore, the individuals' external and internal loyalties, their attitudes towards common goals and their decision-making capacity matter. Formal groups and units are created in organisations in order to achieve the production goals of the organisation. Informal groups are created to cater for social interactions, friendships and contacts. If the work group is large, subgroups may occur, i.e. smaller groups within the larger group. Secondary groups are characterised by sporadic contact, for example in a workplace. Member groups are groups the individual is a member of. A work unit is by definition a member group, but so are memberships in a sports club, political party or a chain of clothing stores. A reference group is a group to which the individual compares themselves, strives to adjust and be acknowledged. Furthermore, individuals are part of a primary group, which has a limited number of members who have close, personal and stable relationships with each other, for example the individual's close friends and family.

In all groups, dynamic and ongoing processes balance the individuals' egos with what they must do to solve their mutual task. The term group dynamics refers to how individuals in a group act and interact in order to execute their mutual task, for instance solving a problem, making a decision or producing something. This demands interaction between the members of the group and it is a continuous process (Figure 14). The different phases and processes of group dynamics, which all groups experience in a continuous cycle, are as follows:

- *The initiatory phase*, characterised by uncertainty, unclear norms and roles. This phase is most apparent in new groups, or if a group faces a new problem to solve.
- *The honeymoon phase*, characterised by a growing mutual appreciation, affinity, team spirit and security in the group, when everyone wants to find their role.
- *The integration phase*, characterised by the crystallisation of roles and subgroups, norms become clearer, communication more coordinated and content more work-oriented. The individuals want to solve the problem, take on different roles and position themselves hierarchically. Some are satisfied

**The mature efficient group.** The group can manage to resolve the task and knows their different roles.

**The conflict phase**
Conflicts occur in all groups and co-operations. It becomes clear when individual differences become apparent and are acknowledged. If conflicts are managed, it can contribute to group unity, and strengthen the group. However, unresolved conflicts risk dividing the group.

**Initiatory phase**
This phase is characterised by uncertainty, unclear norms and roles in the group

**The honeymoon phase**
This phase is characterised by a growing mutual appreciation, affinity, sense of community and security in the group.

**The integration phase**
Different roles crystallize, subgroups and norms become clearer, communication becomes more coordinated and content more work-oriented to achieve the goals. Someone take on the role of a leader. If anyone else wants to be a leader and feels uncomfortable, it becomes apparent that the "honeymoon" of the group is over.

**FIGURE 14** The five-phase cycle in the development of group dynamics and group maturation processes.

with their role, others would have preferred a different role or task. The social group hierarchy attributes the individual to being superior or inferior, having more or less knowledge, ability, etc. These attributes are based on different characteristics, norms being one of them. The higher position an individual has in the group hierarchy, the more power, authority and status in the group. Furthermore, individuals who have higher positions in the group hierarchy often tend to take initiatives in group interaction. It not only affects how many people want to come into contact with the individual in the group but also what possibilities the individual has to interact with other groups. At the same time, all members usually try to assert their own power in a specific area and their position in the group, which tends to result in the group members trying to establish their own territory.
- *The conflict phase* sooner or later occurs in all groups but can be more or less apparent depending on the group members and the task to resolve. Conflicts can contribute to some individuals switching roles in the group. Furthermore, issues which group members have not wanted to address during the honeymoon and integration phases may surface, which can result in consensus on how to solve the task. Accordingly, conflicts can initiate development and be positive, unless they remain unsolved if the group members position themselves and eventually break up the group. There are different types of conflicts. *Conflicts of power* are usually based on different perceptions, for example regarding how a task should best be solved. *Role conflicts*

are based on different opinions regarding who should execute and take responsibility for what. *Conflicts of interest* are based on different interests, goals and priorities. *Value conflicts* are based on different standpoints, concerning, for example, morals, ethics, politics and concepts of religion. *Behavioural conflicts* are based on how individuals treat each other, for example if someone holds negative attitudes, harasses, acts selfishly or is a "know-it-all". *Pseudo conflicts* are based on frustration and underlying problems that no one has the power, possibility or ability to resolve. This results in conflicts concerning other, often insignificant things, in order to let off steam, which can risk causing general dissatisfaction in the group. If the group proceeds strengthened from the conflict, it results in the increased maturity and wisdom of the group, "the mature group", where everyone knows their roles and can manage to resolve the problem together. However, when the group receives a new task, if a new member joins the group, or if a current member leaves the group, the process starts over with the initiatory phase.

## Managing Groups

Problems can emerge in the organisation of work, groups and the social work environment, these problems can, in turn, have negative impacts on individuals in the workplace. Two concepts used when identifying problems in the social work environment that matter to management are *bad conditions* and *bad work* [15]. Bad conditions in the workplace can, for example, be a poor physical work environment and poor material conditions to work, which, in turn, increases the tendency for tension and anxiety in work. This can, in turn, increase the risk of individuals making mistakes in their work tasks and the occurrence of accidents. Both of these can result in bad work: the experience of personal shortcomings among employees and managers, social tension in the workplace, scape goat mentality and bullying. The problems that bad work and bad conditions represent are often magnified and emphasised in the workplace, which can cause disturbances in the work flow and production, for example if conflicts have occurred between different work shifts and work units. Worrying about taking on another work unit's tasks when one's own work shift begins, without knowing how carefully and well the previous work unit has executed their tasks, what has been executed and how it will affect one's own work results can contribute to mental demands and tear, and ill-health caused by high demands and low control.

When introducing new work models and reorganisations, it is crucially important to consider the social conditions and informal organisation in the workplace as much as the technological and formal organisation [15,16]. Socially and technologically alienated work results in poor well-being, ill-health and production loss. The entire work team, in other words, individuals working in all shifts, units and individual employees, should understand the significance of each other's work tasks. Social balance in the work team and a well-functioning formal, technological organisation can only be achieved if relationships in the work team are well-integrated and socially balanced.

## CATEGORISING

When categorising and discussing certain aspects of homogenous groups, generalisations regarding the characteristics of the participants are common. Categorising can be positive since it is a way of finding out which groups may need particular measures. Even if stereotypical perceptions of the categorised individuals' characteristics occur, the individual employee's abilities and needs risk being forgotten. Age is used in the workplace to manage employees and their possibility of, for example, receiving a higher salary, career and competence development. This is based on attitudes and perceptions of age and its significance regarding work efforts and knowledge. Age is sometimes considered a legitimate reason not to hire or to let go of employees. Particularly women appear to be perceived as being of the wrong age. They can face difficulties in finding new employment when they are considered to be of childbearing age. When women are no longer of childbearing age and have entered menopause, they are often considered too old to be employed. Furthermore, women are sometimes perceived as seniors at a younger age than men. A survey conducted in the UK with more than 1,600 participating managers showed that the managers considered a female employee to be senior at 48 years of age on average, and a male employee to be senior at 51 years of age on average [181]. In a study with more than 900 participating managers, 41 percent of respondents regarded it as important to keep employees in the organisation until 65 years of age, and 14 percent stated that they wanted to keep employees in the organisation until 66 years of age or older [107]. However, a deeper analysis showed that the attitude towards keeping employees in the organisation until an older age was affected by when the managers themselves wanted to retire. A statistically significantly larger proportion of managers, who wanted to work until an older age themselves, considered it important for the organisation to keep their employees until an older age, compared to managers who themselves wanted to retire earlier.

## POWER AND MASTER SUPPRESSION TECHNIQUES

In all social environments, power is an important concept and tool in management. In working life, power is a significant factor to manage work units and individual employees to make them execute tasks and act in accordance with the goals of the power holder in the workplace. Power can take different forms [182]:

- *Reward power* which is based on the possibility of promotion, increased salary, acknowledgement and pleasant treatment
- *Coercive power* which is based on the possibility to cause negative consequences
- *Legitimate power* which is based on an individual's position in a social structure
- *Referent power* which is based on people identifying with or admiring the power holder

- *Expert power* which is based on an individual's access to knowledge and experience
- *Information power* which is based on an individual's access to valuable and desired information
- *Connection power* which is based on an individual's connections to influential people

Individuals can exercise power through different master suppression techniques [183]. The five master suppression techniques are as follows:

1. *Making invisible*: silencing or marginalising individuals through ignoring them. No one listens to what you have to say. Others can repeat what you just said, and everyone listens and praises them like they came up with the idea. Other examples are riffling through papers, whispering or yawning when it is your turn to speak.
2. *Ridiculing*: describing someone's argument or person as silly and unimportant. You are ridiculed and people make jokes at your expense. An example is to make remarks about your looks in front of a group. It can also be that someone ridicules what you have to say, despite it being an important message you want to convey and expect others to take seriously. You are perceived as grumpy if you remark on this and are expected to accept the joke made at your expense.
3. *Withholding information*: excluding someone or marginalising their role by withholding important information. For example, a decision you should have been part of making at a meeting is made in an informal situation where you are not included instead, like in a pub or at a golf course. Co-workers may even let you understand that they have had a meeting and express that it was a shame you were absent.
4. *Double binding*: whatever you do is a failure. You have to make a choice but are belittled and punished regardless of what choice you make. If you are absent from the meeting, you are a bad co-worker who is unreliable and does not prioritise work. If you participate in the meeting, despite having an ill child at home, you are a bad parent. If you are very careful in your work tasks, you are told you get nothing done or that you are too slow. If you work quickly, you are told you are sloppy and irresponsible.
5. *Heaping blame and putting to shame*: making someone feel embarrassed because of their characteristics or telling them that something is their own fault. Blame and shame is a combination of ridicule and double bind. Even if you have not been informed of a meeting, you are told that you should have found out about it. No one listens to what you say in the meeting, which makes you feel that you have expressed yourself unclearly, stupidly or uninterestingly. If you propose a coffee break, you are told not to slack off. If you say something when someone pauses to think, you are accused of interrupting them. You are repeatedly interrupted in the middle of a sentence and left with a feeling that what you have to say is unimportant. You are criticised for not following (informal) rules that have never been expressed and may even have been made up to embarrass you.

Everyone uses master suppression techniques, many people have a favourite they often use in order to gain and assert power. However, the fact that everyone uses master suppression techniques every now and then does not mean that it is an acceptable behaviour. Antidotes to master suppression techniques are questioning, respecting, asserting reasonable norms, making visible and allowing space for everyone, promoting healthy communication, being honest and demanding honesty in return, informing and breaking negative patterns.

## Managers and Leaders

A manager is the formal power holder in the workplace, and the title manager refers to a formal position in the organisation. A leader is a person who makes everyone strive to achieve a common goal. However, a manager is not necessarily a good leader who can engage their employees. Similarly, a good leader does not necessarily have the formal title of manager in the workplace. *Management* is a concept that entails the work tasks of the manager and leader when they manage and divide work, i.e. when they plan, organise, recruit, manage, lead and control an organisation or group to achieve a certain goal. There are different approaches or styles for how managers work, which are somewhat incorrectly known as leadership styles. Some categorised approaches and styles often used to describe a manager's leadership are the pacesetting leader; the authoritative, commanding leader; the visionary leader; the democratic leader; the coaching leader and the empathetic, laissez-faire leader. However, managers and leaders often use several of these different approaches and leadership styles in their work, depending on the employee and the task.

*The pacesetting leader* strives for high standards and sets high goals and expectations for themselves, the organisation and their employees. They want to achieve results quickly, therefore, it can be a beneficial leadership style for short-term deadlines. However, the pacesetting leader may risk expecting an excessive pace for their employees in the long run, which can inhibit innovation and contribute to insecurity and a heated, competitive atmosphere between employees. The risk of scape goat mentality increases if deadlines are not met. Furthermore, the pacesetting leader can risk losing faith in their employees if they find it difficult to live up to the high expectations. Moreover, this leadership style risks causing stress reactions, ill-health and sickness absence among employees to a greater extent, unless they are provided with sufficient possibilities for recuperation.

*The authoritative, commanding leader* is perhaps mainly associated with the army, police force and emergency or surgery departments in health care. The authoritative, commanding leader demands compliance. This leadership style is needed and effective in situations that require clear leadership, for example in a crisis situation. In these situations, the authoritative leader provides a sense of security to rely on. However, this leadership style may also be needed in certain situations where other options fail, for example when the leader must manage troublesome employees in an authoritative way. However, an authoritative and commanding leader inhibits long-term creativity, spontaneity and innovation development in the organisation.

*The visionary leader* perceives a far-off organisational goal, wants to be creative and paints picture of how the common goal should be achieved together with their employees as a team. However, the steps to achieve the goal are unclear.

The visionary leader points to the end goal but does not know what the process of achieving it should look like. Employees have freedom under responsibility in their tasks and are allowed to decide on how to solve them themselves. The visionary leader can inspire creativity and independence, for example in uninspired teams. This increases the possibility of spontaneity and innovations to change and develop the organisation. The visionary leader is very useful in reorganisations when the organisation faces a crisis or needs to change tack. However, if employees are not mature enough to manage their own work and need clarity and directions, this can be a disadvantageous leadership style and create more insecurity. Accordingly, this leadership style is not beneficial in situations where employees need more support or clear directions and guidelines on how to execute their work tasks.

*The democratic leader* strives for consensus in the team and in the organisation. Everyone in the workplace should be included, speak their mind and come to an agreement. The democratic leadership style is beneficial if employees are very independent and experts within their fields. Furthermore, it requires that everyone has access to the same and equal amount of information regarding the organisation and to achieve the organisation's production and budget goals. Moreover, this leadership style is appropriate to make everyone feel included in situations where it is important that all employees understand and support the organisational goals. The democratic leadership style will not work if the manager has much more knowledge and information than they convey to their employees, it will create confusion and insecurity instead. Furthermore, the democratic leadership style is inappropriate if rapid decisions and emergency actions are needed in order to save the organisation or to achieve goals. In these cases, long discussions and attempts to achieve consensus in the team will not work.

*The coaching leader* focuses on developing their employees, their skills and best qualities in order to achieve the individual goals of the employee and the common goals of the organisation. The coaching leader is like a mentor and sounding board for their employees and strives to identify and develop the strengths and weaknesses of each individual. However, this requires that the individual employee has goals and is aware of them. Therefore, this leadership style works best with motivated employees who enjoy their work tasks. In order for the coaching leader to achieve organisational goals, they must also be good at manipulating and adjusting the employees' individual goals to align with the organisational goals, in order for both sides to benefit from the results. If an employee lacks motivation or interests that align with the organisational goals, it is very difficult and a waste of resources to try to coach them. These cases need clearer leadership and directions regarding tasks to be executed.

*The empathetic laissez-faire leader* wants everyone to enjoy themselves and promotes relationships between people in the workplace. The empathetic leader cares for their employees to have the emotional support they may need. The empathetic leader is most appropriate and outstanding when an individual employee, or the entire team, is going through tough times. However, the relationships risk getting in the way and diverting focus from clear directions and uncomfortable decisions to achieve the organisational goals, which can result in decreased productivity in the work unit. The laissez-faire leader often avoids conflict and lets the work unit and

# The Social Work Environment 131

employees manage their own work, which enables informal leaders and insecurity regarding what goals are expected to be achieved at work.

Regardless of leadership style or mix of leadership styles, managers and leaders who have the ability to perceive their employees' individual needs, wishes and emotions are described as better leaders. Furthermore, there is a positive correlation between empathy and leadership success. An important factor is that the employee is acknowledged and appreciated for their individuality in the social work environment. Leaders who perceive their employees as individual subjects, rather than interchangeable objects, provide better conditions to create a sustainable social work environment that promotes health. Leaders who are focused on their employees show better results in production, efficiency and well-being among employees, compared to leaders who are focused on production and who mainly perceive their employees as replaceable cogs in the organisation and production. The three main components of good leadership are as follows:

- The manager understands their responsibility for their employees, and for employees' experience of their work as valuable.
- The manager takes responsibility for plans, priorities and boundaries related to the employees' work tasks and work description. Furthermore, the manager should be able to make employees follow the plan, even if work tasks are executed with freedom under responsibility (i.e. that the employee is trusted to take responsibility for and prioritise their own work tasks and working hours, provided that they execute what they are expected to).
- The manager takes responsibility to create and maintain the employees' ability to execute their work tasks.

## THE NEED OF AN AGE CONSCIOUS AND SITUATIONAL LEADERSHIP

Overall, the support of managers and employers is crucial for their employees' possibility of working. A young or recently recruited employee needs an instructive leader in order to experience a sense of security and well-being and to experience inclusion and support in the social work environment to develop their employability. However, an older and experienced employee generally needs more support and appreciation for taking their own initiatives and independency in order to experience well-being and develop their employability. The experience of leadership in the organisation, whether employees feel acknowledged and included, and the level of social support in the workplace affect whether employees want to participate in the social work environment. Bad leadership and a lack of inclusion, autonomy and social support have shown to correlate with lower levels of work ability.

In age conscious and situational leadership according to the SwAge™ model, it is important to take biological, chronological, social and mental age in relation to the nine spheres for actions for the ability and willingness to work into account. The manager and leader must use different methods and measures in their instructions and support in order to manage the employees' functional diversity in the work situation regarding, for example, physical work demands, demand and control, working time, financial incentives, the personal social environment, participation, motivation, possibilities of utilising skills and learning regardless of age. *Age-conscious leadership* and *age management* are described as tools to increase employees' possible ability and willingness to stay in the workplace, and for a healthy and sustainable working life for all ages [1,10,11,80,107,108,116,149].

Different sectors and lines of business appear to hold different attitudes to employees' age, partly due to the production content, but also due to the organisational culture and historical traditions. Organisations characterised by knowledge, like universities, traditional craft organisations and organisations where the experience of customer, client and patient treatment is significant, are examples of organisations where attitudes towards senior employees usually are positive. However, attitudes towards senior employees are usually not as positive in organisations that demand high physical capacity, speed, agility and strength. Examples of such organisations could be in competitive sports, wood cutting, construction, etc. and in sectors characterised by rapid technology development, such as IT companies and electro-technology industries, where the latest, updated knowledge is desired. Some organisations can be described as "neophiles", in other words, they have an extreme strive for youth and renewal [184]. Ageing can be particularly difficult in these types of organisations. However, the willingness to remain in the workforce increases if the employer holds a positive attitude towards and values senior employees' experiences and knowledge [1,11,107,108,149].

Bad leadership and insufficient support in the social work environment are associated with lower levels of work ability and less willingness to stay in the workplace. Few organisations and companies take practical measures to increase the possibility of an extended working life, despite the fact that many managers state the increased number of elderly citizens in the population creates significant challenges in society. It appears that above all, the managers' own retirement plans, budgeting and required production efficiency affect their attitudes towards measures for senior employees. Furthermore, the attitudes vary depending on the senior employee's occupation. Mainly employees with special skills are offered continued employment and measures to facilitate an extended working life until an older age [6,107,108,172]. This should change, managers and leaders should take all nine determinant areas (Figure 1), and how these associate with different age definitions, into account in order to promote and enable a healthy and sustainable working life for all ages.

## SOCIAL AND INSTRUMENTAL SUPPORT AND INCLUSION IN THE WORKPLACE

No human is an island, individuals interact and often need support from other individuals in their daily lives.

Social or emotional support is a concept usually described as the accessibility to support from other people, which at least fulfils basic social needs through interactions with other people. Instrumental support refers to support from practical resources and access to appropriate facilities, for example the possibility of using correct materials, equipment, technology, sufficient staffing and supported knowledge. The sphere for action *Relationships, social support and inclusion* in the

# The Social Work Environment

SwAge™ model includes the management of social and instrumental support in all nine determinant areas for a healthy and sustainable working life for all ages. In other words, making sure that the social and instrumental support is sufficient based on individual employees' diagnoses and functional diversity; physical work environment, mental work environment, recuperation; personal finance; personal social situation; social work situation; motivation, stimulation, work satisfaction; knowledge and competence development. Besides, digitisation has contributed to new opportunities for long-distance relationships and long-distance communication, which has affected and changed the social context of work in many workplaces. Part of this is that employees' relationships with customers, clients, patients, students, etc. are increasingly done via digital technology. Another part is that workplace presence has increasingly turned into workplace separation, and relationships with colleagues, technical and administrative support, the manager, etc. take place increasingly remotely via digital technology in more and more workplaces. Through the increasing digitisation in working life, the demands on the manager and the leadership have also changed. The manager must now increasingly lead the work and provide support and feedback to employees remotely and via digital technology.

The experience of support and a sense of security with managers, co-workers and other people in the workplace and in the social work environment is of great significance to reducing and coping with stress. Experiencing happiness enjoying the time spent at work, and knowing who is trustworthy and supportive through emotional and instrumental support in the workplace are factors of significance for a sustainable working life. Experiencing appreciation and satisfaction with the workplace and occupation are very important factors for well-being and long-term health. The *buffering hypothesis* illustrates how individuals' need for support from other people can work as a buffer against demands and stress [185]. Social support is an efficient buffer against stress and increases resistance to time pressure and excessive demands, which, in turn, has proven to decrease the risk of psychosomatic symptoms such as pain in the neck, shoulders and back, headache and anxiety. For example, individuals need support from co-workers, managers and superiors. Opportunities to talk to and exchange thoughts with others strengthen the sense of inclusion and community in the workplace, for many people good friendships are what make them experience well-being at work. Feedback on results from the manager, customers, patients and co-workers is crucial for motivation and performance. Appreciation and acknowledgement from superiors is an important part of social support.

The process of how work demands are experienced as work stress, and cause health effects, has been described in a figure by Le Blance and others, based on a summary of theories and empirical studies [186] (Figure 15). Different types of work demands, i.e. stressors at work, can result in different types of stress reactions, demands and impacts on health. The relation between experienced stressors at work and what stress reactions they result in is affected by the individuals' own resources to cope with stress, together with situational resources, in the form of social and instrumental support, as a function to cope with stress.

Sociologically speaking, two mechanisms explain why work matters to the individuals' existence. The first mechanism is necessity, in other words, the possibility of physical survival.

**FIGURE 15** The process of experiencing work stressors: the influence of personal resources and supportive social resources at work, resulting in a plausible health effect out of a stress reaction [186].

The other mechanism can be perceived as the level of social interaction in the individual's environment, i.e. work as one of society's normative, sanctioned duties which the individual should fulfil in order to live a righteous and ethical life. Therefore, breaking this norm and not participating in the workforce can be negative for individuals in several ways. For example, individuals, especially men with high status at work, who retire prematurely have shown deteriorated mental health after leaving work for retirement [6,187]. Leaving working life, which has been the source of experiencing high status in a community through the social work environment, decreased these individuals' social status in life and affected their life satisfaction negatively. However, individuals can also have completely opposed experiences, for example being included or excluded from the social community in the workplace. Experiencing a lack of support and exclusion from a social community generally decreases the willingness to remain in the workplace and increases the wish to leave the workplace [6,15,188,189].

## THE SOCIAL EFFECTS OF INFORMAL ORGANISATIONS IN WORKING LIFE

Individuals' willingness, possibility and ability often depend on the type of motivation they experience for different tasks. It is not a question of what provides stimulation at work, but rather that the experience of stimulation at work is important for the willingness to remain in the workplace. When measuring employees' work ability, the manager's attitudes and support towards the employee not only proved to have practical significance to well-being but also appeared to have a significant correlation to high levels of work ability [13,190]. Furthermore, increased social support provided increased work ability for individuals who had previously had lower levels of work ability. At the same time, situations with a lack of social support, acknowledgement and appreciation at work correlated to decreased work ability.

Early theories show that being acknowledged for one's work efforts and in the social interplay in the work unit affects work, i.e. *the informal organisation*. Motivation and work satisfaction has been the subject of several studies, as early as in the beginning of the 20th century. For example, experiences were conducted in the Western Electric Company's factory in Hawthorn, Chicago [189]. One of the experiments has been called the lightning study, since it measured productivity in different light conditions and concluded that if employees were acknowledged in their work situation, it resulted in increased work performance and decreased sickness absence. The effect of attention and experience of acknowledgement increases work satisfaction and is therefore usually referred to as the Hawthorn effect (read more in Section *Health Promotion and Organisational Productivity*). Another observed conclusion from the experiments conducted at Western Electric's factory was the significance of the informal organisation and its social interaction for the management and production outcome of the formal organisation. All employees in the work units were assumed to follow the informal hierarchy in the work unit, which only partly coincided with the formal hierarchy. People who stood out did not help others, told others to the management which caused implications, did not participate in social activities during or after working hours, were insufficiently or excessively productive compared to group standards, were punished, for example through social isolation. The informal organisation affected production and work flow to at least the same degree as the formal organisation in a work organisation [15,189]. A socially ineffective structure risks causing decreased work ethics, counteract recuperation and increase staff turnover. Not experiencing well-being and experiencing work as tedious cause indifference to the quality of work, unconcern to take responsibility and initiative, high absence and occurrence of psychosomatic symptoms. However, the greater influence an individual has on their work, the greater their engagement is. It is fun to work, develop and manage work tasks when one has the possibility to decide in what way and when they should be executed, provided that there are sufficient resources and that one has the necessary knowledge. Positive social interactions counteract exclusion and social isolation in the workplace and increase participation, a sense of community and the motivation to remain in the workplace. In order to prevent social isolation in the work situation, co-workers may, for example, create small informal groups of two to five individuals who help each other out with work tasks. However, some people are excluded from these informal groups, for example lone wolves, people who do not help others, people with poor work ethics or a bad reputation, risk takers or people who are bullies or want to think that they are better than others. Furthermore, it is more difficult for people with lack energy, for example due to older age, physical issues or mental symptoms, to be included in these informal groups. However, it is possible to counteract exclusion and social isolation through participating in spontaneous activities, for example playing board games with co-workers in the workplace or in leisure time, practicing sports with co-workers or socialising in other ways in activities that are not part of the work tasks. Being included in an informal group is particularly important for people who are already isolated in their work tasks and therefore are not part of a formal group in the workplace. However, informal groups can cause issues when problems occur at work and employees readily try to clear themselves of guilt or wrongdoings in order not to be shown in bad light to other co-workers in the informal groups and to managers. This risk increases when employees try to find a scape goat, which results in *bad work* (read more in Section *Organising Groups*). Senior employees who are no longer able to keep their previous work pace can experience accusations. However, the employee who points out the problem to managers or management, or accuse someone of causing the problem, risk being subjected to social isolation themselves.

People who do not experience well-being in their work situation due to social isolation appear to have a greater tendency for "recuperative absence" [6,95,106]. Some people even consciously hurt themselves at work in order to avoid the workplace and subsequently leave working life in order to avoid social isolation and exclusion in the workplace. However, there are often other reasons for absence from work and staying home, for example to avoid becoming worn out, which, in turn, can cause social isolation. The individual who stays at home does not take their social responsibility or execute their work tasks, which means that their co-workers need to execute their work tasks as well, which often results in the individual being subjected to social isolation and exclusion from the informal group.

If an employee chooses to leave working life permanently, for example through early retirement, due to lack of well-being or social isolation, this can in some ways compensate for the isolation by avoiding it. However, the employee may potentially be unable to contribute as much to their family's finances, which, in turn, contributes to decreased status in their family's and society's perception. If the individual experiences that they are unable to cope with the lack of support in the workplace and leave work, this can simultaneously contribute to the individual feeling maladjusted in relation to societal norms of labour force participation until a certain age. When the individual can no longer cope in a social work environment with a lack of support but at the same time loses their status as a labour force participant, the risk of mental symptoms and feelings of inferiority increase [6].

## DISREGARD, BULLYING AND DISCRIMINATION

Individuals can be stigmatised, disregarded and discriminated against on different grounds. A social work environment characterised by bullying can be a significant reason for the development of ill mental health, exit from the workplace and premature retirement. Individuals who are subjected to bullying often develop anxiety, depression, physical symptoms and sometimes even suicidal thoughts (read more in (3) *Mental Work Environment*, Section *Stress*). Because of this, it is important to prevent bullying in workplaces and in society. At the same time, being subjected to bullying is stigmatised, and some people are ashamed of admitting to the fact that they are bullied. The bullied individual rather expresses that they experience other issues in their work situation, for example high stress levels and no resources to cope with stress, a lack of control in work tasks and excessive demands from the surroundings, to be the cause of their mental symptoms, while in reality the bullying is the main cause.

The social identity an individual is attributed to probably affects the status, inclusion and sense of belonging they experience in their social environment. People continuously, consciously or unconsciously seek identification, a sense of community and belonging. Accordingly, they seek situations and people they recognise, where they feel equal and experience the greatest sense of community. This also entails a certain status. Furthermore, the processes of the brain probably affect what community the individual is part of since the brain facilitates daily processes through sorting and categorising impressions and memories (read more in (9) *Knowledge, Skills and Competence Development*, Section *Memory*), which also applies to all people the individual meets. The brain sorts the people we meet into different categories, for example people with certain attitudes, cultural identity or the hierarchal status of their occupation in the organisation or in society. However, this categorisation must not affect the attitude towards other people, how we treat them or what we expect from them. Categorisation and discussion of homogenous groups, for example a certain age group, must not cause generalisations and stereotypical conceptions of the group [1,2,9,149,166].

The concepts of horizontal and vertical segregation are often used when discussing disregard and segregation. Horizontal and vertical segregation, based on, for example, age, gender, functional diversity, religious views, gender identity or gender expression or ethnicity, probably has great significance to our ability and willingness to work, or if we want to leave work (Figure 16).

*Horizontal segregation* refers to an uneven distribution in the division of positions in a hierarchy, which different individuals have based on different categories. For example, the distribution of women and men between sectors and occupations is not equal, which is usually described as female- or male-dominated sectors and occupations based on whether a majority of female or male employees work there. Furthermore, the horizontal segregation in different occupations contributes to different perceptions of status for a certain occupation. Female-dominated occupations are usually perceived as having a lower status compared to male-dominated

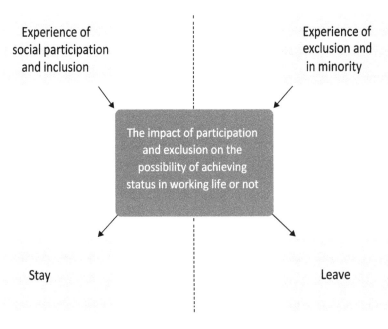

**FIGURE 16** Model of the significance of experiencing inclusion in a social group, alternatively experience of exclusion and in minority and the impact of participation and exclusion on the possibility of achieving status in working life or not, as well as its effect on the willingness to stay in or to leave working life.

**Source: Nilsson [9].**

occupations. Furthermore, individuals employed in female-dominated occupations often do not feel appreciated to the level of their occupational value.

*Vertical segregation* refers to an unequal distribution of men and women in different positions within the same occupation and sector. Women systematically hold lower positions with lower salaries, while men hold managerial positions to a greater extent. Horizontal and vertical segregation can also contribute to individuals who are in the minority experiencing exclusion, which affects the significance of a sustainable working life for all ages, and those who have the ability and willingness to work until an older age. People of gender minority in an occupation dominated by a certain gender often want to retire earlier compared to individuals of the gender in the majority in the same occupation [9,94,106]. For example, studies in health care showed that a larger proportion of male medical practitioners, compared to female medical practitioners, expressed that they could consider staying in working life until after 65 years of age.

Furthermore, female medical practitioners expressed, to a lesser extent than male nurses, that they considered working until after 65 years of age. From an organisational, hierarchical perspective, female medical practitioners hold a higher position than male nurses. However, male dominance is often considered obvious in the profession of medical practitioners, and confusion about gender identity and occupational identity is common. It is not easy to be a woman in a male-dominated profession and vice versa. When women enter traditionally male-dominated professions, thereby breaking the tradition, it can result in them entering a "no-man's-land", being perceived as deviant and outside of traditional gender roles in culture and society.

Similarly, men who enter female-dominated professions are perceived as breaking male gender roles. The motivation for a continued working life may be dented if the individual is continuously perceived as an outsider in the workplace. Men working in traditionally female-dominated professions have shown a wish to retire early, compared to men in other professions, and wish to retire earlier than female employees in the female-dominated professions [9].

Gender-marked professions could cause health problems [9,191,192]. If an occupation is female-dominated, it could have a negative impact on the status of the occupation. Men who decide to remain in female-dominated professions often appear to make career progress more rapidly than women in the profession, since being a man generally entails higher status than being a woman However, career progression must be possible in order for this to occur. In the female-dominated profession of nurse assistants, there is no such tendency. In the female-dominated profession of nurses, more men in general stated that they wanted to work for a slightly longer time. Possibly, the status the men had achieved through the vertical gender segregation, since career progression is possible in another way for nurses, may have contributed to male nurses considering working for an additional couple of years. Some gender theories describe the reason why it is important for men to achieve a higher position in order to want to stay in the workplace. Some gender researchers state that men are "raised" by their society and culture into a masculine homo-social group where they are expected to be dominant and take their tasks as providers more seriously than women [192–196]. If men do not achieve high status or position in the female-dominated profession, or if the profession does not provide these opportunities, they risk being subjected to exclusion through not belonging to the dominant gender in the occupation while simultaneously not maintaining the gender role stereotype, and vice versa, though, due to the critical discussion, it is still far from achieving anything close to gender equality in some sections [194].

However, gender is not the only factor of significance to status, inclusion and tendency to stay in the workplace or wish to leave it. The experience of horizontal and vertical segregation based on, for example, age, functional diversity, religious beliefs, gender identity or gender expression and ethnicity can also be of great significance. Therefore, making working life more integrated and contributing to greater inclusion is an important aspect of enabling a larger amount of people's ability and willingness to participate in working life until an older age.

### ANTI-DISCRIMINATION LAWS

The purpose of the European Commission directives is to prevent discrimination and promote equal rights and possibilities. This is defined in the following directives:

- Directive 2000/43/EC against discrimination on grounds of race and ethnic origin
- Directive 2000/78/EC against discrimination at work on grounds of religion or belief, disability, age or sexual orientation
- Directive 2006/54/EC equal treatment for men and women in matters of employment and occupation
- Directive 2004/113/EC equal treatment for men and women in the access to and supply of goods and services

Anti-discrimination legislation is mandatory. Discrimination can occur in the following ways:

- *Direct discrimination*, which means that someone is disregarded through being treated worse than someone else who would have been in a similar situation if the disregard occurs in relation to the grounds of discrimination such as gender, sexual orientation, religion or belief, ethnicity, functional diversity or disability or age.
- *Indirect discrimination*, which means that someone is disregarded through decisions, criteria or manner that appears to be neutral but that can disregard certain people in particular, unless the decision, criteria or manner has a legitimate purpose, and the means used are appropriate and necessary to achieve the purpose.
- *Inadequate accessibility* for people with functional diversity but with regards to financial and practical conditions.
- *Harassment*, which means a behaviour that violates someone's dignity and that is related to any of the grounds for discrimination.
- *Sexual harassment*, which means a behaviour of sexual character that violates someone's dignity.

Studies show that age discrimination appears to be the most common reason for discrimination in Swedish workplaces, followed by gender discrimination [26,138]. In working life, age is often used as a guide for employees' wages, career opportunities and competence development in the workplace [197]. Stereotypical perceptions based on individuals' social ages can be positive, however, they can also be negative and result in discrimination. For example, several studies have shown that negative perceptions of senior employees generally entail that they have lesser tendencies to change, are ignorant of recent knowledge and technology and are slower compared to younger employees [11,107,108,149,166,198]. A study of managers' attitudes towards their senior employees revealed that 81 percent of managers thought that their senior employees found it difficult to accept changes and re-organisations [107]. For the employees working for these managers, such prejudice can be expressed through fewer opportunities to receive competence development, and participate in new projects and investments [4,6,11,107,199,200]. Furthermore, studies have shown that senior employees have marginalised, disregarded and condescended to younger employees because senior employees feel threatened by being pushed out by younger employees [11]. Younger employees were attributed to being sloppy, irresponsible and not as trustworthy as senior employees. In other words, age discrimination can affect individuals of all ages.

Disregard or discrimination on the grounds of age is described as *ageism*. The definition of ageism or age discrimination is an unreasonable prejudice and discrimination of an individual or group who are attributed to certain characteristics based on their age. Ageism or age discrimination can be directed towards people of all ages and occurs when an individual, based on their age, is wrongly perceived as unsuitable or inappropriate for a certain purpose. Forcing individuals to leave working life at a certain chronological age could be perceived as ageism, however, so would forcing individuals at a high biological age to work without adequate aid [198] Ageism was first described in the Washington Post in 1969. The term was subsequently used by, for example, *The grey panthers* in their struggle for seniors' rights in society. The grey panthers was established by Maggie Kuhn in 1970 as a response to her being forced to retire from the Presbyterian Church at 65 years of age. Besides opposing mandatory retirement at a certain age, the grey panthers has questioned other ageist laws and stereotypical perceptions based on age.

However, the term ageism was not used to a great extent in Europe until the start of the 1990s, when it was used in the context of worry regarding the impact ageism has on premature exits from the labour market. Age discrimination and ageism directed towards employees are statistically related to premature retirement, for men in particular. Because of this, the European Union introduced a directive against age discrimination in 2000 (directive 2000/78/EG). This directive meant that every country in the European Union should introduce legislation against age discrimination or ageism in 2006 at the latest.

## NORMS, NORM BREAKERS AND POWER

People who are discriminated against are often the same people who are marginalised on the grounds of discrimination, and who have less power regarding the cultural norm and power elite in the social group they are situated in when the discrimination occurs. Group norms and finding unifying factors between participants in the group are important for the sense of community. When we meet new people, we often begin to seek mutual interests in order to find a position in the group. If someone says they are from a particular city or have attended a certain school, they are often questioned if they know a person who has attended the same school or is from the same city. In order to remember names and who is who in the group, the brain categorises group members based on their gender, appearance, occupation, social group, etc. in order to facilitate the building of memories. This results in some people being perceived as more similar to ourselves, while other individuals stand out from our own norms and are perceived as norm-breakers. When the time comes for a group project, or going for lunch or a coffee, group members often tend to divide themselves into different norm groups where the individuals recognise themselves and feel secure in the otherwise new, and perhaps partly insecure, situation. Individuals who have the most common characteristics or attributes, and who often are most similar to the leaders, tend to find it easier to blend with the new group and get a higher status, categorise and set the norms in the group, while people who are most far from the most common characteristics tend to have a lower status in the group. Therefore, belonging to the norm and fitting in with the group's expectations result in different power expressions, for example a Caucasian middle-aged man in a leading position is perceived as the norm of the power elite in working life. Women in working life tend to have lower positions compared to men in most positions in the organisational hierarchy. This means that even if women have a socially high-positioned occupation, they will partly be perceived and treated as if they are in a lower position than the men in the social class below them. Above all, people who break norms regarding more than one ground of discrimination run a higher risk of being subjected to disregard or discrimination, since they stand out more from the social and cultural norm in the rest of the group [201]. Different ways of breaking norms overlap, which is described as *intersectionality*. Every norm-breaking characteristic an individual has, in relation to the norm, accumulates in the *intersect*. Therefore, if an individual breaks several norms, it results in a more apparent norm-breaking in relation to the homogenous norm in the group. Consequently, an unemployed, senior, functionally diverse, homosexual woman of colour originating from Ghana will run a higher risk of being subjected to disregard and discrimination when applying for a job in Sweden than compared to a middle-aged, heterosexual, Caucasian man originating from Stockholm. People who break norms tend to not receive the same opportunities from the norm group to be employable and develop their employability through access to support, community or that someone invests in them. Because of this, they may find it more difficult to enter working life as a new employee after unemployment, return from long-term sickness absence or work until an older age. Furthermore, people who break norms have a tendency to want to leave working life if they experience exclusion and a lack of participation in the social work environment. Therefore, they tend not to be as motivated to remain employable in a working life and social work environment where they are perceived as undesired, norm-breaking outsiders.

Age has been described as perhaps the last legitimate reason to avoid recruiting or dismissing an employee [197,198,202], and as previously mentioned, age is one of the most common grounds for discrimination in working life [26,138]. In a study of employees aged 55 years of age and above, 8 percent of female and 3 percent of male respondents expressed that they experienced age discrimination in their work situation [9]. Women are subjected to age discrimination to a greater extent since they, trough the intersect of age, gender and working life, tend to be of the "wrong" age in more professions and more social age groups than men. On the one hand, women may find it more difficult to become employed or included in new projects in the workplace when they are considered to be of childbearing age, since they risk taking time off work due to issues related to pregnancy, to give birth and to care for ill children. On the other hand, women in the social age group when they approach menopause may not have access to measures for maintained employability, since they are already perceived as being on their way out of working life for retirement.

## CONSIDER AND REFLECT

- What is your well-being like in relation to your social work situation and work conditions?
- What is your work-life balance like?
- Reflect on whether efforts and rewards are balanced in your social work environment.
- Reflect on the social and instrumental support in your work situation based on all nine determinant areas of the SwAge™ model.
- Reflect on norm-breakers and intersectionality in your workplace.
- Reflect on whether the attitudes in your workplace mirror that all employees are valuable and important.
- Is there a risk of disregard and discrimination based on the grounds of discrimination in your workplace? If yes, on what grounds of discrimination?
- What are the group dynamics like in your workplace?
- How are conflicts managed in your workplace?
- Are there notorious users of master suppression techniques in your workplace? If yes, what techniques do they use?
- Is any individual or group perceived as an obstacle in the work process and are they disregarded by managers or others in the organisation?
- Do you or anyone else run the risk of suffering from occupational injuries, occupational illness or occupational accidents due to the design of the social work environment?
- Are systematic risk evaluations conducted in your workplace, based on the social work environment and social work situation?
- Reflect on whether you receive sufficient information about what is going on in the organisation in order to feel included.
- What leadership style makes you work best?
- What is your own leadership style?
- What leadership style is used to manage and divide work in your organisation?
- Reflect on how the SwAge™ model can be used as a tool of reflection and reference for management and leadership.
- Reflect on what you can do in the future to promote good social support and a sense of community in your social work environment.

# Organisational Perspective and Action Proposals that Matter to Social Inclusion, Support and Sense of Community

The balance between the personal social environment and the experience of the social work environment is very important for the ability and willingness to work, for employability, and relate to social age (Figure 13).

Furthermore, the social work environment includes job lock, which can entail a risk of decreased employability and increased ill-health. Job locks can have different characteristics and take place in different scenarios. An employee who enjoys their workplace and occupation may have physical symptoms which cause them to find it more difficult to cope with their work tasks. At the same time, these issues may make it difficult to find new employment, and the individual may find themselves locked in their current work situation. Furthermore, some individuals do not dare or believe that they have insufficient competence to leave their current workplace. Staying in a problematic work situation and not daring or being able to leave and approach a new workplace or a new occupation can risk resulting in an experienced loss of control at work and cause the individual to develop low self-esteem. Individuals with insufficient employability risk being more difficult to relocate and may experience difficulties in finding new employment. Therefore, if a larger amount of people should be able to work until an older age, it is important to contribute to the employees' competence and health, as well as to ensure that there are work tasks for individuals who are not fully productive in the labour market.

## ORGANISING GROUPS

The term organisation derives from the Greek word *organon* (instrument, tool, organ) and refers to planning in order to achieve a goal, management of different parts, internal cooperation or a larger association. The organisation develops through organising and through people in the organisation doing their tasks. A work organisation can be defined as a social structure and coordination of people with different roles, set by cooperation, work division, specialisation and defined goals. Organisations need stability over time, material and immaterial assets, and a surrounding world to relate to in some part. Fundamentally, the organisation does not exist without its individual members and their efforts, when an individual leaves or another individual enters, the organisation changes and becomes a new organisation. This can reflect the approach that the whole is greater than the sum of the parts, but also that the individual parts, in other words, the individuals, do not always have an overview of the entire organisation. Another approach is that the organisation is nothing but a structure of coordinated roles, regardless of who or what individual will take on which role. Furthermore, an individual can have several, or all, roles in the organisation. In the latter approach to the organisation, the individual is replaceable. Furthermore, norms regarding how the individual is expected to be in order to fit in with the organisation emerge. People who question the norm, or who are not part of or fit the norm risk being perceived as norm-breakers.

## LEADERSHIP FOR EMPLOYEESHIP – SOCIAL INNOVATION AND SUPPORT THROUGH NEED-, USER-, AND PERSON-CENTRED DEVELOPMENT

A group or a company needs coordination and active cooperation to achieve something that the individuals would not be able to do on their own. Both cooperative partners and critics are needed in an innovative and creative organisational climate. Coordination and cooperation are in principle necessary in order to find solutions, make things happen and achieve success. Communication, trust and common goals are needed in order to cooperate. Teamwork between individuals is key to efficient cooperation. There must not always be consensus in the group, and everyone must not agree with each other in all aspects in order to cooperate. On the contrary, a critical voice is valuable in most processes and can contribute to new solutions. Even conflicts are natural and contribute to development, unless they escalate, slow down proceedings or become permanent, unsolvable issues. Insufficient communication, misunderstandings, competition, organisational instability and ambiguity can also contribute to problems with cooperation. Cooperation and coordination must be prioritised in the organisation to efficiently manage development and different challenges. Challenges on different levels to achieve a healthy and sustainable working life must be identified in order to be utilised or improved. In these matters, one must clarify the type of support and leadership needed for different situations.

In order to raise awareness of possibilities, threats, strengths or problems that must be taken into account, everyone in the workplace must reflect on the nine determinant areas for a sustainable working life and conduct workplace analyses. The manager has the main responsibility to manage the organisation, however, every employee is important. Let every participant use the matrix themselves, to reflect and identify factors that work well and possible challenges, risks and areas to rectify and improve, based on their situation (read more in Part III *Practical Application Based on the SwAge™ Model*, Chapter *A Tool for Systematic Workplace Management and Action Plan for a Sustainable Working Life for All Ages*).

Compile the results from all individual matrixes to a common matrix, then discuss and reflect on the common results in the group. Read the description of how to do this in Section *Employeeship – Increasing Individuals' Experiences of Being Part*

# Organisational Perspective and Action Proposals 145

*of a Greater Whole.* This approach defines what and where problems are present, and what actions are needed to rectify them. Through active employeeship, the conditions for consensus improve, and it creates a common perception of the organisation and the need for action and measures. Furthermore, it can contribute to an increased sense of inclusion and security in the group. However, it is the manager as the leader of the organisation, who has the overall responsibility to:

- "Brainstorm" and try to understand the reality of the problems and possibilities, based on the described risks, problems, possibilities and factors that work well
- Acknowledge all employees, make sure that they experience security and that all employees dare to speak their minds, ask, share their experiences and question within the framework of the discussion or topic
- Describe and discuss what can be done and how it should be achieved
- Plan different actions, measures and activities, make a schedule and divide responsibility
- Implement planned activities, actions and measures
- Follow up, account for what has been done and the purpose of the actions, what the actions have resulted in and how this has affected the situation of individuals, groups (occupation, gender, business field, etc.), the entire organisation and society
- If needed, make additions to the organisational plan

This approach allows the manager as leader of the organisation to gain increased knowledge of the following aspects:

- What kind of support and leadership is needed for different employees, different tasks and measures to prevent risks in the work situation, the work environment and production and to move organisational development and learning forward.
- To identify participants who are the most central and important cooperative partners internally and externally in the workplace. The easiest way to do this is to simply write down all people of importance to work procedures and development and who affect or are affected by the issue. Include both direct and indirect internal relationships in the organisation, for example between the manager and employees, between co-workers, and external relationships to, for example, patients, customers, students, relatives, suppliers or other relevant actors and stakeholders.
- Make an estimated order of participants based on who is most important to the challenge, question, issue or need in question. Include both people who mainly are important as cooperative partners and people who may be opposed or critical. Reflect and try to identify how and why they are cooperative partners, or why they are opposed and critical. Do they support the process or do they need support? What is important to them in relation to the question or issue? Do you need to pay attention to something in particular? What, how and why needs immediate actions or long-term actions in

order to manage and steer the issue in the right direction? Rank them as 1 = Very important, 2 = Important, 3 = Less important, as cooperative partners or as critics.
- Rank, prioritise and identify the need for resources for the action or measure. As a leader of the organisation, the manager has an overview of how plausible the schedule for measures and activities is in relation to other events, tasks and actions that can affect the implementation.
- Evaluate and follow up on results for individuals, work units and production in relation to allotted time, finances and other resources. Evaluate results in relation to goals regarding, for example, sickness absence, retirement, development, learning in the organisation, etc. (Read more in Part *III Practical Application Based on the SwAge™ Model*, Chapter *Evaluations and Follow-up of Actions and Measures*).

## SITUATIONAL AND AGE-CONSCIOUS LEADERSHIP

A situational and age-conscious leadership characterised by instructive and supportive behaviour must be developed in order for employees to feel acknowledged as subjective individuals in a healthy and sustainable working life for all ages [1,6,11,41,107,108,116,200]. Situational leadership means that different leadership styles are needed in different situations, meaning that the leader must be perceptive and flexible enough to adjust and switch their leadership style in the different situations the employees and organisations find themselves in [203,204]. As a leader and manager, one must be perceptive of the employee's needs to execute work tasks. The leadership style cannot be one and the same for all employees, situations, experienced and new employees, work tasks, work procedures, etc. The needed leadership and leadership style is based on the relation between the individual employee's competence in relation to their work tasks, commitment and ability (A) and the support and communication in instructions and leadership (L) that the leader uses to manage and divide work tasks to the employee (Figure 17). With appropriate leadership, the employee is expected to develop autonomy in their work situation, with appropriate competence and engagement in the goals and tasks of the organisation. The four levels of the individual employee's stage of development and maturity of their work ability (A), competence and their engagement in work in relation to execute their work tasks are as follows:

- A1 – The enthusiastic but inexperienced employee, with low competence and with high engagement. An employee who is enthusiastic and eager to learn but currently has little experience, if any at all.
- A2 – The disappointed and passive employee, with low-to-moderate competence and low engagement. An employee who is disappointed and frustrated because they have stagnated and experienced no development.
- A3 – The skilled but cautious employee, with moderate-to-high competence and varying engagement. An employee who is competent but too uncertain to be self-reliant and autonomous in the execution of work tasks.

# Organisational Perspective and Action Proposals

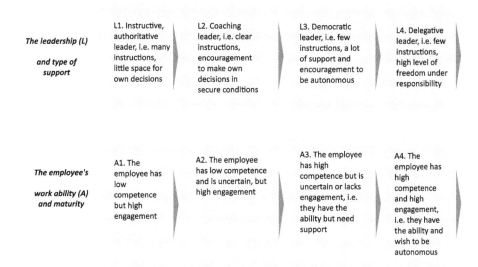

**FIGURE 17** The need for different leadership (L) styles in relation to the employee's stage of development and maturity of their work ability (A), competence, work engagement and willingness to work.

- A4 – The autonomous and result-oriented employee, with high competence and high engagement. An employee who trusts their own competence and has the ability to execute their work tasks, who is autonomous and can cope with working with freedom under responsibility.

In order to meet the employee's (F) needs to perform to their best standards in different situations, the leader must switch between the following four leadership styles:

- L1 – The instructive, authoritative leader who offers many instructions on how to execute work tasks, who shows and explains the frameworks and offers little space for the employee to make their own decisions.
- L2 – The coaching leader who is very instructive and explains how and why, who has a very supportive leadership style that encourages autonomy and individual initiatives.
- L3 – The supportive, democratic leader who gives few instructions but who has a very supportive leadership style, encourages the employee's autonomy and makes decisions together with the employee
- L4 – The delegative leader, who mainly is present in circumstances where the employee is experienced and has a sense of security in the social work environment and in the execution of work tasks. The delegative leader offers few instructions since the employee has expert knowledge. Furthermore, the delegative leader is not particularly supportive but allows autonomy and primarily offers resources for the employee to execute and manage their work tasks with freedom under responsibility.

According to situational leadership, the most appropriate leadership style (L) and type of support which matches the employee's ability (A), the different stages of maturity, is a combination of instructive and authoritative management, which is supportive but allows employees to have their own responsibility (Figure 17).

In order for a situational leadership to work, all nine determinant areas for the ability and willingness to work must be taken into account in relation to the employee's developmental stage, combined with the different age definitions of the SwAge™ model. Considering the nine determinant areas, and the employee's age based on the different age definitions, the situational leader should

- Be perceptive and able to assess the employee's need of instructions and support in their developmental stage related to their work task
- Be flexible and adapt different leadership styles based on the employee's current needs
- Create a sense of trust and security between themselves as manager and the employee in order to reach agreements on what the leader and individual need from each other when they work together, in order to achieve goals, perform to expected standards and be productive

Furthermore, a manager must be aware that the possibility of combining working life with the employee's personal social environment is significant for their ability and willingness to work. All employees have different personal social situations, which are affected by their social age. For example, if an employee is of a social age when it is common and expected to have children, this will affect their working life participation and the social work environment in the workplace. Repeatedly having to stay home from work to care for ill children can put the employee in a dilemma, regarding their social work environment and sometimes even their personal finance. Role conflicts and bad conscience can occur when the employee is forced to choose between caring for their ill child or causing unpleasant situations for their employees, for example if they must work on their day off or take on work tasks that they are not familiar with. At the same time, the employees' children may well be the most important responsibility in their lives. Furthermore, a senior employee may need to care for their personal social environment, for example if they have close relatives in need of their help, or other needs or responsibilities. Employees, primarily women, who experience difficulties in balancing their working life with their family life, often want to reduce their number of working hours. A situational and age-conscious leadership that enables a reduced number of working hours and a lower workplace during certain parts of life provides greater possibilities for the employee to stay in working life for longer. This is because the employees have a greater possibility of combining their personal social life with their work, despite the increased need for recuperation during certain times. For example, measures such as the working time model 80-90-100 can provide increased opportunities and tendencies for the ability and willingness to work until an older age (read more in (5) *Financial Incentives*, Section *Working Time Models to Facilitate an Extended Working Life*), however, so can employments with less demanding work tasks, where the senior employee's knowledge and experiences are utilised. Such employments are often experienced as stimulating, while they simultaneously provide continued participation in the social work environment. The active hours and pace during a day and during the life course are probably significant to our experience of well-being and health. A situational and age-conscious leadership, with reflections based on all nine determinant areas, enables the manager to support and help employees based on their unique situation and abilities. In this way, employees can develop in the long term and achieve their highest performance level, in agreement with the organisation's goals and tasks, and motivate and monitor their own long-term employability.

## COMMUNICATION AS A TOOL TO PREVENT RISKS OF DISREGARD AND DISCRIMINATION IN THE SOCIAL WORK ENVIRONMENT

In an organisation, some of the most important tools are communication and the language used in the division of power, work and coordination. Language is the structure that creates, maintains and is the basis of terms and understanding when humans cooperate. A philosophical term for language as an entirety of coherent expressions, conversations, terms, arguments and statements is the *discourse*, which derives from the French word *discours* (speech, talk) and from the Latin *discúrsus* (running about). A discourse is the specific way of creating terms regarding a certain subject, which are given specific expressions and power positions when communicating [205,206]. In other words, a discourse is created through communication in a certain context which defines power positions and hierarchies in relation to how it is expressed by the sender to the recipient through spoken language, written language or body language. Furthermore, it concerns how and what approaches, arguments and questions arise through this communication, or lack of communication. Who holds power is central to a discourse. When people discuss and describe their statements and actions, they use different discourses as resources in order to strengthen their narrative and as instruments of power [207,208]. Discourses construct the individual's identities in different roles, as well as how the individual uses language in negotiations. When people speak, they use language and expressions from conceptual apparatuses in order to describe contrasts between different poles regarding the discussed subject, as well as to present, motivate and construct their actions as normal and ethical.

The dominant discourse and conceptual apparatus in a workplace contribute to the division of power and to who is allowed to speak their mind in different social environments. The prevalent fundamental assumptions and the conceptual apparatus used in the discourse in the workplace are often not perceivable until someone who thinks differently speaks up [209]. The discourses which control work and the organisation of work contribute to the individuals' experience of their social work environment [6]. The dominant discourse in the workplace sets the boundaries for what is considered to be true, reasonable, desired and relevant to discuss. The dominant discourse in the workplace affects how observances are expressed and interpreted in communication, how the content of communication is assessed and how individuals act as a consequence. Therefore, what, how and why certain factors and conditions are perceived as risks of ill-health in the work environment and in the work situation, and whether employability and an extended working life are important, largely depend on the dominant discourse in the social work environment. The power of the dominant discourse in an organisation risks preventing individuals from breaking norms, even if they suffer. For example, questioning attitudes or work content, drawing attention to excessively high work demands, insufficient possibilities to gain control of one's work situation, an excessively high work pace, poor working hours, stress, insufficient competence, and failure to utilise ergonomic and technological tools. An extreme example is individuals who worked as guards in Nazi concentration camps during the Holocaust in 1933–1945, who, based on the dominant discourse in their social work environment, executed disgusting, abhorrent and violent acts that were part of their work tasks. Things that they would never dream of doing in their personal social environment.

Even if not all workplaces and social work environments are problematic, the dominant discourse in a workplace must be questioned and reflected on, based on whether it is possible to think, speak and act in different ways in order to perceive potential risks and problems. However, it takes great courage to become a "whistle blower", to dare speak up and draw attention to improprieties in order to change strong discourses. The reason is partly that participants in the social work environment are intervened and subjected to the force of the discourse, partly that not everyone has equal ability to speak up due to the dominant discourse. Because of this, it can be difficult to discuss and problematise the discourse in the workplace if the individual does not experience that they have the right to question or speak up, if they do not have the competence or position required to participate in the discussion, according to the dominant discourse [209]. Furthermore, the individual can experience that what they say is irrelevant, since the legitimate and allowed topics to refer to only are set by the dominant discourse. Similarly, the individual can experience that what they want to communicate is misunderstood due to the dominant discourse, according to which other people may be unable or unwilling to include and understand what they say to shed light on the problem.

In workplaces and in organisations, all individuals must be heard and taken seriously, not only employees that the dominant discourse allows to speak and communicate their views, but also individuals who otherwise risk being marginalised in different ways. Furthermore, it is important that everyone feels acknowledged in the organisation and are made justice regarding the resources, time, engagement, skills, knowledge, creativity, etc. that all employees have, and which should be encouraged and utilised in the social work environment.

## Social Support, Security and Strengthening the Social Work Environment

Since human beings fundamentally are social creatures, acknowledgement and social support from the group are vital in order to experience well-being and not to feel threatened, disregarded or excluded. If someone listens and cares for you and what you have to say, it will increase the possibility and feeling of acknowledgement, which, in turn, increases the sense of security and experience of well-being.

The main effects of social support are that it improves health and well-being by catering to the significant human needs of security, social contact, appreciation and inclusion. In a workplace, social support and interpersonal relationships matter to employees' strategies to cope with stress, among other things [6,11,13,95,169,210,211]. Furthermore, social support can contribute to increased motivation and quality of work and increase employees' satisfaction with themselves and their work efforts.

# Organisational Perspective and Action Proposals

The support in question can come from both managers and co-workers, as well as from the surrounding environment outside the workplace. Social support can be divided into the following types [210]:

- *Emotional support* – experience of love, appreciation, care and trust gained through empathy and listening.
- *Appraisal* – feedback and acknowledgement.
- *Informational support* – suggestions, advice, directives and information required to cope with and solve tasks and issues. A way of helping people help themselves.
- *Instrumental support* – practical help, for example sharing of burden, help with work tasks when needed, money, time, adjusted tasks and surroundings.

These different types of social support can be provided not only objectively or subjectively but also in general or focused on a particular issue. Emotional support has been described as the most important in order to cope with work-related stress. Instrumental support is perceived as the least important aspect, even if it is important to decrease demands and the risk of time constraints and an excessively high work pace. All things considered, social support and the experience of humanity in the organisation are significant and one of the most important things we can provide for each other.

However, support must be provided in an appropriate way. Even if intentions are good, it does not necessarily mean that supportive behaviour is experienced as helpful, it depends on the amount of support, whether it comes at the right moment and in what way the social support is provided. Furthermore, the relationship with the person who provides social support and the experience of the situation is crucial. One must not forget that social support is a continuous process from which both parties benefit. In the long run, support should be evenly distributed and provided equally, otherwise, the person who has received help without being able to return the favour may experience that they are in a position of dependency or that their feelings of guilt are too strong to ask for help when needed. However, appropriate support from the right people can be of vital importance in order to decrease work-related stress, improve health and well-being and protect oneself from negative impacts of stress. In order to improve an individual's possibility of receiving social support, every employee should always have at least one supportive person to turn to in the workplace, since there are no guarantees that everyone has a supportive person in their personal life. There are three kinds of social support circles in a workplace: (1) people in close friendships who are key persons in the supportive process, (2) formal, actual support persons who provide practical social support when needed in the workplace, but who are not in contact outside the workplace, (3) external, silent support persons who affect the energy of the supportive process by helping with work when needed, providing moral support and participating in social contexts during breaks and pauses. Together, the three circles constitute a supportive network for the receiver of social support.

When organising, managing and developing the organisation, group or individuals, it is important to reflect on the following:

- Groups – cooperation, group processes, teams, informal groups, subgroups
- Norms and roles – victimisation, harassment and conflicts

- Learning – both individual and collective learning, workplace learning
- Leadership – leadership qualities, leadership styles, power, manipulation and reactions

A healthy social work environment includes the elimination of physical risks and issues and adjusts equipment and premises to people's different physical and mental preconditions to facilitate work [1,7,10,13,43,80,169]. Furthermore, employees should experience a sense of autonomy, inclusion and influence in small and large matters and be able to affect the division of work, work pace and way of working in relation to other people and to the technological system. Work should provide physical, intellectual and cultural stimulation, variation, social interactions, a sense of coherence, learning and both personal and professional development. Physical and mental work demands, workload and challenges should be balanced. The workplace must be characterised by equality, justice, respect, trust, democratic leadership and open communication. Furthermore, employment must provide good opportunities to combine work with a sustainable and rich life outside work and contribute to the experience of well-being and social support. Social support increases the experience of inclusion and sense of community in a group and the experience of security and safety. It is easier to find solutions to problems and obstacles in groups where participants experience these factors.

## INVENTORY OF THE WORKPLACE CULTURE

In workplaces, individuals who otherwise would never have met or wanted to spend time together are forced to share many hours every day in each other's company. Consequently, the social work environment and the workplace culture is an area that demands constant attention in order to prevent subgroups, informal leaders, exclusion, bullying and discrimination. In order to promote a healthy and supportive social work environment, one can obtain an inventory of the workplace culture (Figure 18). Does the workplace culture contribute to:

- Exclusion/discrimination or inclusion?
- Talking about or talking to each other?
- Mistrust or trust?
- Self-assertion or humility?
- Hidden agendas or transparency?
- Notorious use of master suppression techniques or openness and generosity when it comes to errors and mistakes?

In an organisation or company, everyone participating constitutes each other's social work environment. Openness makes a great difference to well-being. A simple way of contributing to social support in the workplace is to greet one's co-workers. Nice co-workers and good relationships in the workplace are significant health factors for most people, accordingly, the organisation should create a culture where it is natural to greet each other every day. Furthermore, trying to take breaks together, eating lunch together and organising other social gatherings together increase the possibility of building good social support between co-workers.

# Organisational Perspective and Action Proposals

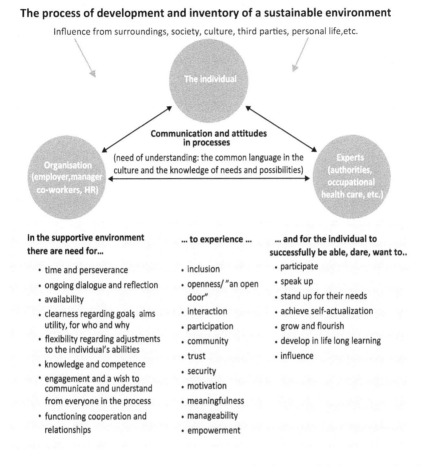

**FIGURE 18** The need and purpose for a sustainable situation and of the inventory of workplace culture and supportive environment.

Additionally, receiving feedback on a regular basis is important to feel acknowledged and appreciated. Therefore, receiving feedback on a regular basis, including both appraisal and constructive criticism, is important. The managers play a key role when it comes to feedback, however, appraisal and constructive criticism between co-workers, and encouragement, acknowledgement and support are factors of importance to the workplace atmosphere. A good atmosphere decreases the risk of scape goat mentality. Furthermore, it is important to be observant of victimisation and bullying in the work unit, both from managers and employees. Bullying and victimisation are severe work environment issues that require rapid actions from managers and management. Victimisation, discrimination, disregard and bullying risk resulting in deep and long-term injuries and illness to the subjected individual. Because of this, it is important to observe such tendencies, for example if anyone appears to be lonely or excluded from conversations. Be wary of the development of a workplace culture involving inappropriate jokes and conduct, it may not be appreciated by everyone, rather the opposite.

If problems are experienced as unsolvable and cannot be resolved, help from outside the organisation is required as fast as possible. A person who criticises their co-workers, organisation or management in a destructive way must know that their behaviour impairs the work environment for everyone. Because of this, it is important not to remain passive when rumours or harassment occur, or when someone does not accept decisions, it is important that the individual is informed, in a calm and collected manner, what is needed and expected of them and that they need to behave more constructively. If problems remain, it may be appropriate for the organisation to help the individual leave the workplace or find a different job where they experience well-being. A good atmosphere increases the sense of security and team spirit in the group and contributes to people daring to make fools of themselves and admit to mistakes and errors, that we all make.

Furthermore, a good atmosphere increases creativity and contributes to everyone's development when group productivity increases. A way to achieve creativity is to utilise differences such as age, ethnicity and gender distribution. It may feel convenient to cooperate with people who are similar to oneself, due to the chance of shared references and values, however, in a group where individuals of different genders, ages, backgrounds, etc. cooperate the input, new ways of thinking, creativity and networking within the group and with other groups often increase – provided that everyone experiences a sense of security. A workplace with employees of different ages, equal gender distribution, ethnic diversity and people of different backgrounds and competence is often more creative and competitive in the long run. Because of this, it can be a good idea to utilise differences when creating work units and teams.

Social activities in the workplace or after working hours, during training days or at conferences make individuals get to know more sides of each other and strengthen their sense of community and unity. Particularly in tough times, when stress and time constraints may inhibit our energy to put our best foot forward, we need to laugh together and let off steam in order to decrease the risk of conflicts. However, it is difficult to begin laughing together at troubles and fight side by side when times are tough if one has not previously felt safe and appreciated. Therefore, showing appreciation and warmth to each other is important for the well-being of the work unit.

In health promotion at work, and in the creation of a person-centred approach for different target groups in the organisation, it is important to reflect on the following:

- Who is of interest/who is included, i.e. target group?
- How voluntary is it, do people have a choice?
- Who is interested in what is best for the individual?
- Who is interested in the individual's motivation and interests?
- What are the possibilities of changing focus from the "ill"/issue to "healthy"/opportunities?
- What different (financial, knowledge, cooperation, etc.) resources are available or needed?

## CONFLICT MANAGEMENT IN THE SOCIAL WORK ENVIRONMENT

Conflicts are essentially natural and always occur in groups, even though they can be very stressful, both for the individuals concerned and for surrounding people. Conflicts

Organisational Perspective and Action Proposals 155

affect individuals' moods, well-being and long-term productivity. Furthermore, it can contribute to individuals' wish to leave the workplace and the organisation. Because of this, conflicts must be dealt with quickly. One person usually needs to manage and guide in order to resolve the conflict. The manager usually has this role in the workplace. Take the following steps to guide and resolve a conflict:

- Take inventory of what the conflict actually concerns. Every individual involved should be allowed to describe the problem based on their perspective, one at a time. Distinguish facts from values, norms from norm-breaking. The individual holding the dialogue should try to remain as neutral as possible and not give advice.
- Describe the issue which has caused the conflict, based on the information you have collected and what the situation is like for the different parties.
- Analyse the issue, individuals and motives for the conflict. Examine different options, risks and consequences. At this stage, the nine determinant areas of the SwAge™ model can be a useful tool to distinguish different areas of importance to the conflict.
- Assess how serious the conflict is or may become, and what alternative solutions are available to solve the conflict. Present several options, describe pros and cons and evaluate different possible future scenarios based on the different options.
- Make a decision and turn the decision into an action plan to solve the conflict and prevent new conflicts in the situation.
- Talk to everyone involved and establish the decision with everyone involved, to make sure that they understand the reasoning and why the chosen solution works best for the organisation. In this way, they can accept and respect the decision even if it entails a loss of prestige and power for someone.
- Follow up on how the situation develops and if anything should be adjusted or changed.

## EXERCISE TO INCREASE UNDERSTANDING IN ORDER TO WORK TOGETHER

A way to increase understanding of the significance and needs of everyone in the work unit, and as a way to decrease the risk of conflict, is to do exercises and inspire reflection. For example, a useful exercise is to divide the work unit into pairs. Both people in a pair hold the other person's right hand with their thumb turned upwards. After this, the task is to touch the other person's right shoulder with their right hand, as many times as possible in a minute. Some pairs will compete against each other to become the one holding the power and win. Others quickly find a solution together and understand that if they cooperate in the pair and take turns, they will both succeed. Several people will probably have an eye-opener they hopefully will remember and make use of in the daily social work environment of the organisation.

## EXERCISE FOR SOCIAL SUPPORT AND EMPATHY IN THE WORK UNIT

Everyone feels troubled at times, and people need social support and an empathetic, supportive social environment in many situations. Perhaps an individual does not want to, and it is probably not appropriate, to share all their life troubles in the workplace and with the employer. However, in order to increase understanding of what people go through, increase empathy for each other in the work unit and to promote the possibilities of good social support, an inventory of challenges and needs of support in working life, the work environment and work situation can be a possible exercise. The exercise should not be confused with coaching conversations or be expected to have any therapeutic purposes. Make a joint decision on how much time should be allocated for each step of the exercise (a suggestion is about two to five minutes for each person and step of the exercise).

- Begin the exercise by dividing the group into smaller groups of two to three persons. The group members take turns interviewing each other, in other words, the first person interviews the second person and then the third, about how they feel, cope and perceive challenges in their professional life, work environment, work situation and employability. Use the nine determinant areas of the SwAge™ model as a checklist in the interview.
- The interview ends with the interviewer making a quick summary of what they have perceived as challenges based on the nine determinant areas, and the interviewee gives feedback on whether they think it is a correct assessment or not. The purpose of this is to evaluate whether the reflection has been correctly perceived and to get a clearer picture of each other's challenges.
- Each person writes down and phrases the interviewee's perceived challenges and needs of support in order to find solutions and measures, draw a mind map or write in bullet points for example.
- Take turns telling each other what you have perceived, identified and phrased when it comes to the need for different kinds of support (read more in *Social support, security and strengthening the social work environment*). Take notes and rephrase your description if needed, based on how the interviewee reacts, whether they agree or disagree with your description.
- Every participant in the group reflects and contemplates the issues, using the SwAge™ model as support. Think of two to five ideas for possible solutions and support in order to contribute to resolving the issues and challenges. Write these down and try to visualise what internal and external possibilities, strengths, threats and issues can come from these suggestions as well.
- Present the suggestions and ideas to each other and explain your thoughts and reflections.
- Reflect on the experiences of the exercise, and how to utilise the results together in the large group.

## CONSIDER AND REFLECT

- What is the social support in your social work environment like between:
  - management and managers?
  - management and employees?
  - managers and employees, and between employees and managers?
  - employees and employees?
- How does the organisation work to create unity and inclusion?
- Does the organisation promote a sense of community and security in work units, where employees dare to admit to their errors and mistakes?

# Organisational Perspective and Action Proposals

- Does the organisation promote a sense of community and security in work units, where employees dare to show their abilities and skills?
- Reflect on how you can develop and increase social support, the sense of community, inclusion and self-esteem and appreciation among employees in the social work environment.
- Does the organisation have a situational, age-conscious leadership based on the different parts of the SwAge™ model, where employees are acknowledged and receive support based on their situation, age, opportunities, abilities, needs, etc.?
- Does the organisation actively work to eliminate risks of ambiguity in leadership and informal leaders?
- Reflect on and identify the dominant discourses in the workplace, organisation and surrounding society that risk contributing to people having unequal possibilities of speaking up and being heard.
- Reflect on whether it is possible to facilitate employees' social integration, inclusion in the work unit and possibilities of getting to know each other in (and possibly outside) the workplace.
- How can work be organised in order to enable the possibility of taking pauses and breaks together, and to acknowledge good initiatives, for example organising social activities to increase the sense of community?
- How can issues in the social work environment and approaches in work and the work unit be discussed? For example, would the case method (read more in *Methods for Competence Development and Transfer of Tacit Knowledge*) work as a way of making everyone participate, contribute and be acknowledged for their experiences and knowledge?
- How do the organisation, management and employees work for zero tolerance against discrimination and disregard in the workplace?
- Reflect on whether everyone in the workplace is informed of the equality plan, and whether follow-up is continuous to make sure the equality plan is up to date.
- How can information be communicated to make sure that everyone has good knowledge of and feels included in what is happening in and around the organisation? (However, make sure the amount of information is appropriate, so employees do not have to pay attention to information that is irrelevant to them).
- Reflect on whether everyone in the organisation appears to understand and be aware of the equal importance of every employee's work tasks, regardless of occupation and hierarchical position, to the daily operations, productivity and achievement of goals in the organisation.
- Does the social work environment and work schedule work well in relation to:
    - the employee's personal social environment and needs?
    - coping with personal social relationships?
    - being active and enjoying leisure activities?
    - the possibility of changing work shifts or taking a day off when needed to support a relative, or to participate in family and leisure activities?
- Reflect on the organisation's measures to promote employees' employability, based on their social work environment and their personal social environment.

# Societal Perspective and Action Proposals that Matter to Social Inclusion, Support and Sense of Community

The action sphere *Social inclusion, support and sense of community* in the SwAge™ model includes determinant areas the personal social environment and inclusion and participation in the social work environment (Figure 13). Society should promote inclusive, supportive and safe social situations. An inclusive society is a sound, supportive environment to promote better health and well-being among citizens. An individual who experiences inclusion in the community of a group, in which both the group and the individual fit the norms of society, wants to contribute to society in different ways to a greater extent. A lack of support and trust, exclusion and discrimination results in a risk of subgroups, informal leaders and conflicts. Being a norm-breaker and speaking up to people who exclude, disregard and discriminate can be a way to cope with the situation. However, this increases the risk of social division in society, if individuals perceive themselves and others as belonging to different groups, "us" and "them", when they strive to identify with a community and find a sense of security. In subgroups, participants are expected to act based on the dominant group norms, set by the leader and power elite. How subgroups in society, in the personal social environment and the social work environment perceive function affects how and whether individuals want to participate in working life or not [1,2,106].

## SOCIETY AS A SUPPORTIVE ENVIRONMENT FOR SOCIAL INCLUSION AND PARTICIPATION

The term supportive environment was developed during the WHO conference in Ottawa in 1986, as a key strategy for health promotion in different arenas, not least in working life. However, it was not until the third WHO conference in Sundsvall in 1991 that a practical, useful definition was developed, focusing on "creating supportive environments for health" [212]. Supportive environments for health shift the focus, partly from focusing on illness to focusing on health, partly from health promotion focusing on risks, to possibilities of health promotion in different everyday arenas, including where people live, their local society, their home, where they play, learn, work and love. Supportive environment for health is a strategic term that refers to creating favourable conditions for the development of good health [213,214]. Supportive environments for health focus on equality based on different grounds of discrimination and highlight preconditions and possibilities, rather than guilt and shame based on normative concepts. Inclusion in a social environment, with social, political, financial, cultural and existential dimensions, contributes to sustainable development, with implications for established work routines, resources, role division, methods and division of responsibility for needs and issues. In other words, it includes shared responsibility on the macro, meso and micro levels in society. WHO policy framework has been developed by 53 countries and includes different prioritised areas for measures and cooperation between the different levels. Interdisciplinary cooperation between different sectors is stated to be significant to improve public health efficiency and to create socially supportive environments for health promotion in different arenas. An Asian initiative, a vision of an inclusive society for all ages that optimises opportunities in longevity [215].

## HOW SOCIETY AFFECTS DISREGARD AND DISCRIMINATION IN WORKING LIFE

Society is responsible for legislation and regulations to decrease disregard and exclusion in society. For example, the purpose of anti-discrimination laws is to prevent discrimination and promote equal rights and possibilities for everyone in society. The grounds of discrimination included in the EU directives to counteract discrimination are: gender, sexual orientation, religion or belief, ethnicity, functional diversity or disability, and age, and are previously described in this chapter in the Section *Anti-discrimination Laws*. Society is responsible for supporting and promoting norms and a societal culture where no one should experience disregard or discrimination. All employers should actively prevent disregard and discrimination in the workplace, for example by introducing equality plans, promoting equal gender distribution, examining salaries and preventing sexual harassment. In the workplace, the employer is responsible to ensure that no one is discriminated against based on the grounds of discrimination when it comes to employment terms, work conditions, salary, recruitment, promotion, education and competence development or when it comes to parenthood.

Age is a concept subject to the same risk of inequality as socioeconomic class, gender and ethnicity. In working life, younger people can be perceived as less trustworthy. In some societies, senior people are treated with respect and honour, not least in working life. However, in other societies, senior individuals are perceived as weak and as a burden to society. There is a risk of social inequality if different age groups are provided with different resources, power and status. There is an ageing elite in Western societies, the middle-aged, while very young and elderly people, who left the labour force many years ago, are relatively powerless. Retired individuals are perceived as less profitable to recruit and as less employable in society. The reason for this is that they, based on belonging to the social age group of the former labour force and current pensioners, are attributed to being more ill, slower, less dynamic and ill-disposed to change. However, younger individuals who enter working life for the first time risk being, based on their social age group, perceived as less trustworthy, undisciplined, rash and more inclined to do drugs. Middle-aged individuals constitute the age elite since this social age group, according to the norm, is expected to be responsible for production, profitability and growth in society. At the same time, individuals of this age are expected to reproduce and raise future productive and well-behaved citizens. Because of this, individuals in this social age risk experiencing role conflicts and being torn between their personal social environment and their social work environment as especially great, stressful and demanding. Furthermore, in this age group, a larger amount of people are on sickness absence due to stress and exhaustion symptoms [66]. In order for working life to be sustainable until an older age, society should contribute to a more diverse working life, where individuals' personal social situation, social work situation and needs in different social ages are supported and accepted.

How can society be supportive of these aspects? A measure could be to increase knowledge regarding the significance of diversity in society and in organisations, as well as inclusion and exclusion based on social age categories. Employers, managers, leaders, HR staff and union representatives need knowledge of ageing in working life and determinant areas for a sustainable working life. Based on participation, and situational and age-conscious leadership, stigmatised perceptions of employees, based on stereotypical norms, can change in society [1,106–108,216]. In a study with almost 12,000 participating employees, age appeared to be the most common reason for the experience of discrimination in working life [8]. The attitude in society, among employers and recruiters, of there being a general "expiration date" on employees when they approach the age of retirement, must be changed in order to allow and enable a larger amount of people to participate in working life, to change careers and workplaces throughout working life, and until an older age. Anti-discrimination legislation makes this clear, and compliance with legislation should be facilitated. Leadership and work environment should be mandatory parts of courses in different university educations where students will have leading positions. This may increase understanding and knowledge of healthy and sustainable workplaces for people of all genders and of all ages, of different grounds of discrimination and of the societal norm to include rather than exclude. Another important measure in society is to disseminate information on how to prevent discrimination and promote equal rights and opportunities. In workplaces, this primarily concerns examining risks of discrimination and harassment, analysing reasons for identified risks and obstacles, and to following up on and evaluating measures in order to decrease the risk of discrimination, based on the grounds of discrimination. The nine determinant areas of the SwAge™ model can be used in this inventory and examination of possible discrimination. Read more in Part III *Practical Application Based on the SwAge™ Model.*

## HOW SOCIETY AFFECTS WORK-LIFE BALANCE

The possibility of decreasing the number of working hours when needed and wanted, to spend more time with family, leisure activities or voluntary work, while still working, contributes to a larger amount of people wanting to work until an older age

[3,4,8,11,26,95,105,138,211]. At the same time, some people must work full-time to better fit in with the social work community in the workplace. This is regulated through working time regulations on the societal level. In order to facilitate individuals' sense of security, coping and support, both in their personal social and work social environment, it must be possible to work full-time or part-time, depending on needs in different social ages.

Work schedules must take the individual's personal social and social work needs and community in and outside the workplace and in different social ages into account, in order to prevent employees developing ill-health due to excessive demands, or that they retire prematurely because they wish to have time for other things in life than just working.

## CONSIDER AND REFLECT

- Reflect on what measures related to the action sphere of *Social inclusion, support and sense of community* can be taken on the societal level in order to improve the conditions for employability:
  - on the organisational level
  - on the individual level
- What significance does society, societal culture and societal development have on the employees' experience of:
  - their personal social environment
  - their social work environment
- Reflect on what measures can be taken on the societal level in order to improve the conditions for inclusion, supportive and secure social contexts:
  - on the organisational level
  - on the individual level
- In what ways can measures on the societal level contribute to employers, management and managers, who are utmost responsible for:
  - having sufficient knowledge of and interest in leadership
  - allocating time in the organisation to develop leadership and provide better conditions for support and inclusion in the organisation
- Reflect on what significance society has for the knowledge of and compliance with laws and regulations related to the action sphere *Social inclusion, support and sense of community*

# Sphere D. The Execution of Work Tasks and Activities

Experiencing motivation and stimulation and having appropriate skills, knowledge and competence are factors of importance for the ability and willingness to work, and for employability, and are associated with cognitive age [1–3,5–8] (Figure 19). People need activities and routines in their everyday lives to experience well-being. It is important that employees experience motivation and stimulation in tasks and that the task feels meaningful and part of a larger whole, in order to be willing to execute tasks. However, skills, knowledge and competence are also needed for the ability to execute tasks. Because of this, the execution of work tasks is associated with cognitive age in many ways, since it requires a certain cognitive maturity, ability, motivation and function to execute work tasks and activities. Individuals more or less consciously consider the execution of work tasks and activities and thereby whether they can and want to work, or if they should leave their workplace, for example for a new occupation or for retirement (read more in the Section *To Stay or to Leave?*)

---

**Examples of Considerations of Significance for the Decision to Stay in the Workplace**
- If I experience motivation, meaningfulness and satisfaction in work tasks and activities.
- If I experience stimulation and the possibility of being creative and developing as an individual in the workplace.
- If I have the possibility of executing work tasks I consider to be the core of my work and that are part of my work description and profession.
- If I identify with the possibility of executing my work tasks and my profession in the workplace.
- If I have appropriate skills, knowledge and competence to execute my work tasks. If I have the opportunity to receive new knowledge, develop my competence and experience challenges, stimulation and development in my work and profession.

*(Continued)*

**Examples of Considerations of Significance for the Decision to Leave the Workplace**
- If I do not experience stimulation, motivation or meaningfulness in my work tasks and activities. If I do not experience my work tasks and activities as desirable.
- If I do not experience that my work tasks are, or have never been, interesting or that they mostly are routine.
- If my competence, knowledge and skills are excessive for the work tasks and activities in my profession.
- If I experience that my skills and competence are insufficient to execute my work tasks and activities.
- If I do not have the possibility of developing or receiving new knowledge and competence in my workplace.
- If I rather want to spend my time with other tasks and activities than the ones that are part of my profession.
- If I, by leaving work, will have better opportunities of putting my competence, knowledge, skills and abilities to use in tasks and activities somewhere else.

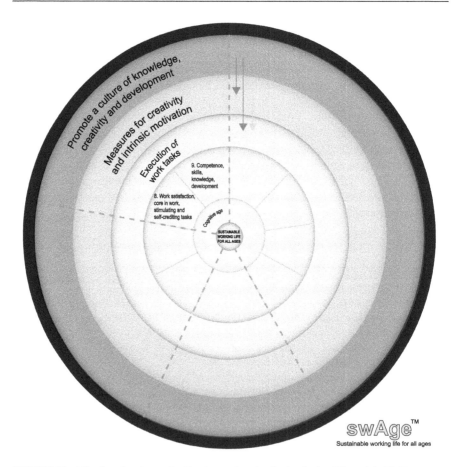

**FIGURE 19** The fourth sphere of reflection and action in the SwAge™ model, (D) *Execution of work tasks and activities*, which includes the determinant areas: (8) Work satisfaction, motivation, stimulation and the core of work, (9) Knowledge, skills and competence development.

# Area 8. Work Satisfaction, Motivation, Stimulation and the Core of Work

The core of work tasks and activities, i.e. the content of work tasks and activities, and the experience of work satisfaction, motivation and stimulation in work tasks are significant to whether an individual states a willingness to work [1–3,8] (Figure 19). Regardless of occupation, professional status and occupational possibilities, it is important that individuals experience motivation and incentives to be willing to work and develop their employability [1,5,8,80,108].

## MEANINGFULNESS AND STIMULATION OR FUTILITY IN THE WORK SITUATION

There is a difference between facilities in the work environment and motivating factors in the work situation [46,73,148]. Work motivation is an important factor for working life participation, and there are many theories regarding how to motivate people to work. However, in order for individuals to both be able to and want to work, both facilities and motivational factors must be favourable. The fundamental factors for *both the ability and willingness* to work are, according to the SwAge™ model, the determinant areas: sufficiently good health, personal finance supported by salary and benefits and a supportive personal social environment [1,2,4–7,80,137]. Furthermore, the facilities, or the "hardware" of work, that must be good for the ability to work are mainly the physical work environment, the mental work environment, working hours, work pace and the possibility of recuperation, knowledge, skills, competence and the opportunity of competence development in work tasks. Motivational factors, or the "software" that makes an individual *want* to work, are inclusion and participation in the social work environment and the experience of motivation and stimulation with the content of work tasks and activities.

In each situation and action, the individual makes a cognitive assessment of his capacity to perform a task. Self-efficacy is a concept developed by Albert Bandura and refers to the individual's confidence in their own ability to cope with an action in a particular situation [217]. Self-efficacy is about belief in one's own ability in a specific action, task or activity and not in one's own ability in general. Self-efficacy can vary between lower and higher confidence in that ability, depending on the situation the individual is in and what the result should be. Self-efficacy thus does not refer to a personality trait but is determined by a dynamic process, which changes with experience and develops over time. How an individual experiences the degree of their self-efficacy affects whether the individual will take the initiative for action, what level of effort the individual has when he goes in to perform the action and how well the individual is able to continue trying to cope with the action and the task of the individual encounter resistance. The individual's confidence in his own ability is influenced, among other things, by previous experiences of failures and successes. An individual's development of self-efficacy for a certain action or task is influenced by enactive mastery experience, vicarious experience, social persuasion, as well as individual physiological and affective states at the time. Seeing others' performance and actions in a situation, or taking part in influence and persuasion from significant others regarding how the individual should act, can also affect the individual's own self-efficacy through indirect experience. The degree of self-efficacy affects personal change and choices in life.

Receiving attention and acknowledgement for work efforts affects employees' productivity [189]. However, some claim that people are fundamentally lazy, unwilling to work and avoid working as much as they can. *Theory X* perceives people as unwilling to take responsibility, and that people must be forced, controlled, managed or threatened to work [128]. This way of motivating individuals to work is also described in the traditional division of work, for example, based on or characterised by Taylorism, i.e. Fredric Winston Taylor's *scientific management* [126]. According to Taylor, scientific management in companies and organisations can make work and production more efficient and rational. If the management would adopt a holistic perspective of the organisation or company, it would enable an overview of work efficiency, and which employee could execute a particular work task the fastest and with best results. In this way, the employees did not have to take responsibility for knowledge of the best way to execute work tasks and plan the execution of work tasks. Instead, the employees should concentrate on their speciality in work. Work and time studies were carefully conducted in the workplace to measure capacity and to show which individual employee was the most skilled and efficient at executing a particular work task. The purpose of this way of organising work was to achieve higher efficiency, fewer errors and increased productivity. However, at the same time, it appeared that this approach resulted in negative effects through increased ill-health and decreased work satisfaction among employees, among other things, due to *alienation* in work. Work became monotonous and tedious, employees had to repeat the same work tasks time and time again. The employee became an exchangeable cog in the production machinery and no longer experienced the satisfaction of planning, launching, executing and finishing the work tasks they were responsible for. The organisation of work changed, for example, previously small autonomous groups where employees divided their own work tasks, took care of each other and trusted each other, changed and moved responsibility to the management, which resulted in decreased employee autonomy, specialised work tasks, a decreased sense of community and worsened communication between employees. Employees could no longer perceive their own significance in the entire production flow of the organisation, but only their own limited part and work tasks.

Alienation risk causes decreased work satisfaction and work motivation. Robert Blauner [218] describes four reasons for alienation in work:

- *Powerlessness* – lack of control and autonomy to monitor one's own work
- *Meaninglessness* – only executing a limited, specialised work task makes it difficult for the employee to grasp the entire work flow and what significance their work efforts have for the end result
- *Isolation* – partly if one does not belong to a work unit, partly if the work task is not connected to a context in the surrounding society and culture
- *Self-estrangement* – standardised work affects the employee's possibility of expressing their autonomy, potential and personality negatively, which results in a decreased sense of self-respect and status and an increased desire to escape the situation.

A too strict division of work can make work monotonous and tedious, resulting in people not wanting to work. Furthermore, it can result in learned helplessness, since autonomy has been eliminated. The experience of being locked in a work situation with uninspiring work tasks can cause stress reactions and remain in the body as somatic symptoms and physical pain. Therefore, this type of work division risks counteracting the organisation's goals of increased efficiency and productivity after a while. Individuals who find themselves in a less motivating and satisfying work situation, who feel less loyalty towards the organisation or who cannot speak up to make a difference often choose to exit the workplace as soon as possible [219]. Reorganisations can make individuals experience insecurity and anxiety regarding their own situation and future, which affects work satisfaction. In a study with participating employees aged 55 years of age and older, 23 percent stated that they intended to leave the workplace for retirement as soon as possible, due to negative changes in their profession, or because the core of their work tasks had changed into something they did not consider meaningful and motivating [8].

Group participation in the workplace, positive social interactions and a sense of community are particularly important for people who feel isolated in their work tasks. In this way, they can increase their satisfaction with the work situation, thereby avoiding alienation. The level and type of alienation differ between different sectors and professions since they contain different opportunities for learning, career development and responsibility. Furthermore, it differs depending on whether the employee is allowed to and has the competence to control the technology used in work, or whether they lack this possibility and knowledge and rather feel controlled by the technology. The greater influence an individual has on their work situation, the greater their engagement, which also increases the ability and willingness to keep working.

Work motivation appears to be reinforced in the long run if people are presupposed to have a wish to grow and develop. *Theory Y* presupposes that the organisation of work tasks and work content affect whether people want to work, take responsibility and develop their creative abilities [128]. The most important driving force and motivation to work is the experience of satisfaction in work tasks and activities, self-actualisation and control of one's own situation. Therefore, it is important that the employee's own goals are integrated with the goals of the organisation. When the employee works to achieve their own goals of satisfaction and executes interesting and stimulating activities and tasks in their work, they simultaneously work to achieve the goals of the organisation. The possibility for and need of the employees' personal expressions, knowledge and initiatives vary and differ depending on sector and occupation, however, very few people want to remain in a working life characterised by uninspiring work tasks. It is important for employees that their professional skills are put to use and that they have the possibility of doing good work in order to experience satisfaction in their work situation. However, it is not a question of what particular aspect the employee experiences as stimulating in their work situation, but rather the fact that they experience stimulation in their work situation that is significant for the willingness to stay in the workplace.

## BUILDING A WALL OR BUILDING A WINDMILL

There is a Chinese proverb which goes something along the lines of: "when the wind of change blows, some people build walls, while others build windmills".

How does the wind of change in society and in working life affect employees? To exemplify, we can reflect on the case of a senior nurse assistant, Greta, whom I met and interviewed in a research project. Greta was on long-term sickness absence and experienced that working life had caused her ill-health.

---

The straw that broke the camel's back was the increased time constraints in work, with additional little work procedures that should be done, even if nothing else was removed. Documentation, ordering laundry service and food through a computer system with frequent malfunctions, rather than using pen and paper, were added. According to Greta, this system caused problems and took a lot of extra time rather than increased work efficiency. She had entered the care profession because she wanted to provide care for and talk to people, not to work with technology and computers. Working life, the rhythm and the professional role had changed without anyone asking Greta. The computer errors she did not understand or could cope with made her feel inferior. Furthermore, constantly having to ask a co-worker for help to execute these tasks made her feel like an obstacle and a hindrance for others and for younger staff who did not experience the technology as a problem. In her experience, the additional work tasks and technological issues took time and energy to the point where she did not have sufficient time to execute her work tasks to high standards. Furthermore, Greta experienced that her learned professional skills and long experience were not valued or desired in the new organisation of work, where patients were treated rapidly and efficiently, like on a conveyor belt. The stress resulted in constant tiredness and the diagnosis of fatigue syndrome. The wind of change blew Greta away from working life. Why did Greta blow away while others adjusted and found shelter or even continued their career after 65 years of age? In these questions, one must consider that there are no absolute boundaries between working life and personal life. A relevant factor is that Greta's husband was ill and had left working life early and that her only child, a daughter who was divorced with three small children, suffered from breast cancer. In the end, all this worry was too much for Greta. She was reported ill and left working life prematurely with sickness benefits.

# Work Satisfaction, Motivation, Stimulation and the Core of Work

At the same time, other people cower when everything becomes too much. Afterwards, they rise up, dust themselves off and move on in their own way in order to cope. This can be exemplified in the case of the teacher Hans, who taught Swedish to immigrants in Sweden, and whom I interviewed for a research project.

---

When the headmaster of the school tried to establish new routines, teams of teachers and common workspaces, Hans built a wall. He participated in mandatory meetings in order to avoid conflict with the headmaster. In other aspects, Hans continued to execute his work in the same way he had always done. He taught the students he liked with joy, while he spent less attention, interest and engagement on students who showed no interest in studies or in learning anything during the lessons. He sought comfort in jazz music and his musician friends as often as possible, situations he experienced well-being in and drew energy from.

---

A third group appears to draw energy from the hardships they have suffered. They utilise the experience and can use it as a stepping stone to reach "a higher level" in life. These are individuals who perceive, emphasise and enhance opportunities and the satisfaction of these, whereas others only see hardships. An example is people who, after a traumatic event, life-threatening accidents, severe illness or crises, see life as a gift to change their life for the better. Per is a man whom I met and interviewed for a research project.

---

Per experienced a high level of well-being through his work. He felt that his work tasks were stimulating and motivating and enjoyed finding solutions to them. To him and in his life, it provided additional value to resolve complicated issues that arose, and he felt it made him stronger afterwards. Per suffered from cancer when he was 66 years of age and was ill to the point where he had to spend Christmas and New Year's Eve in a separate room in the oncology department of the university hospital. Some people would have left work immediately, but to Per work was a source of meaningfulness and motivation in life. His wife knew this. She fetched Per's papers, folders and computer and brought them to his separate room at the hospital where he was treated for his severe cancer. Subsequently, Per spent his days calling customers and working from his hospital bed. In the interview, he said that he experienced his work to be so meaningful that it had made him survive cancer. When I met Per, he was 72 years old and still working. He had stopped doing parts of work tasks he did not enjoy and spent his time doing the work tasks he drew stimulation and energy from. Per explained that he experienced executing these tasks so rewarding that he did not know when he would quit working. The work tasks were an important part of Per's life, and as he said: activities, routines and tasks are required in order to feel good in life. The fact that he had developed his experiences through stimulating activities and tasks meant that it was just a bonus to receive a salary for his work efforts.

Many people have both the ability and willingness to work, some even want to work until 70 years of age or older. What primarily characterises this group is their experience of meaningfulness in work, and the possibility to use their professional skills and knowledge in work tasks [1,6]. These individuals experience stimulation in their professional role, are proud of their work to a greater extent and appear to have a creative outlet through their work. Furthermore, in most cases, they have better self-rated health and well-being, compared to other groups, even if they have chronic illness.

## THE INDIVIDUAL'S EXPERIENCE OF COMPREHENSIBILITY, MANAGEABILITY AND MEANINGFULNESS IN THE WORK SITUATION

The difference between whether individuals metaphorically blow away, build a wall or build windmills relates to salutogenesis. *Salutogenesis* refers to an individual's ability of experiencing comprehensibility, manageability and meaningfulness in different life situations [220]. Other concepts that border on this are coping, empowerment, locus of control, learned helplessness and self-efficacy. *Resilience* is another relevant concept referring to different individuals' resistance and ability to adjust in exposed situations [221]. It can simply and metaphorically be described as the ability to remain and ward off stressful work situations, discrimination, bullying, changes in work situations, work loss, retirement and life traumas, for example, and when needed, to crouch in the wind without breaking. Individuals' resilience can decrease due to different traumas, for example, not only in cases of illness, injuries and ageing but also in reorganisations of work where the individual's supportive structures change and perhaps even disappear.

There are some significant predictors of high resilience. These predictors can be summarised as:

- The organism's *experience of stressors*, the problem or agent which activates the resilience process and creates a disturbance in the homeostasis of the individual, the work unit, the organisation, the company or society. The perception of and reaction to the stressor varies, depending on the context of the experience and the organism's previous experiences.
- *Design of the external environmental conditions*, which includes the balance between risk factors and protective factors in the surrounding environment. For example, demand/control, clear structure/lack of structure, effort/reward, inclusion/exclusion, sound social support or poor social support through leadership, co-workers, etc. in the social work environment or in the personal social environment.
- *Interaction processes and mutual influence between the organism and stressors*. The organism tries, more or less passively or actively, to understand, master and cope with stressors, agents and demanding issues in the environment in order to create a more preventive situation, in other words, to "build a wall".

- *Internal/external abilities to cope with stressors, agents and issues*, i.e. the physical, cognitive, emotional, behavioural or spiritual strengths needed to succeed in different tasks, cultures or environments.
- *Adjustability, endurance and elasticity* to, in a short-term or long-term perspective, cope with the process of being exposed to stressors, agents and issues, and being able to bounce back to one's comfort zone. The organism usually learns this best through gradual exposure, where it successfully adjusts step by step before it faces the next challenge.
- *Previous experiences of positive and successful results of adjustment, endurance and elasticity*, regardless of the type of stressor, risk or traumatic experience. This means that the organism learns to cope with life events and has a greater chance of good salutogenesis, i.e. that it manages to "build a wall" or "build a windmill" rather than "blow away" when it faces stressors or stressful events later in life.

Altogether, salutogenesis, empowerment, sense of coherence, locus of control, learned helplessness and self-efficacy concern the experience of mastering and coping with one's existence (read more in (1) *Diagnoses, self-rated health and functional diversity* and (3) *Mental work environment*). In turn, these experiences are fundamental to the individual's actions and strategies when they are exposed to or find themselves in demanding situations [76–79,209,222]. If a person with abilities that relate to salutogenesis find themselves in an unhealthy (work) environment, are subjected to big changes, new technology, have functional diversity or are part of what is sometimes defined as exposed groups (seniors, women, ethnic minorities, etc.), they do not perceive their situation as an obstacle, but rather use their creativity to change their situation into something they can draw energy from in order to move on. Others, who may not have equally good abilities related to salutogenesis and creativity, stay, feel unwell and develop illness in their current situation, in other words, their health and well-being deteriorate. Satisfaction with daily activities, high levels of well-being and good quality of life promote better experienced health. Having access to several meaningful areas in life decreases vulnerability and increases the possibility of creating and maintaining contexts in life when changes and losses occur. Meaningful activities and engagement in work tasks and society make people feel valued and desired [223] (read more in (3) *Mental work environment* and (7) *Social work environment*). If the attitude towards the employee is positive in the workplace, and the individual is treated as wise and experienced, this contributes to an increased willingness to stay in the workplace [6,106,166]. However, if the organisation, company or manager displays a lack of interest in the employee and their execution of work tasks, it results in bitterness with working life and work tasks, and a wish to leave the workplace [6,106,162]. Furthermore, feeling satisfied with one's daily work tasks and feeling acknowledged in the work situation have been mentioned as significant factors for the willingness to extend working life [1,5,6,11,224]. Financial incentives are not the primary force motivating people to work unless they are in a financially exposed situation [2,225]. Individuals who themselves choose to continue working often feel satisfied with their work tasks and are loyal to their organisation. It is preferable if senior employees approaching the age of retirement continue working because they want to, rather than because they do not have any other financial options. Different types of reemployment after retirement are often connected to stimulating work tasks that contribute to general life satisfaction [1,6,226]. Furthermore, individuals with longer time horizons, who count on living until a very old age, are more positively inclined towards postponed retirement in general [225].

## SATISFACTION OF NEEDS IS FUNDAMENTAL TO WORK MOTIVATION

Motivation in general concerns expectations of satisfaction, and so does work motivation. Individuals are motivated by conditions being improved, or at least maintained. However, we often seek quick satisfaction rather than long-term satisfaction. This behaviour is based on evolution, our limbic system disregards the prefrontal cortex in the brain to pressure us to make decisions and take actions that give greater, immediate satisfaction. The brain urges us to at least eat a little piece of chocolate and a couple of crisps, since the fast-releasing carbohydrates provide pleasure and instant energy. Choosing options that are healthier for us in the long run requires impulse control and a sense of purpose, for example, getting up from the couch and going for a run rather than eating another piece of chocolate. The present moment is prioritised and easier to be motivated by in relation to what eventually may happen in ten years' time. This affects the decisions individuals make to improve their health in the long run, and how the individual acts to maintain their competence and ability on a high level, to ward off possible future threats of changes in the labour markets, re-organisations and terminations. In other words, the measures individuals take to maintain their long-term employability are affected by the motivation for immediate satisfaction and the motivation for planning long-term satisfaction that may not even become a reality.

Motivation is driven by satisfying needs – when one need has been satisfied we seek to satisfy the next in a constantly ongoing process. The psychologist Abraham H. Maslow studied motivation among individuals who expressed having a high level of well-being and health. He considered these individuals to be more mature and to have achieved a higher level of self-actualisation, to the point where they almost appeared to belong to another type of human being. Maslow thought that these people had a lot to teach others, who did not consider it as obvious to help and prioritise oneself. Maslow's studies of human nature led him, among other things, to the central thesis that human beings have an innate tendency to drive themselves to higher levels of well-being, self-actualisation and material wealth [188]. Maslow developed the hierarchy of needs as an explanatory model for how people prioritise their needs, which in his model is a gradual hierarchy. According to Maslow, an individual moves back and forth in the hierarchy, it is not a matter of simply moving upwards in the hierarchy. The gradual shape of the model shows that the needs of the lower levels must be satisfied before the higher goals become important to the individual. The levels in Maslow's hierarchy of needs are the following:

1. *Physiological needs* refer to fundamental needs, like hunger, thirst, sleep and sex, which form the basis for the individual's motivation. Unless these needs are satisfied, the individual cannot move on to the next level, or they

will risk being affected by illness, irritation, pain and discomfort. Because of this, individuals strive to satisfy their fundamental needs as quickly as possible.
2. *Safety needs*, also called security needs, refer to physical protection against violence, accidents and injuries to the body.
3. *Needs of love and belonging* refer to social needs, which include human interaction with friends and family. Human beings are herd animals, even if the individual lives on their own, people generally need a certain level of acknowledgement in order to minimise feelings of loneliness.
4. *Esteem needs* refer to the individual's need for self-esteem, and their need to be respected by others for the individual they are, with both negative and positive characteristics.
5. *Needs of self-actualisation* refer to the individual's need to freely develop knowledge and possibilities in order to reach their full potential.

Maslow's studies of human nature led him, among other things, to the central thesis that human beings have an innate tendency to drive themselves to higher levels of health, creativity and self-actualisation. He described that people perceive satisfaction as motivation to seek new satisfaction and that this is the reason for people striving to achieve new goals. When a human being has satisfied one need, they seek a higher goal. The first three needs are fundamental to the individual's well-being and conditions to be motivated. The two latter needs are developmental needs, they regard personal development and the possibility to reach one's full capacity. The individual's needs are hierarchically organised and relatively based on their strength. For example, hunger and thirst are fundamental needs that must be satisfied but are uninteresting to the individual once they are satisfied. However, what a person chooses to prioritise is individual, and among other things controlled by innate, learned and situational factors in the hierarchy of needs. An individual moves freely between or remains in any of the levels, based on the need they currently must satisfy.

Different personalities' needs, based on their characteristic traits, are defined in relation to their satisfaction of needs. Individuals have discernible, universal traits that are not culturally dependent. The *five-factor model of personality* discerns five traits that are central to what is sometimes described as different personalities, that different personality tests, for example used in recruitment, often are based on [227]. The five factors are usually mentioned by the acronym OCEAN, based on the first letter of each factor:

- *Openness*: tendency to be creative, curious, imaginative and appreciate art, emotions, unusual ideas, diversity and different experiences. Individuals who score high on this scale can be described as creative, original, imaginative and curious. They are unconventional and have broad interests. Individuals who score low on this scale are described as traditional, down to earth, having few interests and rigid opinions of how things should be.
- *Conscientiousness*: tendency to be self-disciplined and to act dutifully. Individuals who score high on this scale tend to be goal-oriented, organised, persistent and willing to plan first rather than act spontaneously. They are

trustworthy, work hard, have self-discipline, are punctual, keep their surroundings tidy, are ambitious and plan forward. Individuals who score low on this scale have a tendency to be lazy, unthoughtful, unrealistic, careless and lack self-discipline and clear goals.
- *Extraversion*: tendency to seek stimulation, thrills, adventure and other people's company. They are characterised by high energy, positive emotions, self-confidence, interaction, high activity levels and a great capacity to feel joy. Individuals who score high on this scale can be described as social, active, talkative, good judges of character, optimistic, amusing and emotional. Individuals who score low on this scale are described as reserved, calm, keep people at a distance, quiet and want to be left alone.
- *Agreeableness*: tendency to be compassionate, act friendly, adjust and cooperate rather than to be suspicious and have a hostile approach towards other people. Individuals who score high on this scale have a gentle demeanour, are trustworthy, pleasant, good-hearted, helpful, forgiving, believe there is good in all people and tend to be naive. Individuals who score low on this scale are described as cynical, unpleasant, suspicious, manipulative, easily annoyed and find it difficult to work with others.
- *Neuroticism*: individuals who score high on this scale are described as having an increased tendency for emotional instability, to be hypersensitive, sensitive to stress and easy access to unpleasant feelings like anger, anxiety, depression, vulnerability and hypochondria. Individuals who score low on this scale are described as calm, relaxed, secure and pleased with themselves.

What individuals do to satisfy their fundamental needs and find motivation, for example the experience of work motivation, can differ depending on their personality traits. What is mainly relevant and notable regarding these five different types of personality traits is that individuals never only have one type of trait, but different levels of all five personality traits. Furthermore, surroundings, previous experiences and current mood play a part, in other words, the most apparent personality trait is partly situational.

## EFFORT-REWARD BALANCE/IMBALANCE

There are different theories regarding what and how to affect employees' motivation, experience of meaningfulness, stimulation and well-being in work tasks and activities. Effort-reward imbalance (ERI) refers to the balance between efforts and rewards and is a theory often used together with the concept of *performance-based self-esteem*. The theory was developed by Johannes Siegrist, Senior Professor in work stress research at the University of Düsseldorf, in order to examine what contributes to cardiovascular diseases [228]. The fundamental idea of the theory is that the effort put down in work, through physical and mental efforts and working hours – perhaps even overtime – and the work pace the individual must, can and want to work in, must be rewarded in some shape or form in order for the individual to experience satisfaction and acknowledgement for their efforts. Effort and reward are based on the social norm of reciprocity and balance. Reward has explicit parts, like salary, promotion, different career opportunities or other rewards for the effort. However, it is also very important to consider the implicit parts, like acknowledgement of what the individual has performed or produced in their daily work. An imbalance between effort and reward causes strain and stress reactions. A lack of balance can occur if a person experiences that they put a lot of time and energy into their work, in relation to what they receive in the form of salary, acknowledgement or the experience of satisfaction that the effort is meaningful to the entire organisation or in society. There are individual differences in the experience of balance between effort and reward. An important task is to try to consider what is objective, and what is subjective and largely depends on how the individual interprets situations and is affected by them. An individual living with a constant experience of ERI will be dissatisfied with their situation, and in the long run, their engagement and motivation will decrease. Unless the individual chooses to leave or change their situation, it can eventually cause ill-health. The individual has the highest risk of developing ill-health if they experience high stress levels due to over-engagement in their efforts while simultaneously receiving a low reward, compared to stress based only on high effort that results in high reward, or only on low reward resulting from low effort. Individuals with high levels of engagement and a large need for feedback or appreciation run a higher risk of experiencing ERI and therefore have an increased vulnerability. Imbalance in the workplace can, in general, cause illness or issues, especially if the individual has a need to display their competence and receive acknowledgement from the surroundings, and if they have a self-esteem and self-image that highly depends on work tasks and the response the person receives in their work. These are some parts of the previously mentioned concept of performance-based self-esteem. In a long-term perspective, ERI can cause a lack of engagement, since needs are not satisfied, and the individual's employability decreases. Therefore, satisfying needs such as work motivation is an important instrument of power in the organisation of work that managers must consider and handle with care.

## MOTIVATION THROUGH EMPOWERMENT AND NUDGING

*Empowerment* is a concept developed in the US that has been used frequently in management since the 1980s, and which subsequently has been used in organisational development and competence development. Furthermore, the concept of *learned helplessness* was often used in organisational- and behavioural psychology in order to describe what happens when individuals are deprived of the possibility of making their own decisions in life and work tasks. Learned helplessness can be described as the opposite of empowerment. Empowerment refers to an individual experiencing power in their own situation, their work tasks, their surroundings, etc. In other words, the individual should experience power in relation to their own personal, socioeconomic and environmental factors that affect the individual and their situation. In health areas, empowerment can, for example, be used through coaching conversations, to support individuals to take control of their own life and health, and to choose options that promote health or to follow a prescription or treatment. In social services, empowerment can, for example, be used to help individuals to strengthen their self-esteem and refrain from drug addiction, or to resist gang influence. The individual receives written or oral information, and based on their previous knowledge and the new information, the individual makes a decision regarding their situation. The individual's choice must not be the correct decision according to anyone else. What is essential is that the individual experiences empowerment, so they can improve their own situation. In organisations or companies, if employees experience that they cannot affect their work situation, it will often result in a lack of work motivation, negligence in work tasks, bad treatment of patients, students, clients or customers and low efficiency. Because of this, it is in everyone's interest that organisations and companies should create fundamental support for employees to experience empowerment, rather than learned helplessness.

*Nudging* is partly the same as empowerment, though it is a term that has mainly been used in economic research areas. Put simply, nudging refers to a mental nudge in a direction that makes a person choose one option over another [133], in other words, a policy instrument to influence individuals, both in society and in working life. The idea behind the concept is that people usually know what is the right or best thing to do, though we do not always make rational decisions, we rather make many of our decisions on routine and out of old habits. People usually have issues with self-discipline, we give in to temptations, procrastinate or do what everyone else does. We know that it is good to eat healthy, exercise, study or to be ambitious and productive at work, but we do not always act accordingly. Therefore, individuals sometimes need help to make the right decision.

The Dutch airport Schiphol had problems with untidy bathrooms in need of constant cleaning, due to careless use of the urinal which resulted in urine ending up on the floor. In this situation, nudging was used by painting a small fly at the bottom of the urinal. It turned out that the amount of urine ending up outside of the urinal decreased by 80 percent when men had something to aim at. The bathrooms were experienced as cleaner without the need for constant cleaning. Painting a small fly in the urinals was a cost-effective policy instrument that motivated people to act according to a desired behaviour.

The experiment at Schiphol was the starting point of what was to be called nudging when two American researchers decided to examine why the fly had such a great effect. Behavioural economist Richard Thaler and law professor Cass Sunstein [133] described how we all can become better people and make wiser decisions if we just receive a mental nudge in the right direction. Richard Thaler was awarded the Nobel Memorial Prize in Economic Sciences in 2017, among other things for his theories on nudging. Nudging is often discussed in positive terms since it can be used to direct people in making decisions that are beneficial to them and/or society. Another way nudging is applied is when fruits are placed by the checkout in supermarkets, in order to make people choose an apple rather than candy, or to paint footsteps on the pavement leading to a bin in order to decrease littering. Furthermore, nudging can be used to help people change their behaviours, for example to complicate smoking by referring to smoking areas, or to serve food on smaller plates to trick the eye into thinking the individual is eating more than they do. However, in order to make these small nudges from leaders in society or in a workplace work, the individual must experience empowerment and the freedom of choice. Furthermore, it is important that the individual perceives their own effort as meaningful and part of a greater whole in order to be motivated to execute the task well. For example, we accept waste separation if we sympathise with the thought of a better environment, however, acceptance depends on who benefits from the nudge. If we benefit ourselves, we are more likely to accept it compared to if society or the workplace benefits from us putting extra energy into doing something. Furthermore, nudges that result in individual benefits are experienced as less intrusive on the freedom of choice. People who experience that they gain benefits and satisfaction through executing their work tasks are motivated and willing to work, while people who do not experience satisfaction and stimulation through their work tasks are not as motivated and are more inclined to consider leaving their work (read more on the considerations to work related to the spheres for action in Section *To Stay or to Leave?*).

Furthermore, social inclusion and informal groups are strong influences that provide different nudges, they strengthen or weaken empowerment and affect the experience of motivation and meaningfulness in the execution of work tasks. You can read more in Section *Social Effects of the Informal Organisation in Working Life* regarding the influence on motivation and on the execution of work tasks.

## OCCUPATIONAL IDENTITY

Occupational identity is the individual's experience of their occupation as part of and connected to their identity, and it is a significant motivational factor regarding whether people remain in working life or not [6]. Not only are knowledge and skills part of the occupational identity but also are questions concerning social conditions and attitudes. Individuals assimilate certain values and attitudes connected to their occupation and make them their own approach. Therefore, occupational identity becomes a part of people's entire identity. However, individuals' personal values and attitudes regarding the occupation contribute to the design and idea of the occupational identity, the occupational role and the occupational collective. The facade becomes a collective representation of an occupational group and thereby a reality in its own right. Because of this, problems may arise when an individual from one occupational group experiences that they have to execute a task that is not part of their role. The organisational hierarchy plays a part in these situations. For example, work tasks that assistant nurses consider to be good may be considered acceptable by nurses, while medical practitioners may consider that the same tasks are beneath them. When an individual enters these pre-established roles, the individual will, regardless of their original motivation to execute a certain task, or through upholding the facade corresponding to the role, discover that they must do both.

Previous studies show that senior employees are generally more satisfied with their work tasks compared to younger employees [229–231]. One way to explain this is that the senior employee has had a longer time to "find their way" in life and achieve what they want. Another way to explain it is that senior employees are described to, through their merits and tacit knowledge, cope better with work and receive more acknowledgement and appreciation. A third explanation is that material and technological conditions have improved, which facilitates work. A fourth explanation is that senior employees no longer perceive that they have other options than to accept their work situation and be satisfied with it. However, it is clear that people who value and perceive their work as an important part of their identity and their life are more willing to continue working in the work place, and until an older age [3,232,233]. Work motivation and work satisfaction increase through the possibility of meaningful and developing work tasks, and the ability to use experiences, knowledge and competence in work tasks. The experience of being needed and desired affects work motivation. Several people who continue working do so because they have unique and extensive tacit knowledge and are motivated by the experienced acknowledgement and meaningfulness the execution of their work tasks contributes to. Individuals who experience a high level of satisfaction in their occupation and perceive their work tasks as very meaningful can, if they are forced to leave their work, often show deteriorated mental health after retirement, unless they have any equally meaningful tasks to engage in as pensioners. This is probably due to the fact that retirement decreases their possibility of executing something they consider to be meaningful, which contributes to their life satisfaction. The fact that some pensioners want to be re-employed after retirement has been connected to their retirement resulting in insufficiently satisfied needs, based on meaningful, self-actualising activities [6,234]. At the same time, individuals who are tired of their work tasks and do not experience their work content as stimulating or meaningful are more inclined to retire prematurely due to insufficiently satisfied needs through their work tasks.

## CONSIDER AND REFLECT

- Reflect on whether you are generally satisfied with your work tasks and activities, and in your life situation.
- How and when do you experience work motivation?
- How and when do you experience meaningfulness through your work tasks and activities?
- How and when do you experience stimulation through your work tasks and activities?
- Reflect on how and if you experience motivation, stimulation or meaningfulness through your tasks and activities
  - in the workplace.
  - in the work unit.
  - in the organisation or company.
  - in relation to society at large.
- Reflect on whether there are possibilities for employees to have their own areas of responsibility and empowerment in your work place.
- Reflect on whether and how motivation, meaningfulness or stimulation and overstimulation can affect labour force participation, sickness absence and retirement in your workplace and in working life in general.
- Reflect on whether working life is segregated, people who have an outlet for their creativity and experience work as stimulating, work more (perhaps too much), while people who are in a less motivating environment may not want to work to such a large extent.

- Reflect on whether your experience of motivation, stimulation and meaningfulness through work tasks and activities related to your future employability and your ability to work until an older age.
- Reflect on how and what you could do in the future to experience motivation, stimulation and meaningfulness through work tasks and activities.

# Area 9. Knowledge, Skills and Competence Development

Knowledge, skills, competence, intelligence, memory and learning are factors of significance to whether individuals state that they are able to work, to their employability, and relate to cognitive age [1–3,8] (Figure 19). Without these factors, it is difficult to execute work tasks and activities correctly or to execute anything at all at work. The rapid development, technical solutions and digitisation have shown the need for constant learning and updating of knowledge in relation to employability. However, a lot of other skills, knowledge and know-how are important in work and employability. Furthermore, knowledge, skills and the possibility of competence development are closely related to cognitive age, development and maturity.

## PERCEPTION AND COGNITION

Individuals' knowledge and abilities are controlled by their brains. Perception and cognition, the mental processes at work when we interpret sensory impressions, are part of the brain's responsibilities (read more in Sections *"The Sensory System"*, *"The Nervous System"* and *"The Functions of the Nervous System"*).

### Perception

*Perception* refers to an amount of conscious and unconscious processes involved with translating the individual's sensory impressions into meaningful information. Perception is the process that interprets information from the sensory system, for example regarding objects, events, the material we sit or walk on, light, sound, written or spoken words. Perception itself occurs after an impression. Individuals essentially have similar impressions of a stressor, however, the perception differs due to the individuals' different experiences used to interpret the impressions. In other words, perception is what provides meaning to an impression through interpretation. Some processes seek to find patterns in detailed impressions of sensory information, while other processes seek to fit the incoming sensory information into expected patterns and previous knowledge. These unconscious processes are vital in order for us to recognise faces, orient ourselves in the surroundings and estimate distances. When the brain works to sort sensory impressions, different types of illusions sometimes occur. Different animals usually have more abilities related to sensory impressions and perception than human beings, for example the ability to sense magnetic fields, the ability to sense heat at a distance or the ability to perceive electromagnetic waves. All sensory information in the brain can be blocked in order to avoid overload caused by more sensory information than the brain can handle.

# Knowledge, Skills and Competence Development

For example, an injured individual sometimes does not feel any pain during a massive physical trauma, since the excessive amount of sensory information is blocked in the neural pathways, much like a traffic jam during rush hour. Another strategy for the brain to handle excessive incoming sensory information is to pick and choose what information it primarily should focus on when interpreting sensory impressions into perception.

Neurons in the brain disappear and die during the entire adult life, which often causes deteriorated perception, because of this it relates to cognitive ageing. Cell death is mainly caused by illness, lifestyle and what the brain is subjected to, rather than the time that has passed, i.e. chronological age. When neurons die and disappear, the number of contact surfaces between different neurons, where the information exchange takes place through synapses, in other words, the signals between the different neurons decrease. The decreased number of contact surfaces, and thereby the decreased number of possible synapses between neurons, can be the reason for the deteriorated perception and increased reaction time that occurs with increasing age. Reaction time deteriorates because it takes a longer time to activate the processes needed for perception. Many work tasks and performances are measured and valued based on how quickly one executes a task. This also applies to different tests regarding linguistic abilities, logic abilities, spatial abilities, decision-making and certain memory functions. Therefore, perception is of great significance to the execution of work tasks and activities, and until what cognitive age an individual is able to work.

## COGNITION

*Cognition* is a generic term referring to the mental processes in the brain which concerns knowledge, thinking, information, emotions and will. The cognitive processes include thinking, attention, memory, learning, consciousness, decision-making and problem-solving. Additionally, other parts of the brain's processes are part of cognition, like voluntary motor functions and perception of reality, which are partly or entirely learned. Furthermore, sensory perceptions and interpretations that depend on knowledge and experiences are part of cognition, while others that concern sensory impressions are part of what is called perception. Moreover, individuals' voluntary actions are based on cognition.

In other words, the concept of cognition is used to describe the brain's most complex functions located in the cerebral cortex. Cognitive performance affects how the individual functions when approaching and interacting with cognitive challenges in everyday life. Cognitive function presupposes coordination and integration of the brain's processes and affects individuals' attention, perception, concentration, working memory and episodic memory, decision-making, executive processes, problem-solving, languages as well as spatial and motor functions. People's cognitive abilities are crucial to their adjustment to demands in the external environment. However, it is also crucial to how they can change conditions in their environment since individuals' voluntary actions are based on the cognitive abilities of their brains. Furthermore, cognitive ability enables us to acquire experiences and establish memories that help us consciously cope with and master everyday life, working life and other complex situations. An individual who has lived for a long time has been able to acquire extensive amounts of experiences and knowledge, which have been stored in memory and positively affected their cognition. However, an older age simultaneously increases the risk of exposure to factors that can contribute to deteriorated cognition, for example injuries, toxins, long-term stress and various types of illness. Knowledge and continuous development of cognitive abilities are required to maintain employability. Individuals' self-efficacy, self-esteem and self-confidence provide crucial connections between knowledge, understanding, abilities, experiences, personal characteristics and employability.

## COGNITIVE ABILITIES

Cognitive abilities are divided into *fluid cognitive abilities* and *crystallised cognitive abilities* [235]. Fluid cognitive abilities and intelligence are used to resolve unknown problems. Executive and logic functions are used to plan, value, organise and logically initiate solutions to new, previously unknown cognitive challenges. Cognitive flexibility is important, where previous experiences and solutions can be discarded in favour of new thoughts and to find solutions to problems that lack clear instructions, without which we cannot proceed. Creativity is used to create and generate new ideas and to identify new solutions. Furthermore, spatial ability is important to fluid cognitive abilities, i.e. the ability to perceive and orient oneself in different environments, to construct, draw a map and perceive patterns that appear. Fluid cognitive ability depends on the ability to think of and perceive a problem from new perspectives. Crystallised intelligence refers to the ability to use previous experiences and knowledge to resolve cognitive challenges and problems. Crystallised cognitive ability increases with age, since individuals learn and collect more experiences throughout life. This is part of what often is described as wisdom (read more in long-term memory). The semantic memory, which is closely connected to learning languages, word comprehension and facts, increases. Crystallised intelligence is affected not only by educational level and general knowledge but also by relations between sorted concepts and the brain's network of concepts, with a well-developed structure regarding categorised opposites and hierarchical organisation of concepts. Crystallised cognitive ability provides the individual with a pragmatic frame of reference based on habits and accepted norms to refer to when the individual makes decisions and masters recurring everyday problems. In other words, the brain's functions can be divided into perceptive processes to process information, and cognitive structures for knowledge, abilities and experiences (read more in sections *Cognition* and *Perception*). Previous experiences make up a net of associations or a cognitive framework called schema or schemata. Schemata can be likened to the brain's built program or instructions for actions, reactions and expectations in various familiar situations, created by previous experiences. Functions that are based on schemata appear to be stable or develop and improve over the years. Professional skills can be described to be based on schemata. Cognitive structures do not deteriorate with age, however, the perception process of general information processing becomes slower with increasing age. In tests, senior people have been better at handling crystallised

# Knowledge, Skills and Competence Development

intelligence than younger people. The measurable differences between different individuals, regarding the deterioration of crystallised intelligence due to increasing age, can probably be explained by the fact that individuals have trained these abilities to different extents, therefore, they have developed different types and different amounts of schemata.

## INTELLIGENCE

*Intelligence* can be defined as the brain's cognitive processes of flexible and secure information processing and problem-solving. Intelligence derives from the Latin word *intellego*, which translates to understand, comprehend, realise or discern. Intelligence includes not only abstract thinking, relational thinking, learning, adjusting to new situations and efficient learning from experiences but also the ability to intentionally develop one's thinking based on available information. Furthermore, intelligence usually includes the ability to associate, reason, know languages, plan, solve problems, think abstractly, understand ideas and complicated causal links and learn. Intelligence includes logical thinking (logos) and linguistic abilities and expressions (episteme). An individual's intelligence, like most personal characteristics, is set based on a combination of hereditary and environmental factors. However, there are several types of intelligence [236]:

- *Logical and mathematical intelligence* refers to the individual's cognitive ability of logical reasoning and of discerning numerical patterns.
- *Linguistic intelligence* refers to the individual's cognitive ability of using correct words, expressing themselves and influencing their surroundings in order to convey a message.
- *Spatial and visual intelligence* refers to the individual's cognitive ability of perceiving three-dimensional figures and perspectives, metaphorical abilities and to assess, for example how hard to kick a football in order to score a goal.
- *Kinesthetic (bodily) intelligence* refers to the individual's cognitive ability of moving their body through gross and fine motor ability, with help from the proprioceptive sense, for example in a dance, sport, or handicraft.
- *Emotional intelligence* (EQ) refers to the individual's cognitive ability of coping with their own and other people's feelings, having relationships, motivating themselves and being empathetic. Furthermore, it includes the ability of knowing and understanding emotions and emotional knowledge, as well as to reflect on emotions in order to promote emotional and intellectual development. EQ is defined as the individual's ability to recognise their own and other people's emotions, to handle feelings in themselves and in relationships and to reason and reflect on feelings in order to strengthen their own ability. EQ is regarded as important in working life and to employability. For example, job listings often describe the significance of personality, referring to the individual's characteristics, ability to be empathic, social competence and "soft skills". EQ is especially important in occupations where relationships and human interactions are central, for example

to customers, clients, students and patients. Individuals with a high level of EQ are described as motivating themselves and others to achieve higher goals. Furthermore, high EQ contributes to increased possibilities of career success and to building strong relationships.
- *Social intelligence* refers to the individual's cognitive ability of interacting with other individuals, people and animals. It also relates to the individual's competence and understanding of how they affect others and how to behave in order for others to appreciate their company. This type of intelligence is part of what is described as *emotional intelligence* or EQ (emotional quotient).
- *Self-awareness intelligence* refers to the individual's cognitive ability of being aware of how own positive or negative thoughts direct and affect their bodily functions, senses, feelings, desires and emotional states. With help from self-awareness intelligence, an individual can use affirmations to direct and explore their life objectively. This is also a part of EQ.
- *Musical intelligence* refers to the individual's cognitive ability of perceiving, creating and expressing themselves through music.
- *Nature intelligence* refers to the individual's cognitive ability of understanding connections in nature and includes abilities like having a "green thumb" and understanding of animals.
- *Existential intelligence* refers to the individual's cognitive ability of reflecting on large questions and discuss religion and philosophy.

All individuals have, to a greater or lesser extent, cognitive abilities for all types of intelligence. However, one or several types of intelligence are usually more prominent in an individual. Cognitive abilities consist of innate, biologically conditioned talents, learned and practised skills as well as socio-cultural and societal ideals. For example, societies that highly value mathematical, logical and linguistic intelligence promote an educational environment to stimulate the development of these abilities, however, this comes at the expense of other cognitive abilities. Therefore, it is important that all types of cognitive abilities are stimulated and integrated throughout life.

## MEMORY

The individual stores experiences in their memory, which provides the ability to recognise, learn and facilitate problem-solving. Memory makes every individual unique. There are different types of memory: sensory memory, short-term memory, long-term memory, semantic memory and episodic memory [237].

### Sensory Memory

Human beings store memories in a way that part of the perception process, what is called the sensory memory, receives impulses from different senses like hearing, vision, touch, taste and smell. The sensory memory functions automatically without conscious awareness, receives unfiltered information and only stores information for a few seconds. The sensory memory distributes part of the information to the short-term memory for continued processing, based on the needs of the brain's cognitive processes. The hippocampus, located in the inner region of the temporal lobe, is part of the limbic system and involved in the creation of new memories. The hippocampus receives impulses from the sensory system and redirects certain sensory impressions and is of great importance to the individual's sense of direction and spatial orientation, especially when finding their way in a previously unknown area. Furthermore, the hippocampus has target cells and receptors for several important hormones like cortisol, thyroid hormones, neurotransmitters and sex hormones. Additionally, the hippocampus is of great significance to the hypothalamic–pituitary–adrenal axis, in other words, the system of hormone glands and hormones that normally constitute the body's response to stressors and the individual's fight or flight response. No memories are stored in hippocampus itself, however, it is involved in the creation of memories and connects association pathways between different parts of the brain, for example it is significant to the transfer of information from the short-term to the long-term memory.

### SHORT-TERM MEMORY

*Short-term memory* is also known as working memory. The short-term memory stores information that the individual needs to handle their ongoing and current situation. Short-term memory can be based on different impressions, that is, visual short-term memory is based on visual impressions, auditory short-term memory is based on sounds and short-term motor memory or muscle memory is based on the body's movements. An example of information stored in short-term memory is when an individual remembers a telephone number in the short time span it takes to press the buttons. In general, it is difficult to store more than six to eight things for more than 30 seconds in order not to overload the short-term memory. However, different people have different abilities to store information in short-term memory and have different abilities to store information input from different senses. The function of short-term memory is to keep information fresh during a current activity, unused information will disappear. However, since short-term memory is quite limited, individuals use different methods to keep information fresh. An example of this is *chunking*, in other words, the process of dividing and compressing large amounts of information into smaller parts, for example through mind maps, keywords or summarised headlines that have logical connections to the individual. An example of chunking is when we divide combinations of words or numbers into smaller parts, that is, 1776195020161989 becomes 17 76 19 50 20 16 19 89. If there is a need to store information in short-term memory for future use, the information is transferred to the long-term memory. The impulses, impressions and information that are not transferred for storage in the long-term memory are forgotten.

### LONG-TERM MEMORY

*Long-term memory* is the part of memory where information is transferred from short-term memory for long-term storage. Whether information is stored in the long-term memory or not depends on how important the information is and to what extent it is processed. If we often come across the same information, or if we have repeated information in short-term memory, it is often perceived as more important, for example a telephone number or numerical code we often use. Knowledge is stored for long periods of time in the long-term memory, without the individual having to make an effort. Long-term memory is based on hierarchical memory models. There is no limit to how much information can be stored. Appropriate clues are needed to retrieve information stored in long-term memory and to recall a certain memory model in the long-term memory. Information from the long-term memory appears when the individual finds themselves in a similar situation, or when information is needed to solve a new or similar problem. However, with increased amounts of similar memories over time, the information from different memories risks being mixed up or combined with other experiences. Some information stored in long-term memory cannot be recalled, for example events in early childhood when the perception was not able to interpret sensory impressions through previous experiences, because of this the information has mainly been stored as unconscious sensory impulses. Long-term memory can be divided into *explicit* and *implicit* memories.

The *implicit long-term memory* is the *procedural memory*. Information on how the individual should brush their teeth, make themselves understood by speaking a certain language, dance the waltz, make coffee, cross the street, ride a bicycle, drive a car and other know-how is stored here. Procedural memories are well-practised, automatised memories and learned reflexes that have become permanent in the body. These memories are so deeply unconscious that they do not even seem to disappear in cases of memory disorders such as dementia.

The *explicit long-term memory* is the factual memory and consists of two parts: the semantic memory and the episodic memory. The *semantic memory* is a memory of general knowledge which stores knowledge, facts and accounts from books, education and other people. The semantic memory is usually detached from the context, time and place in which the memory was created. For example, the semantic memory is used when recalling the multiplication table, Spanish verbs and the names of rivers in Asia. In the *episodic memory*, experiences are stored as episodic events from our lived life, which are often connected to the time or place the memory was created. Episodic memory is primarily affected early on in memory disorders. In these cases, the episodic memory storage is no longer renewed, since a memory disorder such as dementia affects the ability to create new memories first. Because of this, the affected individual mainly forgets events that have occurred after the disorder develops.

Semantic and episodic memories can be connected if facts in our semantic memory are associated with events in memories of our lived lives, in other words, they are associated with events that we have experienced and that are stored in the episodic memory. The possibility of easily retrieving these memories from the long-term memory increases, since the connections create more contact surfaces in the neurons when several memories are activated at once. Through activating more senses like smell, taste, hearing, vision and touch in active learning, the possibilities of storing knowledge in long-term memory and remembering the knowledge increases. In pedagogy, initiatives are promoted to facilitate learning and improving memory by enabling several different memory storages at once by activating sensory

# Knowledge, Skills and Competence Development

impressions in the episodic memory, connected to obtaining knowledge through the semantic memory. The collected experiences in the episodic and semantic memory create what we usually call wisdom (read more in *Crystallised intelligence*). In cases of injuries and disorders that have caused disturbances in memory functions, the semantic memory deficiencies can be compensated for a long time with a bit of help from the surroundings and small clues that can make the person suffering from an injury or disorder find the right track and manage to execute a task.

The formal capacity for long-term memory and learning in working life is not extensively complicated by cognitive changes related to ageing. However, increasing age contributes to slower processing of the individual's memories. Illness and lifestyle mainly affect the individual's physiological changes, which can cause memory to deteriorate with increasing age, and for cognitive processes of general information processing to become slower. This is mainly apparent in new or more complex tasks, if previous experiences, previous strategies or habitual thought patterns and thought loops no longer work. It takes a longer time for senior individuals to cognitively react to, structure and organise information in an area that is completely new to them, or if the information is conveyed in a completely new way, compared to a familiar area or way of obtaining information [24,238–240]. Therefore, experiences of processing information in a specific, practised way appear to increase reaction time and new information takes a longer time to store. However, organising incoming information in memory appears to be better and easier for an experienced individual with previous experiences of similar tasks who has already developed a system for this. Many individuals have, through extensive experience, learned to facilitate the functions of their short-term memory. For example, experiments have shown that older secretaries are at least as quick as younger secretaries in writing a text, since the older secretaries' experience of reading a text and organising their work was more efficient, while the younger secretaries' faster movements enabled more keystrokes per minute [239]. The possibility of learning new skills with increasing age is primarily affected by whether seniors have the possibility of competence development regardless of their age. Time restraints, as well as the senior employee's belief in their ability to learn new things and develop, are also of great significance, and it is important to use learning methods that connect new knowledge to previous knowledge and life experience. Therefore, in a learning situation, it is important to provide sufficient time, so that the information is conveyed in a way that the senior employee is used to and that they have a greater possibility of affecting the arrangement. Extended time for learning is especially important if the senior person has not been able to expand and modify their schemata through advancing in their own or in a connected occupational area for a long time.

## KNOWLEDGE

The Greek philosopher Plato (428 B.C.–348 B.C.) and his disciple Aristotle (384 B.C.–322 B.C.) described that knowledge could be divided into theoretical knowledge and practical knowledge, i.e. a priori knowledge (sense-derived knowledge and theoretical deduction) and a posteriori knowledge (experiential knowledge, empirically and inductively based on our perception, after which we reach a conclusion), respectively.

The theory of knowledge, also called epistemology, is still often based on Plato's and Aristotle's definitions of knowledge. They called for the questioning of previous knowledge since reflection and doubt are the beginnings of knowledge. They meant that knowledge is what occurs in the intersection between systematically examined facts regarding the conditions and status of things, and the thoughts, ideas and conceptions that the individual has good reasons to believe in. In most cases, it is impossible to have a full understanding of all facts regarding a subject, therefore, individuals must often accept that collected facts are not always complete in order to find a solution to problems. Because of this, one must often take context, probability, previous experiences, similarities to other facts, hypotheses and probable explanations into account in order to find solutions to problems and to develop knowledge.

Competence and knowledge are required to participate in working life, to get a job and to be able to execute one's work tasks. The individual's competence is a combination of their abilities, expertise, tacit knowledge and experiences. Furthermore, competence, combined with the individual's driving forces, is telling of how well an individual fits different occupations, roles and work tasks. Knowledge is a learned, theoretical ability to understand, reproduce and apply information and ideas. It may concern facts, rules, connections, causations, explanatory models, concept definitions, approaches or innovations. Knowledge is often socially, historically and culturally constructed and is a source of power. Someone decides what is knowledge and what is "fake news", how knowledge should be produced, who has the right to take part in it, how and who should administer previous knowledge and whether it is allowed to create new knowledge with regards to previous knowledge and experiences.

Knowledge is sometimes divided into four different categories: knowledge, cognitive abilities, practical abilities and a category constituted of attitudes, emotions, values, ethics and motivation. However, the Bologna model of higher education builds on six types of knowledge divided into three sections: knowledge and understanding, skills and abilities as well as critical awareness and approach. These six types of knowledge are found in the objectives for higher education examinations and educational programmes in Europe.

In the Bologna model, knowledge is defined as a sort of fundamental knowledge and facts combined with understanding. Skills are separated from knowledge and can be both cognitive and practical. Sometimes the term ability is added to skills. The third main category contains more ideological and emotional aspects of knowledge formation. This includes attitudes, values, approaches, ethics and emotions.

Additionally, knowledge can be divided based on how it is obtained, in other words, if one utilises knowledge through objective channels or if it is subjectively obtained. Knowledge obtained through objective channels refers to knowledge from education and objective information. It is the understanding of a subject and disciplinary knowledge, for example basic facts, concepts, theories, perspectives and perceptions of a specific discipline or subject. A person can obtain knowledge through memorising and internalising information that has been communicated by others, for example through education. Furthermore, knowledge can be obtained subjectively, through the individual's trial and error or learning by doing, or through imitating experienced and knowledgeable people in the surroundings. A person can develop new knowledge themselves through drawing conclusions from practical experiences and discover new knowledge through synthesising and combining previous knowledge, or through systematic research using scientific methods. It is in this way, through trying things out, that a small child learns their first abilities.

The subjectively developed, fundamental abilities, together with what is called practical knowledge and tacit knowledge, are often referred to as generic abilities and competence which, for example, regard critical reflection, evaluating information or comparing different standpoints [241,242]. Key competence, tacit knowledge or practical knowledge in working life and professions is the knowledge that occurs when an employee actively reflects on their work. Therefore, the experience in itself is not the most important, but rather the ability to process experiences and turn these into actions. Practical and tacit knowledge enables the employees to have a flexible approach to their work, and for them to know how to act in a certain situation, often like a learned, conditioned reflex (as opposed to innate, unconditioned muscle reflexes), without actively considering how to act. Practical and tacit knowledge is personal and cannot be obtained through education. Practical knowledge is developed through life experiences and cannot be learned and studied like knowledge obtained from a book. Experienced nurses', teachers', attorneys' or engineers' professional knowledge can partly be described as practical knowledge or tacit knowledge. Theoretical and practical knowledge, abilities, evaluations and approaches obtained from education and work are combined through their application in working life and create a synthesis of knowledge integrated into a pattern of actions. Practical knowledge means that actions, both bodily and thought actions, can be executed immediately and without active thought since the knowledge almost has become conditioned reflexes. Examples of practical and experiential knowledge are the care tasks executed by nurse assistants when they pat a patient on the back to make them comfortable, give the patient something good to eat, fluff the patient's pillow or do something else to make the individual patient or patients calm and comfortable. Patients feeling calm and not stressed is very important in different treatments and for healing processes. This type of knowledge may not even be considered to be knowledge or valued compared to formal education. However, tacit knowledge gained from many years of experience in a profession or workplace can be very valuable to the organisation and to production. Furthermore, it can entail a certain power compared to younger generations, since it is obtained through years of experience in a profession, which contributes to more experienced employees having a slightly better position in the hierarchy of the organisation and company. However, some senior employees can experience their situation to be threatened, for example when new technology causes some of the tacit knowledge to become obsolete and less valuable.

In working life, the understanding of a subject and disciplinary knowledge are of great importance. Employers often want employees to have a degree with relevant, subject-specific competence, knowledge and understanding, however, generic abilities are also required. The generic abilities often requested are good planning skills, coordination and organisational skills, the ability to work under stress, the ability to cope with stress, the ability to schedule time, the ability to adapt, flexibility, independence, responsibility, the ability to make decisions, the ability to work in a team, the ability to deal with other people, good oral and written communication skills in different languages and the ability to use new technology, as well as imagination and creativity. Many of these factors are included in what is known as tacit knowledge, i.e. qualities, knowledge and skills an individual cannot obtain through an education. A model to learn tacit knowledge is sometimes described as the relationship between a master and an apprentice [241]. This means that a person who is to learn a profession follows and learns from actual situations in the profession, a sort of imitation of actions. The master can explain what tasks should be executed and why, though learning is often situational and personal. The expert knowledge of one master may be different from other masters. Tacit knowledge is always individual, unique and personal.

## LEARNING AND EMPLOYABILITY

Appropriate education is fundamental to employability and significant to fit in today's working life. Therefore, an individual's educational level, competence, and the possibility of developing and using new knowledge are significant factors to remain employable and to extend their working life. Not least, the rapid development, technical solutions and digitisation have shown the need for constant learning and updating of knowledge. Furthermore, the individual's possibility of utilising their professional skills and competence is important for their willingness to remain in working life. People with good employability often continue working, even after the general age of retirement, compared to people with poorer employability [1,10]. Moreover, not being allowed to use one's professional skills and knowledge is a significant reason for employees leaving the workplace. Continued development of knowledge, abilities and education is often a prerequisite for the ability to continue one's working life and remain employable.

Experienced self-esteem affects how individuals think, feel, motivate themselves and act [217]. Being in control of one's own abilities, knowledge, development and experiences is important for the experience of self-esteem, which is a significant factor when it comes to employability. It encourages the individual to make an effort and to remain motivated so that the individual can successfully achieve their goals.

At the same time, individuals' performance levels are affected by how the individuals experience themselves. It is important to believe in one's own ability in order to succeed and to display this belief to the surrounding environment. Individuals who believe that they are able to do what is necessary are much more inclined to succeed compared to individuals who lack self-confidence. An important part of this is whether the employee experiences that their knowledge is considered valuable by others and if they feel acknowledged and respected in the workplace. If this is the case, the individual's self-esteem, pride, motivation to develop and willingness to remain in the workplace increase. However, some employees opt out of competence development in favour of primarily tending to daily routines if conflicts between ordinary work tasks and competence development arise. One of the purposes of education is to equip the participants with intellectual tools and belief in their own abilities to continue educating themselves through lifelong learning. This is facilitated if the individual's surroundings are supportive, in order for the individual to experience self-confidence when developing and strengthening their abilities. Furthermore, learning is facilitated if the environment encourages the individual to evaluate and reflect on their experiences and self-efficacy when learning.

In strong hierarchical organisational models and linear systems where employees' competence includes and builds on each other's competence, the individual competence in every occupational group risks being forgotten and not sufficiently acknowledged, especially for individuals in the lower levels of the hierarchy. Accordingly, the knowledge and competence of occupational groups with lower status in the hierarchy risk being disregarded. At the same time, seniors and ill people with "wrong" education, competence and skills risk being pushed out of the labour market, due to technological development, increased competition and organisational trends. As previously mentioned, people who have experienced being pushed into retirement display lower levels of mental well-being. Furthermore, if retirement is initiated because the employee's competence is no longer needed, this will also have negative impacts on the individual's mental health. Individuals with longer education display better self-rated health, and their health does not deteriorate as quickly with increasing age as for people with shorter education. Additionally, people with longer education often work until an older age.

### Senior Individuals and Competence Development

Senior employees' level of education, competence, possibility of competence development, and ability to utilize their knowledge is significant to their retirement planning. A problem with life-long learning throughout working life is that senior employees tend to be disregarded and given fewer opportunities to participate in competence development [1,8,11,13,26,106–108,149,200].

If senior individuals obtain new information connected to their previous experiences and knowledge, their ability to use the new knowledge is better than younger individuals' ability to use new knowledge [239,243,244]. Furthermore, general knowledge and generic knowledge improve with increasing age among some individuals [24]. However, in order to facilitate stress-free learning of new knowledge among senior individuals, it is important to develop methods to connect learning to previous knowledge and life experiences. Furthermore, it is important that employees have sufficient time to obtain new knowledge and connect it to their previous experiences, and that they have the possibility to take breaks and recuperate.

### Cognitive Impairments and Education

Some individuals in working life suffer cognitive impairments due to illness or injury. Cognitive impairments can cause difficulties in interactions, communication or in executing practical actions. In turn, this can cause learning difficulties, difficulties participating in activities and lesser possibilities of participating in different areas of life. Cognitive impairments can cause a limited ability to, for example:

- Focus attentively and follow instructions regarding how something should be executed, organise things or activity, remember plans, solve problems that suddenly occur and make decisions on what should be done and take actions in a situation

- Plan and gain an overview of time in both short-term and long-term perspectives, understand time concepts and how much time different things take, and know when something should happen
- Find a way to places and know where something is located.

Individuals who have been subjected to severe stress, for example in working life, can develop mental fatigue and cognitive impairments. Furthermore, chronic illnesses, for example dementia, MS, hormonal diseases and cardiovascular diseases, can affect cognitive abilities. The internal biological environment in the brain changes with cognitive ageing, and individuals must cope with age-related changes in their brain and nervous system. The individual's own assessment of their cognitive abilities is significant to their experience of their health and functional ability.

Cognitive impairments can limit the ability to execute daily activities. In these cases, cognitive support and aids can facilitate a functioning everyday life. One thing that can significantly facilitate life is if information is presented in a way that is familiar or easily accessible to the individual, in order to make the information more available for the brain to register. In order to make it easier for individuals with cognitive impairments to obtain information, useful examples can be to reduce the amount of text, avoid long sentences and complicated words, make headlines clearer and break down the text into paragraphs, bullet lists and pictures. On a practical note, a good support can be not only to create a fundamental structure and routine for activities but also to eat and sleep at specific times. Furthermore, routines can be facilitated through daily, weekly and monthly schedules.

## CONSIDER AND REFLECT

- How do you experience your knowledge, competence and abilities in relation to your work tasks and activities in the workplace?
- Reflect on your ability and willingness to work in your current occupation in relation to your knowledge, competence and abilities.
- Reflect on whether your work tasks and activities fully utilise your knowledge, competence and abilities or if anything could be changed.
- Reflect on whether you need to develop your competence and abilities in order to be able to continue executing your work tasks and activities and to maintain your employability.
- Reflect on whether your ability to work would improve if you had the possibility of developing your competence and knowledge.
- Do you use knowledge, competence and abilities in your leisure time that you would like to be able to use in a profession?
- What are your possibilities of developing your knowledge and competence at work, would you like to know and develop specific areas further?
- Do you experience that your competence development at work is sufficient in order to remain employable?
- Reflect on whether increased rotation between different work tasks and activities in your workplace would facilitate your future employability.

# Knowledge, Skills and Competence Development

- Reflect on whether all employees in your workplace, regardless of age, have the same opportunity of obtaining new knowledge and participating in new projects.
- Reflect on whether a new education or change of occupation, work tasks or activities would increase your employability and contribute to your ability of working until an older age.
- Reflect on how and what you can do in the future for your competence, knowledge, abilities and development in relation to the work tasks and activities you want to execute in your work.

# Organisational Perspective and Action Proposals that Matter to the Execution of Work Tasks and Activities

The execution of work tasks is significant for the ability and willingness to work, as well as for employability, and relates to cognitive age (Figure 19). It lies in the interest of both the individual's and the organisation's that the individual is able to execute their work tasks and activities in a satisfying manner and remain employable. Employers and management should be clear in their demands and support of individuals' execution of work tasks, competence and competence development, throughout working life. The work tasks and work situation in the workplace should be designed to make the individual experience motivation, stimulation, and meaningfulness through work. Furthermore, the organisation or company affects the individual's possibility to gain appropriate abilities, skills and knowledge to execute their work task or activity. Moreover, stimulation, motivation and the possibility of exercising memory and other cognitive functions affect individuals' cognitive ageing.

## DESIGN OF THE EXECUTION OF WORK TASKS – A HISTORICAL OVERVIEW OF EFFICIENCY AND MEANINGFULNESS IN WORKING LIFE

Working life is constantly changing and developing to maintain productivity. Development in working life, for example technological advancements and constant development of new it-systems, requires constant development and new knowledge for employees too. At the same time, it is important that the development of technology and organisation of work should be adjusted to the needs of the individual and the work task, rather than the other way around.

Historically, organisations, companies and managements have constantly strived to increase efficiency in the execution of work tasks and activities in workplaces. Adam Smith introduced the division of labour in order to decrease production time and transport back in 1770. Frederick Taylor wrote *The Principles of Scientific Management* in 1913, which concerns the replaceable human being needing monitoring and detailed instructions in order to function in production. Time studies were conducted for each work task, and the most efficient employee was placed to execute the appropriate work task and activity. Henry Ford introduced the conveyor belt in

1913 in order to increase production efficiency in his factories. However, the human relations school with a perspective of organisational psychology was introduced in the 1930s [127,189]; this approach was critical towards Taylorism's perception of human beings as "machines". Scientists who conducted the classic Hawthorn experiments in the 1920s and 1930s perceived human beings' internal abilities and driving forces to experience pleasure through work, as well as to learn and develop (read more in the Section *The Social Effects of the Informal Organisations in Working Life*). That individuals' needs of acknowledgement, security and affinity are more important to productivity, work ethics and motivation than the physical conditions in the workplace. Moreover, that informal groups in the workplace affect and direct the individual employee's productivity and attitude towards their work through social control. Furthermore, knowledge increased even if it did not only help eliminate dissatisfaction and ill-health due to hygiene factors in the physical work environment, in order to increase productivity it must be combined with other factors that create work satisfaction, engagement and motivation [147]. From the 1950s until the 1970s, social systems, leadership and relations became more important in organisations. Organisational models include employees' need for personal development, and organisations should support this in order to improve the employees' work efforts [188]. In the 1960s, the organisational model sociotechnical systems theory was introduced, which included analyses of both the social and technical systems, focusing on the work level in the organisations in order to achieve greater structural changes in the work organisation. This could, for example, be achieved through self-monitoring groups, flexible flows, employee teamwork and deepening and development of work. Productivity was expected to increase when employees created a work organisation themselves, based on their occupational role and work rotation [245]. In the 1970s, the general development moved towards flatter organisations. Employees were to monitor themselves and work according to common norms and values. Managers were to be more supportive rather than just controlling and managing. Group dynamics became important.

Modern organisational models were developed in the 1980s and 1990s. The trend moved back towards new time studies, key figures and standardised work procedures. The models focused on rationalising and decreasing time waste as well as producing and delivering "just in time" in order to decrease interest expenses and accrued revenue. Furthermore, active production without storage costs through production to order became more important. *Lean production* and *Toyota production system* were introduced in many workplaces. Keywords were to streamline, identify and eliminate "waste" from the organisation, and above all to ensure that no overproduction, unnecessary transports, unnecessary steps for processing or reworking, storage or unnecessary movements would occur, as well as to decrease waiting time, inspections and reparations.

*New public management* (NPM), inspired by the private sector, was introduced to many areas of the public sector [246]. The focus was on result control, where the management's task was to introduce a clear distribution of responsibility, decentralisation and division into smaller parts. For example, the organisation was streamlined through renting services from the public sector and from the private sector. This increased competition, both within the public sector and towards private options. The keywords in this approach were cost efficiency and economical use of resources, better utilisation of means, explicit goal standards, success indicators, measurable objectives, and focus on actual results through rewards connected to execution – results were regarded as more important than the procedure. This approach has been heavily criticised in the following years. Criticism has mainly concerned that it is impossible to apply models from the production industry in organisations that primarily work with people in care occupations. Human beings are not objects, and there must be greater flexibility and space in organisations that concern people's care and treatment, otherwise, problems may arise. People have emotions and do not always act similarly or behave as expected. This results in things taking different amounts of time, or demand greater or other resources and measures than the plan states. Situations will occur when a client, customer, patient or student does not follow the work plan and pattern, these situations risk causing stress and frustration, both to the employees who provide care and to the individuals who receive care, much like trying to fit a square peg in a round hole. Self-monitoring, empowerment, participation and influence in working life have historically proved to be the most effective ways of motivating and stimulating work satisfaction, work motivation and productivity. Self-monitoring, empowerment, positive division of work and work climate, stimulation through work tasks, a good sense of community at work and an appropriate workload result in a greater amount of people having the ability and willingness to work.

## SATISFACTION IN WORKING LIFE

Individuals spend a large part of their waking hours and their lives in their workplace, executing work tasks. Because of this, it is important to experience work satisfaction, meaningfulness, stimulation and motivation to execute work tasks and activities in the workplace. Part of this is understanding and perceiving the whole of the organisation, the company and the workplace. It is of significant relevance to perceive and understand how one's own work efforts, commitment, time and energy are important parts of the entire organisation, among other things to experience motivation to execute one's daily work tasks and activities. Furthermore, it is important that employees do not experience alienation from the end product in the organisation, that they do not perceive themselves as just an interchangeable cog in the machinery of the organisation, but rather that they experience their own efforts and work tasks as important contributions to the organisation, company, municipality, region or authority where they work, and to society. The employee must experience pride in their own work and how it is executed and perceive the meaningfulness of the work tasks. No one experiences well-being through having a job that feels meaningless every day.

An individual's motivation increases with an increased understanding of their work tasks which are significant for the entire work unit, for co-workers and for the aims of the activities in the workplace. Moreover, how they are significant not only for the productivity, profit, goals and success of the organisation or company but also for the maintenance of societal functions, general welfare and for the future.

An individual's motivation also increases if they have their own areas of responsibility in the workplace. Managers and management can contribute to employees' well-being with their work tasks in the workplace if they make clear the overall goals of the organisation, and how individual employees' work tasks contribute to achieving these. Individuals who experience satisfaction, meaningfulness and stimulation in their work tasks and in their profession want to remain working in the workplace and work until an older age to a greater extent. Some people even identify so strongly with the execution of their work tasks that they have a hard time imagining ever leaving working life. They experience that the possibilities their work tasks provide, perhaps problem-solving or a possibility of using their abilities, skills and creativity, are so rewarding that they cannot live without them. They perceive their work and work tasks as a very important part of their life, identity and well-being. Furthermore, they pay much attention to the fact that their work efforts make a difference in the workplace, in the organisation or company and in society, perhaps even in the world. They may simply experience less stimulation and meaningfulness if they do not have the possibility of executing their work tasks.

At the same time, dissatisfaction with work tasks causes issues for the productivity of the organisation and for the individual's well-being. An individual may find it tedious to execute the exact same work task year after year, or perhaps they are not aware of how their efforts contribute to someone else's well-being, or to the productivity and increased revenue of the organisation. Perhaps an individual does not perceive how their role and work tasks contribute to other people's ability of executing their work tasks, and that it is an important cog, without which the machinery of the organisation would be worse or perhaps even stagnate. Perhaps the core of the work and work task, what once made the individual choose their profession, has changed over time. The content of the individual's work and activities may not even be what they imagined when they applied for the job, or perhaps changes and reorganisations have resulted in the individual having other work tasks that do not correspond to what they really want to do and execute. It can also be because administration takes longer time and now constitutes the majority of work tasks, causing the individual to experience that there is insufficient time to execute their "actual" work tasks, experienced as the core of their work and profession, with pride. Perhaps technological issues make the individual unable to execute work tasks as they want, or to a standard that makes them proud. All of these factors can contribute to the employee wanting to leave work, the workplace and do something completely different. The individual may look for a position in another workplace, or quit working altogether and retire if they have the possibility of executing more fun activities and tasks outside of working life.

## EMPLOYABILITY IN A CHANGING WORKING LIFE – RESILIENCE

Work motivation and work satisfaction are significant factors for a sustainable working life, for individuals' ability and willingness to remain employable and for employees' execution of work tasks and activities. Therefore, the organisation, company and management must manage and enable the employees' work situation, provide good conditions for this and consider the relation to the employees' cognitive age.

Coping strategies to handle different situations are described in (3) *Mental work environment*. Processing stimuli from different stressors include sequences of events in the central nervous system (CNS) (read also the first paragraph in Chapter *Area 1. Diagnoses, Self-rated Health and Functional Diversity* and Section *The Nervous System*). When the individual faces insoluble problems, the sequences of events in the brain are disturbed and become lasting loops of somatic, cognitive and endocrine adaptive responses in the body, which risk causing illness and injuries in the organism. Because of this, there is a risk of developing ill-health in a situation characterised by a lack of meaningfulness, manageability and comprehensibility of one's own significance in relation to the surroundings. Therefore, the work situation should be examined to find measures that increase the employees' sense of meaningfulness, manageability and comprehensibility of their work tasks in relation to the organisation (read more in Part III *Practical Application Based on the SwAge™ Model*).

As previously mentioned, some societies, organisations, companies and individuals have the ability to "build walls" when they face problems and hardships, while others are damaged and "blown away". Furthermore, some have the ability to utilise the problems in order to change their situation and "build windmills" in order to turn the problem into something positive (read more in the Section *Building a Wall or Building a Windmill*). This relates to the concept of *resilience*, which derives from natural science and concerns whether different organisms survive changes in the ecosystem. In other words, it is the capacity of a system to cope with changes and adapt its' development, regardless of whether it is an individual, an organisation, a company, a forest, a city or a national economy [221]. If resilience is good or poor actually depends on coping processes, partly through resistance (i.e. the ability to build a "wall" against stressors), and partly through adaptability to new circumstances (i.e. the ability to build a "windmill"). People who cannot cope with changes run the risk of being "blown away". Good resilience refers to resistance and a buffer for influence and change, to adaptability when the system loses its' equilibrium and the ability to turn problems and disturbances into possibilities, renewal and innovative thinking.

Organisations and companies that work with diversity often spread the risk and have a greater possibility of remaining unaffected by changes and tend to find it easier to reorganise when disturbances occur. Furthermore, good organisational resilience touches on organisations' and companies' understanding of people's need for knowledge and competence in development and diversity, since employees must have access to and possibilities to learn, as well as other resources that enable them to remain employable. In relation to the demographic development, societies, organisations, companies and individuals must have good resilience. Countries and societies must have the ability to reorganise their systems when a larger amount of people live until an older age. The ability to readjust and adapt to new circumstances in organisations and companies is significant for the ability to face new challenges due to the risk of qualified labour shortages and due to the larger amount of senior employees. Individuals must be able to readjust and adapt to new organisational models, to new technology and to working until an older age (read more about the concepts of coping, locus of control, sense of coherence (SOC), resilience).

Good resilience contributes to good SOC the experience of life as manageable, meaningful and comprehensible. Attempting to process different stressors as manageable, comprehensible and meaningful, can cause them to be assessed as less threatening and as normal phenomena and stimuli. In this way, they do not disturb the neat system sequences of self-regulatory processes in the brain that the individual cannot voluntarily control. Furthermore, there is a connection between CNS and the immune system. People with high SOC find it easier to mobilise physical, mental, material, social and cognitive resources to cope with stressors in their situation and to turn their situation into something comprehensible, manageable and meaningful. Therefore, behavioural interventions that contribute to the experience of an inner locus of control, increased security, relaxation, happiness and other positive thoughts can boost the immune system and prevent illness and injuries and positively influence the CNS and the endocrine systems. Individuals who experience an inner locus of control experience their situation as comprehensible, manageable and meaningful, experience SOC and often find it easier to use a problem-focused coping strategy. They will automatically introduce structures and find appropriate resources based on the nine determinant areas for a sustainable working life found in the SwAge™ model (Figure 1). People who do not have the same frame of reference risk using an emotional coping strategy to a greater extent, thereby forgetting or finding it more difficult to look for appropriate resources. In order to compensate, these individuals need more and clearer acknowledgement and acceptance in order to perform at work. However, the frame of reference used to assess and react to stressors is not static in the primary and secondary assessments when choosing the most successful coping strategy (read more in Section *Coping*). Because of this, it is important that workplaces increase the possibility for support, feedback and empowerment through age-conscious and situational leadership. With appropriate support, individuals who often use an emotional coping strategy can develop and build their inner frame of reference concerning how to best act to proceed from crises and stressful situations, through developing and adapting to a more problem-focused coping strategy.

In order to increase employees' motivation, managers and leaders can explain the significance of their work tasks and work efforts in a greater context. Perceiving one's own work tasks in relation to the work unit, the organisation or company and society, the conditions to experience meaningfulness, work satisfaction and stimulation when executing work tasks increase. Furthermore, in order to make employees feel proud and motivated, the managers and leaders can describe and increase understanding of the historical background of the work tasks, the significance of achieving the goals of the organisation and how the individual employees' work tasks and work efforts are significant in the work unit and in relation to competitors or similar organisations. Additionally, managers and leaders can ensure that everyone understands what roles the employees can play in the quest of realising a vision of the future, however, at the same time it is important to make sure that sufficient resources are available, and that work demands are balanced with the possibility of executing work tasks. The individual employee, employer and manager can determine which resources are needed to handle a situation using the nine determinant areas for a sustainable working life.

# EMPLOYEESHIP – INCREASING INDIVIDUALS' EXPERIENCES OF BEING PART OF A GREATER WHOLE

There is a saying that goes "together we are stronger", this is not least applicable to work situations. Promoting employeeship in the learning organisation (read more in the *Learning Organisation*) can increase individuals' experience of being part of a greater whole. Furthermore, promoting employeeship can contribute to increased motivation, stimulation, competence development, learning, employability, the possibility of efficient teams and meetings, and the success of the organisation. In organisations and companies, empathy, understanding and co-creation are prerequisites for a creative climate, motivation, stimulation and development. A permissive and creative culture of development is needed. Employeeship presupposes a dialogue between the manager and employees, between employees, meetings with information transfer between the manager and employee, organisational development and follow-up of goals, results and budgets. The dialogue must contain the possibility for reflection, both individually and in a group setting, in order to increase the possibility of influence and participation in the decision-making process. Because of this, it is important to create space with regards to time and place for interdisciplinary skills and dialogue in the work unit, where every individual's observations, reflections, experiences and use of, for example, the work environment, tasks, activities and work situation are perceived as occasions for co-creation, commitment and employeeship. Such an occasion, employeeship for work environment and development, can, for example, be a regular subject on the agenda in workplace meetings. Furthermore, this would be a suitable occasion if anything concerning production, production processes, the organisation and competence development needs to be discussed. This can be implemented in reality by using the following points:

1. *Reflection on needs and development:* Begin reflecting individually and then together in the group.
2. *Teamwork and cooperation:* Based on what has emerged, several ideas and solutions can be developed in cooperation and dialogue in order to cater to the needs connected to the challenge and the task. Perhaps the background and context of different possibilities and challenges should be examined. Perhaps reflection is required regarding the internal, organisational environment as well as the external, surrounding environment and its influence on the tasks to execute. What external factors that affect us can be utilised, or affect what, how and why we should execute something? Create understanding, both for the individual's and for the group's needs, experiences and everyday life, for example through dialogue, observations and/or interviews. The interaction in this context is an opportunity for teamwork in an affirmative environment and work climate in the divergent phase of idea generation. There are no bad suggestions, only more or less good suggestions. Examples of suggestions and initiatives to proceed with can be better flow in work tasks, keeping the storage room in order, routines to use safety equipment, assessment of computer ergonomics, handling keys to medicine cabinets, expectations of being on call out of working hours, security and activities to promote well-being in

the work unit, developing new leadership, technological solutions, developing education etc. based on identified needs.
   a. Before the joint reflection, employees should begin by writing a description of what they perceive to be different strengths, opportunities, risks or problems that must be considered and that are affected by each of the four spheres for action in the SwAge™ model. Write down identified factors, for example in a bulleted list, using the matrix for workplace analysis based on the SwAge™ model (read more in Part III *Practical Application Based on the SwAge™ Model*, Section *A Tool for Systematic Workplace Management and Action Plan for a Sustainable Working Life for All Ages*).
   b. Discuss and reflect on different strengths, opportunities, risks or problems that must be considered in the work situation together in the work unit, based on each of the four spheres for action in the SwAge™ model. This is the convergent phase of phrasing a problem, where every individual initiative should be taken into account. Before the joint reflection and discussion, the individual employees' reflections may be compiled in a joint matrix which has to be distributed in time before the meeting.
3. *Goals:* Agree on one or a couple of suggestions for measures or solutions that are useful, accessible and realistic to proceed with.
   How does every individual perceive the task, goal and what should be done?
   a. Prioritise actions based on needs, not on how difficult the implementation would be. Visualise what the different solutions could look like together in the group. Reflect on how communication, cooperation and co-creation occur and what may need to be developed between internal and external parties during the entire process and during different, specific parts of designing the solutions and measures.
   b. Enter a phase where suggestions should be developed and thought through to the point where they can be tested and implemented. Develop the action proposals in detail, so that they fit the designated task, activity, organisation, conditions as well as the type of solution and measure in question. Perhaps it is appropriate to begin with a pilot experiment in order for the organisation and work unit to learn more about how solutions and measures can work in the actual context. After this, make use of the outcome, learn and develop the fundamental idea. The joint assessment can also be implemented a measure immediately.
   c. Goals should be clearly phrased to ensure that everyone has a joint understanding of what they mean. What are the quantitative and qualitative values? If individuals or the group feel ambivalent, insecure, scared and uncertain regarding goals, their own and other people's abilities and knowledge, or needs and interests, in the implementation of a measure, this can affect motivation and stimulation and inhibit the work effort. In order to succeed, individuals in the group must mutually depend on each other, the group and the manager. Make sure that the required information and knowledge are distributed. Reflect, identify and discuss the impact and influence the individual's work efforts

have on other people and their work, as well as on grounds for loyalty, personal responsibility, need for commitment and why it is important to emphasise that work tasks must be executed on time. A group attitude characterised by the sense that "if one of us fails we all fail, but if one of us succeeds we can all succeed together" creates a greater tendency for individual employees to feel safe in the group, to offer help and support and to dare to ask for help.
4. *Implementation phase:* Make a schedule and divide tasks, decide how, who and what should be done, and time for follow-up. Individuals and the group need to experience empowerment and support in order to feel safe, to take on challenges without worry, and to be able to perform. Therefore, everyone needs to understand their role and what, why, how and which work task they are expected to execute. Individuals need to accept the role, area of responsibility and tasks they are given and must have or achieve the qualifications needed to execute the task, for example abilities, knowledge, resources, instrumental support and social support.
5. *Put the plan into action!*
6. *Evaluate:* Reflect, discuss, analyse and evaluate what internal and external strengths, possibilities, risks, problems, weaknesses and consequences the tested or implemented action or measure has had on the individual and group levels. Make sure the evaluation takes place together with the group in the decided follow-up time. Reflect, control and evaluate whether the goal was achieved or not in relation to resources and performance (read more in Part III, *Practical Application Based on the SwAge™ Model*, Chapter *Tools for Evaluation and Follow-up of Actions and Measures*).
7. *Reorient and continue development:* Improve and maintain the constantly ongoing development process in the learning organisation through starting over, that is make a description of, for example, possibilities, threats, strengths or problems that need to be taken into account based on every individual initiative (read item 1 above).

Implementing innovations, measures, improvements and developing an organisation is not a linear process, it happens through loops. Some parts of measures change over time, one may need to take a step back, while other measures must be repeated in order to make the routines part of the organisation and not to be forgotten, because they have not been prioritised due to time restraints, savings or because the person who introduced the initiative has left the organisation. Communication and information are significant factors to make everyone feel included and understand in what phase of a measure or innovation the organisation and work unit is in, when decisions have been made to make something permanent, or why the organisation did not choose to proceed with and make permanent a certain measure or innovation. External communication and information regarding the ongoing or achieved goals and development is important. Furthermore, this information is important to make the group and individuals experience pride and the significance of their own and the group's efforts, work and measures form a greater societal perspective. Moreover, it is important to communicate ongoing measures and achieved goals to spread the word of good examples to other work units, organisations and in society.

## ORGANISATION OF WORK IN RELATION TO EMPLOYEES' ABILITIES AND MOTIVATION

It is partly possible to perceive trends in work organisation and management. For example, organisational principles controlled by technology have been developed and succeeded in companies that produce consumer goods. These principles have subsequently been applied in service companies and care organisations.

Due to increased global competition, many organisations have introduced organisational models to streamline production and decrease costs and time in the production flow, for example *lean production* and *NPM*. This poses a problem, since issues in working life and the work environment that were present decades ago which later organisational models eliminated, have reoccurred when work is organised to increase static work tasks, time for learning and reflection has decreased and there is a decreased interest in employees' entire competence and employability. People with some type of illness or impairment are more inclined to be pushed out of working life when times are tough. Because of this, the new organisational models appear to be especially problematic if employees working life should have the ability and willingness to work until an older age.

Organisational models from the business sector, based on examples from profit-making commercial companies, have also been tested in the public sector in many countries, a concept called NPM. These organisational models have been heavily criticised since quantity often has been prioritised above quality in contact-based and service professions [95,247–250]. In order to find "waste", and to achieve better financial efficiency, the employees' work tasks have been broken down to measure how much time each work procedure takes, in order to minimise time spent. Furthermore, the shift from focusing on quality to focusing on quantity has increased the work pace.

The employer and manager have the main responsibility for decisions concerning the most efficient way to execute work tasks, accordingly, decisions are made on the macro level in the organisation. Managers and officials are responsible to turn decisions into practice based on these decisions. After this, the organisation of work and work tasks is designed and managed in detail to specify exactly how work tasks should be executed. Furthermore, prioritisations are based on political decisions and different laws and legislations in the public sector. What has been decided on the macro and meso level, concerning prioritisation of employees' work tasks and time spent on each task, is to be executed by the employee on the micro level in the organisation. Therefore, the organisation of work affects the employees' mental work environment and possibility to execute work tasks [233,251–253]. Even if the basis for managing employees is the same, differences in production, regarding whether quantity or quality is most important, can entail significant differences in what is followed up on and emphasised in work management. Critics claim that the interactions between people in the service sector and in contact-based professions are significantly based on the quality of interactions and that these cannot be quantified, measured or streamlined in the same way that the production of consumer goods and sales can. People bring their emotions with them in interactions, and the interactions themselves also entail emotions, which affect the quantity and quality of the outcome and production. People will not let themselves be controlled or turned on and off, like a control button on a machine in production, in order to improve the quantity and quality of production. In order to avoid gaps between the employee's possibilities to execute their work tasks, the management who makes decisions must listen to the employees who execute the actual work tasks before making prioritisations and changes. The most important factors to ensure good work are well-functioning organisational coordination: to prioritise the tasks that both the employee and the individual the employee interacts with when executing work tasks consider to be equally important, and a sound climate in the work unit. Furthermore, contact-based and service professions require time to acknowledge and interact with patients, clients, students, etc. as unique individuals. These are important organisational factors for a healthy and sustainable organisation and work unit.

## THE LEARNING ORGANISATION

Learning in the organisation increases competence, abilities, knowledge and opportunities to develop in the organisation or company, and for the individual employee. The learning organisation can utilise both adaptive learning and developmental (creative) learning [254]. *Adaptive learning* means that the individual employee, through their learning, should adapt to the organisation. The individual learns something based on a given task, without changing the task, goal or conditions. This learning refers to following given instructions and mastering situations on routine.

*Developmental learning*, however, means that the individual employee should develop themselves and the organisation through their learning. It is allowed to question the task, goal or conditions. The individual has greater responsibility for larger areas and processes and has an overview of the entire production. The individual employee is presumed to take responsibility, reflect, interpret and phrase problems, tasks, etc.

It is usually important to be a learning organisation, considering the rapid changes in today's society. We need to learn new things in order to cope with the rapid changes of today and tomorrow, concerning technology, business models, customer needs and ways to work. This means that the organisational culture must be supportive if the employee should dare to take risks, experiment, learn from their mistakes and proceed quickly. Because of this, organisations and companies should develop a learning organisation, characterised by a mentality that employees can create and learn together with others, both in their own department, in other departments, and outside the organisation and company. The management should convey the need for a learning culture, pave the way through their own actions, and acknowledge initiatives that support learning. Additionally, the organisation and roles should be designed to provide autonomy and resources to employees to design their work and experiment. In order to support a learning organisation, one can, for example, work with the organisational culture, promote the possibility of creativity, encourage new ways of thinking and employees' ability to learn. Furthermore, in order to support the individual's autonomy, experimentation and learning it is important that they have the opportunity to practice their ability to give and receive feedback, not only from the manager, but from co-workers, customers, clients, patients, students, and their entire network.

# Organisational Perspective and Action Proposals

A frequently mentioned factor is the digitisation of learning. Technology has dramatically increased the accessibility, reach and efficiency of learning. For example, web-based courses have increased accessibility to education for everyone, since employees can attend a course or watch lectures when it fits their schedule, other work tasks and working hours. Not utilising technology and simply relying on one's work experience, the closest co-workers' experiences and a single education is not enough. At the same time, technology is nothing but an enabler. A culture of learning and development is a prerequisite for learning in working life, in other words, a learning organisation is characterised by an organisational culture that encourages learning, supports and provides sufficient resources in changes and re-organisations. Learning is a fundamental part of the organisation and the daily work through having a way of working to learn in the learning organisation. This requires a management that supports managers to make learning a part of work and to develop the employees' continuous learning, in the same way, managers who are able and want to develop employees' learning, and who enable and develop employees' creativity and ability to learn independently, are needed. This in turn requires sufficient resources, time, technology and support for learning, coordinated measures to promote learning and resources through, for example, expert support, a digital environment for learning, apps or a course library. However, this also requires that employees are allowed to, able to and want to reflect, question and learn. The composition and atmosphere of a work unit can be an important factor. An individual can experience a sense of security and inclusion in a work unit, however, they can also experience insecurity, discomfort and fear. Experiencing exclusion, disregard, discrimination and fear decreases the possibilities to learn, both for the individual and for the group. However, if the individual experiences social support and a sense of security in the group and network, this can facilitate the individual's development and ability to cope with external stress [255]. When a group works together to achieve a common goal, both the sense of community in the group and the chances to achieve success increase. However, it is natural that the group has a leader who creates an environment for joint actions to achieve a goal, and who inspires group members to achieve the common goals. Well-functioning communication is important for the leadership to work well. Communication is significant in order to inform everyone of the organisation's common goals and the group's tasks, to drive processes in the work unit forward, and to give feedback and listen to group members. Thoughts, emotions, experiences, needs and attitudes can be shared in well-functioning communication, and conflicts that always occur in group processes can be resolved.

The manager has a key role in the ongoing learning of the organisation and is the person who conveys the learning culture and affects whether the employee wants to, is allowed to and is able to learn. Managers enable employees' independent, ongoing learning and develop a learning organisation. The manager is important for feedback regarding how work and work tasks proceed. In order to develop, employees need to create new experiences. It is the manager's task to find challenges, through questions, discussions and observations, that enable the employee to create new experiences, develop and grow in their professional role, for example through challenging projects or that the employee exchanges work tasks with someone else. Furthermore, it is the manager's task to coach employees before, during and after the implementation of new work tasks. The manager should make their expectations clear, make frequent follow-ups and coach employees to the approaches and habits required to be a learning employee. The manager must be strong in their own ongoing learning, pave the way for employees, be supportive and help employees find possibilities in their further learning and development. The manager influences whether planning, communication, follow-up and discussions take place in open or closed settings. The manager is often the most important person to instil transparency, teamwork and common competence development through experiences, facts, and the development of tacit knowledge for everyone in the work unit. Because of this, it is important that the manager and management provide sufficient resources for learning and support to use technology and to benefit more from learning. Furthermore, the manager themselves must be strong in their ongoing learning and experiment, dare to take risks and fail, in order to set an example for their employees. Through this approach, the manager sets the norm for learning in the organisation for the employees: a workplace environment where even managers are open to the fact that they can fail, and what they have learned through failing. We learn from work experiences, information and knowledge flow, from our personal network with co-workers and managers, but also from customers, clients, patients, students and other people we meet through work. Furthermore, the knowledge we have gained in our personal life and leisure time is also learning which can be utilised at work.

## LEARNING PROCESSES IN ORGANISATIONS AND COMPANIES

The learning organisation has different learning processes [256]. *Single-loop learning* means that the employee executes their assigned work tasks based on the prevailing values in the organisation. They create their own action strategies to execute their work tasks and learn to execute assigned tasks better without questioning the prevailing values in the workplace. *Double-loop learning* means that the employee, based on their work tasks and own strategies developed to execute work tasks, learns to reflect and question prevailing structures and values. Furthermore, they change their own and the organisation's prevailing values when needed through this type of learning. *Deutero-learning* occurs when the organisation has learned to execute both single-loop and double-loop learning, consciously aims to be a consistently learning organisation, and creates appropriate environments, processes and resources for learning. Deutero-learning is the fundamental insight of the organisational knowledge gap between what is known and what must be known, or that the organisation, despite good performance standards, discovers new areas that employees should know. In other words, deutero-learning is primarily learning through self-reflection of one's own organisation, and it is a prerequisite to deeply establish double-loop learning in the workplace. Furthermore, it is needed in order for organisations and companies to learn to be consistently observant of and have strategies and methods to execute both single-loop and double-loop learning.

However, the starting point for a learning organisational culture is always that the employee is allowed to, able to and wants to learn. The employee is responsible for their maintained employability through planning their own development, understanding what skills are required, how they can develop, what methods are available, and where to find knowledge. However, employees need support to do this, since it is difficult for individual employees to keep learning in working life without support and feedback from managers and management. In other words, there is a double responsibility for employees' employability, that is, measures are needed from both the employer and the employee if the employee is to remain employable through their entire working life and until an older age. Organisations and society are in a constant state of change, therefore the employer must provide possibilities for competence development, for example, to be able to keep up with the technological development [13,169,196]. However, employees must play their part by participating in competence development and using new systems offered in order to remain employable and not be "blown away" in the next re-organisation. Learning must be made part of work in order to be the most efficient for employees' interweaved factual knowledge, tacit knowledge, experiences and ability development. If learning is part of work, the work activity and work task, the individual can learn exactly what is needed in the context of the work task, and in small steps which fit the employee's available time.

In order to plan personal development of employability in relation to learning, the employer can encourage the employee to (read more in Part III *Practical Application Based on the SwAge™ Model*):

- plan and reflect on their experiences in a way that develops their employment, related skills and self-awareness
- reflect and evaluate how their transferable skills can be used in work tasks, for new work tasks and in their own career
- make realistic and appropriate plans for education and career, based on their new insights regarding their employability
- reflect on and evaluate their employment potential and ability to plan and cope with their future

Reflection is a key concept in all learning processes to increase the possibility of storing knowledge in long-term memory. Furthermore, reflection is a key concept in the learning organisation. It is significant for everyone in the workplace to consider and reflect on "what did I want to achieve today?", "how did it go?", "what did I do well?", "what do I need to improve?", "what have I learned?". Reflection increases the possibility of active double-loop learning.

## METHODS FOR COMPETENCE DEVELOPMENT AND TRANSFER OF TACIT KNOWLEDGE

Increased cognitive age often amounts to extensive generic knowledge and experiences from life, a long time in a profession or a work task. This knowledge is personal and therefore difficult to transfer to younger and recently recruited employees in the organisation since it is difficult to learn from reading or studying. However, there are different methods to attempt to transfer tacit knowledge and experience-based knowledge between employees in a workplace.

*Case method* is a pedagogical approach to capture and transfer generic knowledge between participants in a work group. The method was developed in different law schools and universities in the 1920s, for example Harvard Law School and Harvard Business School. The case method aims to make the participants face realistic problems in life: an event, a scenario or a dilemma in the organisation, from their everyday life in the workplace, in the work unit or in a work task. It is often the participants themselves who describe and decide on a case, which is experienced as relevant to discuss in the case meeting. This can, for example, be achieved by every participant writing down a case they want to discuss, thereafter the group makes a joint decision of which case to discuss. However, a person designated to lead the discussion can also introduce a case in the case meeting. The question or issue in a case has no correct answer. It is the role of the person who leads the case discussion to ensure that everyone is heard and to promote the participants' discussion, mutual reflection and individuals' own thinking to make them find different solutions based on their own reflections and experiences from working life and life. For example, different intervention projects have used the case method to transfer experience-based and tacit knowledge between younger and senior employees [12,170,188,200,254,257]. The results of these projects have shown that the case method facilitates and increases the transfer of the senior employees' tacit knowledge and key competencies, which they have gained through working a long time in the profession and in the workplace, to younger and recently recruited employees. For example, the transfer concerns the treatment of different individuals in the work situation, handling products and individuals, and the easiest way to execute complicated tasks, or what work tasks should be prioritised in order to achieve goals.

Furthermore, the projects appeared to increase the transfer of the younger and recently recruited employees' disciplinary knowledge, which they had gained from education and recent knowledge of a subject, to the senior employees, for example regarding new fundamental facts, concepts, theories, perspectives and perceptions in their specific discipline and subject.

*Socratic dialogue*, or thoughtful dialogue, is a way to structure conversations between a "master" and an "apprentice" or in a group discussion regarding information, texts or an abstract issue. In a Socratic dialogue, the participants allow their thoughts to take new turns. The method is named after the Greek philosopher Socrates and aims to guide the participant/participants to develop better hypotheses, through mutual reflection in conversations, and to eliminate disposable hypotheses which lead to contradictions, based on the idea that dialogue creates wisdom. In the dialogue, participants attempt to perceive a problem in new ways in order to find different solutions. The Socratic dialogue builds on teamwork and reciprocity, and encourages self-reflection, for example regarding the individual's work, events, problems in work tasks and different possible solutions to existing problems. Before the conversation, everyone should have taken part in the same information and have had time to think, analyse, interpret, and prepare questions. The Socratic dialogue starts off with an interpretive question that everyone should answer, through which the participants' separate understandings and values become apparent. The participants take turns and mutually examine the information provided before the participants' answers, and everyone is encouraged to support their answers by referring to the information. The participants build on each other's thoughts, and at the same time, they problematise and reconnect to their own experiences from everyday life. Including valuing questions in the conversation provides insight regarding norms and attitudes. An observer is not necessary in the Socratic dialogue, however, an observer can ensure that the aims of the conversation are achieved, that everyone is heard, to summarise important ideas and emphasise interesting oppositions and to provide their perception of the conversation.

*Problem-based learning* (PBL) was developed in the 1960s and is a pedagogical method based on human beings' innate curiosity, willingness and ability to gain knowledge. The learning builds on facts and concepts related to reality-based situations and problems. PBL is particularly common in medical training and education. PBL participants are expected to take responsibility for and pursue their own learning, with support from a basic group and guidance from a leader. Learning is perceived as a creative process where participants build knowledge together. The participants' task is to gather information on a problem from different sources of knowledge, for example scientific articles, fact sheets, books and webpages. After this, the participants discuss the gathered knowledge and facts together. Through this approach, they can compile their mutual knowledge, containing information on the problem from different sources, and different perspectives required to solve the problem. Through group discussions and mutual problem-solving in PBL, generic knowledge can be transferred between group participants, since participants share their knowledge-generation process to solve problems, discuss new and previous ways to solve problems, and disseminate technological solutions between each other.

*Learning by doing* is a concept and approach in pedagogy which regards the environment and process surrounding learning to be significant in order to make the individual want to learn, and through this become aware of the information they should learn [258]. If the individual has a chance to actively try things out and experiment in order to learn, more senses are activated – and a higher level of active perception and cognition increases the possibility of storing information in long-term memory. Subsequently, knowledge can develop into consciousness. Furthermore, if a recently recruited or younger employee has the possibility of solving a problem together with an experienced employee, through learning by doing, alternatively through following and observing the experienced employee, this increases the possibility of transferring tacit knowledge [259].

## CONSIDER AND REFLECT

- Reflect on the organisation's measures to promote employees' employability based on their knowledge, competence, motivation and stimulation in their work.
- What can provide possibilities for motivation and work satisfaction in your organisation?
- What makes employees' first and foremost work with the core of their work tasks, and that it does not fall by the wayside?

- What contributes to employees' experience of balance between their work efforts in work tasks and the acknowledgement and reward they receive in their work situation?
- What are the employees' possibilities of having/taking responsibility for different areas of the organisation like, in order to increase motivation, stimulation, sense of responsibility and pride in the work situation and work tasks?
- How can you introduce rotation, variation and changes in work tasks in order to decrease monotony and increase possibilities of stimulation instead?
- Are there opportunities for employees to learn new things in your workplace?
- Does the competence development in your workplace take the fact that cognitive functions, learning and memory changes with age into account?
- Do all employees, regardless of age, have the opportunity of participating in projects, competence development and education?
- What contributes to professional pride and pride in work tasks and the workplace?
- What contributes to everyone understanding and knowing that everyone's work tasks, regardless of their occupation and position in the hierarchy, are equally important to daily work procedures, productivity and achieving goals in the organisation?
- What measures can be taken in order to increase clarity and understanding of the individuals' own work tasks and their significance to the organisation?
- Reflect on the employees' possibility to use their knowledge, skills and competence in work tasks, in their profession and in the workplace.
- In what way can employees' employability increase through rotation between different work tasks, so that employees know how to execute more than one work task if they need to be relocated in re-organisations, changes or if they need to change workplace and profession?
- Does the organisation provide possibilities to acknowledge and utilise knowledge, skills and competence that the employee has gained in their leisure time?
- Does the organisation provide access to competence development and possibility of gaining new knowledge, regardless of age, and adjusted to general cognitive functions?
- Do senior employees mentor recently recruited and younger employees in order to increase the possibility of transferring tacit knowledge between generations?
- Do the decisions, long-term goals and daily activities in the organisation promote a learning organisation with a possibility of reflection?
- Is it possible to change work tasks, area of competence and career, regardless of age?
- What enables the possibility of some kind of career ladder during the life course, for all occupational groups in the workplace?

# Societal Perspective and Action Proposals that Matter to the Execution of Work Tasks and Activities

The sphere for action *Execution of work tasks* in the SwAge™ model includes the determinant areas of motivation and stimulation through work tasks, and knowledge, skills and competence (Figure 19). Cognitive age and ageing affect the possibility of executing a task, however, this is also affected by motivation and the possibility of exercising the cognitive functions of the brain. Society decides on the framework of possibilities and attitudes, because of this, decision makers in society must actively promote a societal structure that enables knowledge, creativity, development, and cognitive stimulation for a sustainable working life among citizens. Even if approximately 50 percent of the labour force maintains a relatively high level of employability until 65 years of age, approximately 80 percent of the labour force retires at this age. If the intention in society is that a larger amount of people should be able and willing to work until an older age, we must better understand the factors that affect the discussion of individuals' working life and retirement on the individual micro level, and on the organisational meso level, but also the measures that can be taken on the societal macro level. If an extended working life and an increased amount of senior employees are desirable in society, it is important to consider what makes individuals able and willing to work, as well as how they can remain employable in the critical discussion of working life.

## HOW SOCIETY AFFECTS INDIVIDUALS' EXPERIENCE OF THEIR WORK TASKS AND ACTIVITIES

All individuals need to experience that their activities and work tasks are important cogs in the functions of society, for the survival, development and success of the organisation and that their experiences are important production assets. A societal culture that promotes knowledge and competence for everyone influences the citizens' experience of and motivation in different work tasks. Attitudes, norms and hierarchies in society affect the experience of how significant and meaningful the own work task and work effort is to others on the organisational level and on the societal level. In order for society to function, all must be equally important, and this should be made clear, to avoid individuals with a lower position in the hierarchy experiencing themselves, their work tasks and work efforts as less important. For example, a janitor is one of the most important employees for an organisation to function, in health care, for example, the janitor's work enables the execution of surgeries and decreases waiting time, therefore they must be acknowledged for the significance of their work efforts. Unless the baker carefully follows a recipe, the bakery will not sell any bread and the company can go bankrupt. The manager in the organisation should lead and divide work tasks to ensure that the organisation and company achieve their goals, maintain productivity and keep their budget, and needs to be encouraged to cope with their responsibilities, experience stimulation and find creative solutions. Everyone needs to feel that they are needed and acknowledged in order to be motivated to work. Being able to use one's competence and knowledge in work is an important aspect of this experience. Furthermore, all of this is important for the willingness and ability of working until an older age. An individual no longer having access to competence development, or no longer being part of new projects in the workplace due to their age is victimisation. An innovative societal attitude should utilise individuals' experiences, skills and knowledge as production assets – approaching the age of retirement can bring new possibilities and income-generating activities. Some seniors change activities, start new companies, maintain their business or increase activity in a dormant company when they retire. With the support of a pension, some individuals dare to approach a new career and develop a special interest in working life, or in an area that has previously been a hobby. For example, by facilitating the launch of small businesses, regardless of age, new companies and innovations can be developed, and perhaps even create new jobs for other people in society.

## POSSIBILITIES OF UPDATING KNOWLEDGE

Society faces constant changes. Experiences and generic knowledge are always important and enriching. However, the constant changes in society also mean that competence and knowledge are partly perishable and must continuously be put in new perspectives, renewed and updated. At the same time, some professions change over time to the point that the core of the work tasks no longer correspond to what the individual is engaged in and once chose to work with. Some educations develop, causing the previous examination and authorisation, regardless of tacit and practical knowledge, to no longer fulfil the qualification of previous work tasks. In these professions, knowledge must be promoted to make use of the employees' knowledge and experiences, and to make sure that the employees feel valued and needed, rather than disregarded and unwanted. Therefore, society needs to support or provide possibilities for citizens to educate themselves throughout their working lives. Knowledge is crucial to a society. It is significant for individuals to strive to survive and improve their lives, both individually and collectively. Knowledge, learning and wisdom are concepts that describe the ability to understand, reproduce and utilise information and ideas. Furthermore, knowledge is in many ways connected to the dissemination of information, since knowledge is no longer useful if it is kept secret. Knowledge is dependent on the survival of the information and knowledge, and it can be disseminated further regardless of whether the original carrier of knowledge passes.

A fundamental prerequisite for the individual's possibility and willingness to participate in working life is based on norms in society. Are senior individuals and their knowledge and experiences considered important or outdated in society? Does the individual have a profession where their competencies are desired, or have developments caused their competence to be outdated? Access to competence development and education throughout life is important. Furthermore, it is important that the level of education and pedagogical tools facilitate learning, based on the cognitive age of the individual. For example, senior individuals' memory generally functions differently than the memory of children and younger individuals. Episodic memory, logical ability and verbal speed begin to deteriorate at 45 years of age [260,261]. Therefore, it is important to consider cognitive age when planning education and competence development. Otherwise, the senior individual can experience being less talented rather than feeling stimulated. Society has an important task in promoting a healthy culture for knowledge, creativity, development and life-long learning. Because of this, society must provide education and schools with the primary purpose of disseminating knowledge from experts to novices, as well as from the older generation to the younger generation and vice versa.

## POSSIBILITIES OF CHANGING AND ADJUSTING WORK CONTENT TO THE INDIVIDUALS' CONDITIONS

Today, the work content and work tasks of many professions are very complex and contain several steps and parts, some of which can make work very stressful. Too flat organisational structures mean that everyone should do everything, which has partly resulted in the responsibility for work tasks rubbing off on all employees. At the same time, many employees describe that they want to leave their professional role if they no longer feel comfortable with the work content, or if excessive time restraints and fragmented focus in the work situation occur [6,95,135,172]. Senior employees who are no longer able to cope with high levels of stress look forward to retirement. The rubbed-off responsibility for work tasks can also affect senior employees by making it more difficult for the employer to find work tasks for bridge employment. Moreover, this applies to finding work tasks in rehabilitation after long-term sickness absence, or when recruiting an individual who is not able to fully perform immediately. Because of these organisational structures, some individuals risk being prematurely pushed out of the labour market, but also because of low demands in the labour market, a lack of interest in the individual, their potential illness, injury or other special needs [14,80,18]. Therefore, society must introduce measures to reinstate simpler, less complex or limiting steps for every work task, in order to facilitate individuals' ability to cope with work tasks in which they do not have to be fully performing due to injury, illness or simply the possibility of working in a calmer and less stressful pace [1,5,6,172]. Additionally, society must contribute to the possibility of changing careers throughout the entire working life, even when the individual approaches the age of retirement. Making sure that individuals can rotate between different work tasks and be given different areas of responsibility is not only stimulating and motivating, it also makes individuals better prepared and maintains employability in reorganisations, changes in production and business closures. Society can contribute to the individual by facilitating studies on different educational levels, offering the possibility of knowledge development and by financing studies through the possibility of maintenance grants until an older age. In some professions, the possibility of changing careers can happen through apprenticeship systems and observation. Furthermore, a change of careers can happen if the individual's authority is upgraded, if their tacit knowledge and competence development in the organisation or company count as a supplement for preferment.

## CONSIDER AND REFLECT

- Reflect on the measures that can be taken on the societal level, in relation to the sphere for action *Execution of work tasks*, to increase the prerequisites or employability
  - on the organisational level.
  - on the individual level.
- What significance does society, the societal culture and societal development have on the employee's experience of their
  - work satisfaction, stimulation and motivation to work?
  - competence, skills, knowledge and development?
- Reflect on different societies' and societal cultures' influence on individuals' education, knowledge, competence and life-long learning.
- Reflect on what influence society has on employers' and managers' use and development of employees' skills, knowledge and competence.
- In what way can the societal level contribute to employers, management and managers, who are utmost responsible for the organisation, the adjustment of work tasks and activities to individuals' cognitive abilities and conditions?
- Reflect on what influence society and the societal culture have on whether and how workplaces make use of employees' creativity and intrinsic motivation to work.

# Part II

# Age in Relation to a Sustainable Working Life

We participate in working life for large parts of our life. During this time we all age.

However, age has previously not been paid great attention in research of health and working life. This is considered to a greater extent today due to global demographic development. The challenges of these changes entail an increased need to understand and to create workplaces and society with possibilities of a secure and healthy life, where the population is able to remain socially and financially active until an older age [102]. Furthermore, this challenge and opportunity is especially interesting regarding the approach towards senior employees and how to make a larger amount of people able and willing to work until an older age. Whether and how employees believe that they can and want to extend their working lives is a complex and interdisciplinary issue which involves several different research disciplines.

## THE DEMOGRAPHIC SITUATION

The proportion of senior citizens continuously increases in large parts of the industrial world. This is, among other things, due to better living standards, more secure workplaces and better possibilities to cure illness and injuries. The average life expectancy is above 80 years of age in more than every third country on earth. The ageing of the population is expected to be very fast in countries like Greece, Korea, Poland, Portugal, Slovakia, Slovenia and Spain, while Japan and Italy are already among the countries with the oldest populations [102]. In the year 2050, more than 33 percent of men and 38 percent of women are estimated to be 60 years of age or older in Europe, and the median age in the 27 countries in the EU is estimated to be 47.9 years of age in 2060. People generally live for a longer time as pensioners today compared to a generation ago. At the same time, the proportion of individuals of working age (20–64 years of age) decreases in relation to the proportion of individuals above 65 years of age. The estimated population of working age in the OECD countries (the Organisation for Economic Co-operation and Development) will, on average, decrease by 10 percent until the year 2060, by approximately 0.26 percent each year that is. However, these statistics differ between countries, and in countries like Greece, Japan, Korea, Latvia, Lithuania, and Poland, the proportion of the working-age population is estimated to decrease by 35 percent or more. With a larger amount of pensioners, who no longer contribute to the number of worked hours in the economy, the support rate will increase for people active in the labour force.

The estimation of the *old age support rate* is made by dividing the number of citizens of working age (20–64 years of age) by the number of citizens aged 65 years of age and above. In 2080, the amount of people above 65 years of age is estimated to be 58 per 100 people of working age in the OECD countries [103] (Table 1). At the same time, the population of working age should also contribute to other areas of society, for example by providing for children and adolescents, who have not yet entered working life, and for individuals aged 20–64 who, for different reasons, do not participate in working life [103].

The rapid ageing of the population contributes to increased pressure on the pension systems in different countries – since they must maintain an economically sustainable system with sufficient pensions. Furthermore, the financial crises that have occurred in several countries have contributed to high levels of public debt and a limited space for managing the economy. Therefore, there is an increased risk of greater inequality in society and between countries, caused by slow growth and low interest rates. This in turn poses new challenges to already strained pension systems in several countries. However, low interest rates result in both challenges and opportunities, since low interest rates generally result in low interest rates for the countries' public debt. Nevertheless, challenges in ageing societies can also result in increased benefit pay-outs, which in turn can contribute to higher taxes, slower salary growth, higher rates of unemployment and decreased pension pay-outs. At times, attitudes in society state that senior employees should leave working life and give younger and unemployed individuals a chance to enter the labour market. However, since the labour market and economy are not static over time, a younger individual cannot automatically take on a senior individual's work when they exit working life for retirement. Furthermore, a larger amount of people participating in the labour force until an older age is expected to increase purchasing power and the demand for goods and services. However, a larger amount of senior citizens will probably increase the need for measures to promote health and healthcare, since health risks increase with the deterioration of organs and senses. Already today, senior citizens make up a large proportion of individuals in need of care, therefore, an ageing population risks increasing costs for healthcare in different countries.

In order to achieve sufficient economic sustainability and to maintain pensions, many countries postpone the general retirement age. This is due to the fact that the proportion of the population active in the labour force largely finances the proportion

Age in Relation to a Sustainable Working Life 217

**TABLE 1**
**OECD**

Demographic old-age to working-age ratio: historical and projected values, 1950–2080

| OECD members | 1950 | 1960 | 1990 | 2020 | 2050 | 2080 | | 1950 | 1960 | 1990 | 2020 | 2050 | 2080 |
|---|---|---|---|---|---|---|---|---|---|---|---|---|---|
| Australia | 14.0 | 16.0 | 18.8 | 27.7 | 41.6 | 49.4 | The Netherlands | 13.9 | 16.8 | 20.6 | 34.3 | 53.3 | 62.2 |
| Austria | 17.3 | 21.0 | 24.3 | 31.3 | 56.0 | 60.2 | New Zealand | 16.3 | 17.0 | 19.5 | 28.3 | 43.8 | 57.5 |
| Belgium | 18.1 | 20.3 | 24.8 | 33.1 | 51.3 | 56.8 | Norway | 16.0 | 19.8 | 28.5 | 29.6 | 43.4 | 53.4 |
| Canada | 14.0 | 15.1 | 18.4 | 29.8 | 44.9 | 54.0 | Poland | 9.4 | 10.5 | 17.3 | 30.5 | 60.3 | 68.6 |
| Chile | 7.2 | 7.9 | 10.9 | 19.7 | 44.6 | 67.5 | Portugal | 13.0 | 14.8 | 23.9 | 38.6 | 71.4 | 72.3 |
| Colombia | 7.5 | 7.2 | 8.4 | 15.0 | 36.0 | 64.3 | Slovak Republic | 11.9 | 12.6 | 18.2 | 26.5 | 54.6 | 58.1 |
| Costa Rica | 6.8 | 7.1 | 9.0 | 16.6 | 41.6 | 69.4 | Slovenia | 12.5 | 13.7 | 17.3 | 34.7 | 65.0 | 60.7 |
| Czech Republic | 13.9 | 16.3 | 22.0 | 33.8 | 55.9 | 52.8 | Spain | 12.8 | 14.6 | 23.1 | 32.8 | 78.4 | 74.4 |
| Denmark | 15.6 | 19.0 | 25.9 | 34.9 | 44.6 | 52.4 | Sweden | 16.8 | 20.2 | 30.9 | 35.9 | 45.5 | 53.4 |
| Estonia | 19.3 | 17.7 | 19.7 | 34.9 | 54.9 | 63.2 | Switzerland | 15.8 | 17.6 | 23.6 | 31.3 | 54.4 | 56.7 |
| Finland | 11.9 | 13.5 | 22.0 | 40.1 | 51.4 | 65.0 | Turkey | 6.5 | 7.0 | 9.4 | 15.2 | 37.0 | 58.2 |
| France | 19.5 | 20.8 | 24.0 | 37.3 | 54.5 | 62.2 | The United Kingdom | 17.9 | 20.2 | 26.9 | 32.0 | 47.1 | 55.1 |
| Germany | 16.2 | 19.1 | 23.5 | 36.5 | 58.1 | 59.5 | The United States | 14.2 | 17.3 | 21.6 | 28.4 | 40.4 | 51.1 |
| Greece | 12.4 | 12.2 | 22.9 | 37.8 | 75.0 | 79.7 | OECD | 13.6 | 15.0 | 20.0 | 30.4 | 52.7 | 61.1 |
| Hungary | 13.2 | 15.5 | 22.9 | 33.4 | 52.6 | 55.4 | | | | | | | |
| Iceland | 14.1 | 16.4 | 19.0 | 26.6 | 46.2 | 64.5 | | | | | | | |
| Ireland | 20.9 | 22.8 | 21.6 | 25.0 | 50.6 | 60.0 | Argentina | 7.5 | 10.1 | 17.3 | 20.2 | 30.3 | 45.5 |
| Israel | 7.1 | 9.1 | 17.8 | 23.9 | 31.3 | 39.9 | Brazil | 6.5 | 7.1 | 8.4 | 15.5 | 39.5 | 63.7 |
| Italy | 14.3 | 16.4 | 24.3 | 39.5 | 74.4 | 79.6 | China | 8.5 | 7.6 | 10.2 | 18.5 | 47.5 | 60.6 |
| Japan | 9.9 | 10.4 | 19.3 | 52.0 | 80.7 | 82.9 | India | 6.4 | 6.4 | 7.9 | 11.3 | 22.5 | 40.8 |
| Korea | 6.3 | 7.6 | 8.9 | 22.6 | 78.8 | 94.6 | Indonesia | 8.6 | 7.6 | 7.7 | 10.6 | 27.3 | 41.0 |
| Latvia | 18.1 | 17.7 | 19.9 | 35.5 | 53.0 | 49.9 | Russian Federation | 8.7 | 10.5 | 17.2 | 25.3 | 41.7 | 41.9 |
| Lithuania | 17.5 | 14.0 | 18.4 | 34.7 | 55.7 | 55.7 | Saudi Arabia | 7.5 | 8.4 | 6.1 | 5.3 | 28.2 | 44.8 |
| Luxembourg | 15.8 | 17.6 | 21.1 | 22.3 | 43.8 | 50.1 | South Africa | 8.5 | 8.4 | 8.7 | 9.6 | 17.4 | 26.8 |
| Mexico | 8.0 | 8.3 | 9.6 | 13.2 | 28.9 | 50.9 | EU27 | 14.6 | 16.0 | 21.6 | 33.6 | 56.7 | 62.0 |

*Source:* Pensions at a glance 2021 [102].

of the population not active in the labour force, as well as seniors in society, and contributes to the economy. However, postponing the retirement age has appeared to be one of the more disputed political reforms in many countries, unless measures are taken to promote and enable a sustainable working life for all ages at the same time. Across the globe in general, employees with short-term education, and physically and mentally demanding professions are less inclined to remain employed until an older age, compared to their fellow citizens with long-term education and less physically and mentally demanding professions. Forcing individuals to work without taking measures in working life has not been perceived as a sustainable, long-term strategy, since increased demands and decreased control of one's life situation contribute to ill-health, especially if the work environment is problematic to the individual's health. A better strategy for healthy and sustainable workplaces and for public health is to motivate people's willingness and ability of working until an older age.

## AGE DEFINITIONS AND PERSPECTIVES ON AGE IN WORKING LIFE

Age and ageing in working life are complex to define. We all have an age, therefore age is something that concerns all of us. Age is not least important in relation to entering, participating in, and leaving working life. There are four age definitions, and the perspectives of these are particularly significant to a sustainable working life for all ages and to employability (Figure 1). These four different age definitions are biological age, chronological age, social age and cognitive age (Figure 20). These four age definitions are explained below. Age and ageing must also be clarified in relation to three theoretical levels:

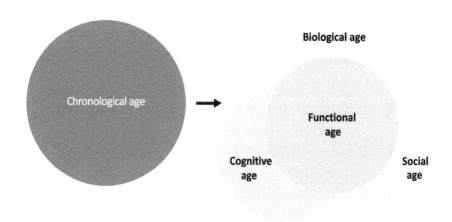

**FIGURE 20**   Age definitions in relation to a sustainable working life for all ages.

- Macro level, the societal level, includes general perceptions in society related to ageing through life and seniors, societal structures, economy, legislation, diversity, development, political and cultural values and approaches related to age and ageing, etc.
- Meso level, the group or organisational level, includes structures in organisations and companies related to norms and values, organisational learning, diversity, development, discrimination and disregard aimed towards particular age groups, economy, re-organisations, age conscious leadership, how different age groups affect each other and group members based on age, etc.
- Micro level, the individual level, includes the individual's experience of their own and the surrounding environment's attitudes towards their age, how body and intellect change throughout life and with age, how this affects the individual's physical and mental state as well as their health, etc.

# Biological Age in Working Life

Biological age is based on the ageing and deterioration of the cells and functions in the body. In biomedical terms, biological age is based on genetic inheritance, which is perceived as the primary biological ageing. The secondary biological ageing is described as biological ageing caused by events and conditions that the individual has been exposed to in their lives – lifestyle, injuries and illness affect our bodies. Biological age is a significant factor for the ability and willingness to work and for employability (Figure 4).

## INDIVIDUAL PERSPECTIVES ON BIOLOGICAL AGE

On a cellular level, the individual's biological age can for example be noticed in the ends of the chromosomes, which are shorter in older individuals than in younger individuals. Telomerase is found in what is called telomeres, in other words, the end of the chromosomes, and contains information in repeated DNA sequences, to ensure that the right genetic material and information is transferred in cell division, in other words, in the replication of the cell. In each cell division, the chromosomes are copied into two exact chromosome copies. However, the entire end of the chromosome, the telomere, is not copied in each cell division. A normal cell can divide approximately 100–200 times before the telomerase is no longer sufficient and no longer protects the cell in replication. When this happens, the cell will no longer divide itself. Since telomeres become shorter for each cell division, chromosomes become somewhat shorter for each cell division, which is why older individuals have shorter chromosomes than younger individuals. Unhealthy demands in working life have been stated to affect the life span (read more in *The Work Environment as a Predictor of Ill-health*). Since work-related physical and mental fatigue is associated with an accelerated shortening rate of the telomeres in the affected individual, their biological ageing accelerates [33].

Individuals' physiology and functions in different biological ages throughout life contribute to different needs for time to rest and recuperate. The fact that a child needs more sleep in different development stages is commonly accepted. However, senior employees in general, based on their biological age, also need more time to rest, a slower work pace and shorter working hours in order to maintain good health and the ability to function.

Signs of ageing are apparent in most bodily functions already before 40 years of age. At 60 years of age, biological signs of ageing show in essentially all bodily functions. For example, the interplay between the brain, sensory organs, circulatory system and inner organs changes. Furthermore, with higher biological age, healing deteriorates and the risk of cancer increases through decreased blood flow and decreased cell renewal. Moreover, the time needed to recuperate and recover from an illness or injury increases.

Age-related changes begin to appear in the circulatory system at 30–40 years of age. The oxygen uptake capacity depends on the capacity and ability of the heart and blood vessels to circulate oxygenated blood to the cells in the organs and tissues of the body and to remove waste products from the cells. Static muscle work decreases the possibility of removing waste products from muscles, which increases the risk of exhaustion. Older age increases the stiffness of blood vessels and the heart which contributes to lower oxygen uptake capacity. Oxygen uptake capacity is estimated to decrease by approximately 1 percent annually. Oxygen uptake capacity matters to the individual's ability to cope with physical work demands. In order to cope with work tasks and to avoid exhaustion after a day at work, an employee must have 30 percent more maximal oxygen uptake capacity than the reference value for different work tasks on average (read more in Section *Bodily Functions that Matter to Work*). This should be monitored to make sure that the reference values for different work tasks are appropriate since oxygen uptake capacity and physical capacity deteriorate with increasing age.

Cardiovascular diseases are sometimes perceived as a part of natural ageing, however, the large proportion of changes in the circulatory systems of the senior Western population cannot solely be attributed to normal age-related changes. The differences in the ageing of cardiovascular systems also depend on heredity and lifestyle.

Measurements have shown that muscle strength in the torso and legs decreases more rapidly with increasing age compared to muscle strength in hands and arms, which remains relatively constant. Furthermore, muscular endurance remains more unaffected with increasing age compared to muscle strength. However, all in all, muscular energy and capacity generally deteriorate with increasing age, since oxygen uptake capacity, muscle strength and muscular endurance work together in muscle work.

Furthermore, biological ageing contributes to the general physiological deterioration of senses such as hearing, sight and touch. In order to perceive sounds and discern speech, these should be approximately 15 decibel stronger than the surrounding sounds. However, many can still find it difficult to hear clearly in noisy environments. Studies show that approximately 5 percent of 35 year olds and 10 percent of 45–50 year olds find it difficult to perceive what words are spoken in environments where many people talk at once. Approximately 20 percent of 70 year olds have been diagnosed with hearing loss [262]. However, the hearing loss that generally occurs with older age is seldom severe to the point where hearing aids are needed before 70–75 years of age.

Sight is affected with increasing age through degenerative processes in the retinas of the eyes. The crystalline lens stiffens and loses transparency, it becomes more difficult for the pupil to change shape. Because of this, many people find it difficult to focus on nearby objects, since the far point moves farther away. Visual acuity is usually measured on a scale from 0 to 2.0, 1.0 corresponds to "normal sight" which can be described as being able to read a standard vehicle registration plate at a 50-meter distance. If the visual acuity is lower than 0.8–0.5 (different rules apply in different countries) in the individual's "best eye", they are not allowed to drive a car. Many people experience that their best corrective visual acuity, which can be achieved through vision aids, decreases. At 70 years of age, approximately 50 percent of people have visual acuity lower than 1.0. The eyes become drier due to decreased tear production, which is primarily a common issue among postmenopausal women. Furthermore, switching focus from objects nearby to objects at a distance and vice versa takes a longer time. This results in difficulties with depth of focus and estimating distances. Furthermore, the crystalline lens and vitreous body in the eye lose transparency, contributing to a need for significantly better lighting and light intensity in order to see clearly. The visual field becomes smaller and peripheral vision decreases. At 75 years of age, the possibility of perceiving things in the peripheral visual field decreases by half compared to the original ability. This makes it more difficult to perceive signs on the roadside and objects approaching from the side, which for example becomes problematic when the individual is in an environment with vehicles and moving objects. Furthermore, night vision deteriorates as well as the ability to switch from light to dark, for example when driving in the dark, thereby increasing the risk of being blinded by lights. Moreover, colour vision generally deteriorates with increasing age, which makes colour-coded information in dim light particularly difficult to perceive.

Furthermore, the irritability of the nervous tissue and reaction ability deteriorate, and reaction time increases with increasing age. Abilities that require several functions, for example balance, also deteriorate. This since the ability to balance includes the functions of touch, strength in muscles and joints, as well as sight. Therefore, deterioration in any of these abilities affects the individual's balance, which can occur at a relatively young age. When all functions deteriorate, the balance deteriorates significantly.

The risk of being subjected to occupational injuries increases with increasing biological age, due to the deterioration of different senses like hearing, sight and touch. Physically and mentally demanding work environments both accelerate biological ageing. Because of this, working life in itself can pose a threat to individuals' health and accelerate biological ageing. However, biological ageing and the deterioration of bodily functions among individuals in an unhealthy work environment have been shown to slow down if they leave the demanding environment of their working lives [263].

As previously described, it is not only general biological ageing that affects the deterioration of bodily functions. Innate diversity, acquired injuries and illnesses during life contribute to differences in the bodily functions that increase in an age group with increasing age. However, individuals' lifestyles provide an explanatory reason for their disposition to develop illness and the possibility to cope with physical work demands. The physiological ageing process can be postponed through the individual's choice of diet and habit of physical exercise. Exercising and strengthening the muscles and joints in the body increases resilience against issues in muscles and joints with increasing age. Ageing should not be regarded as a downward spiral for all functions, since cell renewal continues throughout life, even if it does not continue indefinitely. Physical exercise and activity have proven to improve abilities throughout life since the body particularly renews the functions that are used. Regular physical exercise results in increased well-being, counteracts cardiovascular issues and provides stronger skeleton and sound physical abilities in older age.

## ORGANISATIONAL PERSPECTIVES ON BIOLOGICAL AGE

The biostatistical and organisational perspectives on biological age connect not only to heredity, but also to individuals' lifestyles, injuries and illnesses throughout life [1,2,106]. The development of different body parts and organs, which begins at the moment of conception, turns into biological degeneration with increasing age, and individuals in physically demanding work environments display particularly decreased functions and capacity to work with increased biological age. However, individuals' biological age and physical ageing vary, depending on their occupation as well as their physical and mental demands.

On the organisational level, biological ageing is regarded as something negative and expensive, since it contributes to people becoming worn out, thereby affecting production negatively. It entails a need for, for example, adjusted work pace, working hours and physical work demands. The employer is responsible, through legislation, to find alternative solutions that facilitate the employee's work situation. In cases where the individual no longer is able to cope with their work situation due to ill-health and biological ageing, and alternative work tasks cannot be found, there is a risk that the biological ageing results in termination of employment due to lack of work.

## SOCIETAL PERSPECTIVES ON BIOLOGICAL AGE

An increased middle age in the population of a society is a result of modern developments, people live for a longer time due to improved nourishment and sanitation, medical advances, better healthcare, education and economic welfare. Furthermore, regulations and laws decrease the risk of exposure to accidents and injuries and provide access to security systems in society.

However, we still age. Furthermore, what complicates things further is that it is difficult to understand what factors of biological ageing are caused by external factors, in other words, how an individual's secondary biological ageing relates to their primary biological ageing.

In many countries, the retirement age is proposed to be postponed for all individuals, regardless of profession and work environment. At the same time, research shows that working life affects biological ageing [33]. Furthermore, managers state that health issues related to biological ageing largely cause their employees to leave working life prematurely [107].

# Chronological Age in Working Life

Chronological age refers to passing time and how many years have passed since an individual was born, which relates to the time of labour force participation (Figure 12).

## INDIVIDUAL PERSPECTIVES ON CHRONOLOGICAL AGE

The economy in society affects the need for a labour force and, in turn, this affects between what chronological ages individuals should participate in working life. There are significant differences between individuals' work ability in an age group categorised by chronological age and the differences only increase with increasing age. However, individuals, much like organisations and companies, have subordinated power in relation to regulations and laws in society, their everyday life is regulated by decisions regarding limit values set based on chronological age in the laws, rules, insurance systems and benefits in society. It can be problematic if individuals who have a high level of work ability and want to continue working are forced to exit working life due to their chronological age since the individual's capacity to work is individual and relates to the work task and work environment. However, it can be problematic if a retirement pension is solely based on chronological age, for example, for individuals whose bodies age prematurely and whose bodily functions do not match the average values of their chronological age group. Some individuals, who have not yet reached the chronological age of retirement pension, can be forced to continue working, regardless if they are worn out or suffer from physical or mental ailments and lack work ability.

## ORGANISATIONAL PERSPECTIVES ON CHRONOLOGICAL AGE

The changed demographic structure's impact on the economy has changed and postponed the chronological age span for labour force participation in many countries.

The main focus of organisations and companies is to maintain production and achieve production goals for goods and services. Therefore, it is significant for the survival, competitiveness and production goals for goods and services that organisations and companies consist of employees with sufficient productive abilities. However, organisations must comply with the laws and rules in society that in some regards set a certain chronological age as an "expiration date" for labour force participation. Chronological age is a factor that regulates rules and systems in society, however, it is not a reliable marker in itself for when an individual, based on their abilities in their work situation, should participate in or leave working life. Research has shown that, in general, people's productivity in working life increases until they are 50–55 years of age, after which their productivity slowly decreases [264]. However, this is mainly based on the speed of executing work tasks and muscle strength. A literature review of studies regarding senior employees states that no results show the fact that an older age generally would amount to less professional skills [265]. A potential decline in productivity is not primarily caused by age, but rather by work tasks, work content, work environment and the design of working life, and is very individual. Therefore, chronological age does not mirror individual employees' work ability and possibilities for organisations and companies to achieve their production goals.

Studies from different countries have shown that the chronological age in which employees are regarded as senior employees differs.

## SOCIETAL PERSPECTIVES ON CHRONOLOGICAL AGE

After being born, human beings grow and develop over time. On the basis of societal organisation and systems thinking, there is an interest in classifying individuals' ages based on a general estimation of normal values, in other words, between the chronological ages it is considered normal to participate in working life. Chronological age is one of the factors that mainly classifies when we are allowed to, but also when we should participate in working life. In other words, chronological age forms the basis of the regulation of systems for inclusion and exclusion in working life. This line of reasoning is based on the systems thinking in society, where age is based on the day an individual was born, with a steady cumulative increase of one year for each passed birthday. This reasoning presupposes that all other factors are equal and that there is a normal value of what can be expected of individuals in a certain age group. However, a strict chronological age definition disregards the differences between individuals' ageing, which can, for example, be caused by illness, injuries, work environment and heredity.

To promote international comparability, the working-age population is often defined as all persons aged 15 and older, but this may vary from country to country based on national laws and practices. The maturity of the mental and biological age before 15 years of age is not considered sufficient, and the risk of lasting issues and injuries caused by the demands of working life is considered higher. Some countries also apply an upper age limit for labour force participation, while others do not. For example, in countries with an upper age limit for labour force participation the employer is allowed to terminate employment due to age. This means that it is up to the employer to decide whether the employee has passed their "expiration date".

In countries with age limits on retirement, the chronological age limits when the individual has the right to a public pension. Because of this, the legislation appears to generalise individuals' labour force participation and work ability based on their chronological age. What is interesting is that the demographic structures in several countries appear to contribute to the need of a larger amount of people to remain in the labour force until an older age, in order to increase the number of worked hours in the economies and to maintain the welfare systems [102,266–268]. However, older persons could also be in vulnerable situations [269].

The starting point of public pension and pension systems in social security systems in different countries is to ensure that the individual has insurance against the age effects on the ability to make a living through work. In countries without public pensions, and in past times, older people who were no longer able to make a living, due to illness, injuries and old age, were taken care of by their previous masters or by their younger relatives. However, decisions regarding these arrangements, whether the elderly relative should live in one place or move between younger relatives, the division of support and work with ill elderly relatives did not always proceed without conflict. For example, in Sweden, we find county court sentences from the 18th and 19th centuries that bear witness to how siblings proceeded against each other concerning who should take care of their elderly mother and elderly father. However, when the agricultural society disappeared during industrialisation, care of elderly people became a societal issue. Young and able-bodied people were attracted to new forms of employment in cities and in factories, while older people stayed in rural communities. Municipalities were forced to finance support to poor elderly people. At the same time, diseases such as the Spanish flu claimed thousands of victims. This pandemic primarily affected younger people and caused approximately 3 percent of the Swedish population to die during the start of the last century. Furthermore, many young and able-bodied people in the 17th-century Sweden emigrated, mainly to North America, due to poverty. The Ministry of Foreign Affairs launched a commission in 1907 to examine the emigration. The final report was published in 1913 and described how the flow of young people from Sweden resulted in an ageing population and a decreasing proportion of young and able-bodied people who could help care for their elderly relatives. Elderly people who lacked financial means or who did not have any relatives who wanted or could afford to care for them were referred to spending their last years in municipal poorhouses. The proportion of elderly people in need of care and support from municipalities increased when younger people disappeared, which was very expensive for small, rural municipalities. Because of this, the pension system was introduced.

# Social Age in Working Life

The term social age describes how individuals of different ages function in society, and how our roles, for example, in a family or in a profession, change with increasing age during life. Social age relates to the ability and willingness to work, and to employability (Figure 13).

## INDIVIDUAL PERSPECTIVES ON SOCIAL AGE

Social age can be defined as the age an individual is attributed to by a collective assessment of different life phases. Social age refers to an individual's social identity through different stages in their life. Sometimes the life span is described to consist of four parts: childhood and adolescence, adulthood with reproduction and the phase of working life, the younger senior when working life begins to come to an end and time as a healthy pensioner, and the elderly senior, whose life phase to a larger extent is characterised by a need of care at the end of their life [270]. For example, the individual is attributed to a certain collective identity by belonging to different social age groups. There is an underlying discourse of what is expected and acceptable in different social ages. However, these expectations differ depending on which society, culture and different social contexts the individual finds themselves in.

The concept of social age is at the same risk of inequality as socioeconomic status, gender and ethnicity. Inequality risks causing social differences since different age groups receive different resources, power and status. Middle-aged people constitute the social age elite, while younger and senior individuals are relatively powerless. However, youth is regarded as much more desirable than old age. Senior individuals are sometimes regarded as less desirable to recruit and are often regarded as slower, less dynamic and less inclined to change. The fact that younger middle age is perceived as the social age elite relates to expectations that this age group is responsible for production, profitability and growth in society.

*Intersectionality* is a concept and analytical perspective in social sciences which aims to display what can happen at the intersection of different power structures. We constantly find ourselves in and move between different power structures, and in the areas where they intersect. Every individual finds themselves in, belongs to and positions themselves in several different power structures in different situations. All individuals belong to several groups and categories by being identified as, for example, a woman, man, young person, senior, parent, childless, transgender, cisgender, heterosexual, homosexual, disabled person, Christian, Buddhist, rich or poor. There is an increased risk of exponential exposure in the intersection between different power structures and grounds of discrimination, for example in the intersection between age, ethnicity, socioeconomic status and gender. The groups of men and women do not simply constitute men and women, since every individual also has a certain age, ability, sexuality, ethnicity, etc. In certain contexts, the norm is to be a woman, while in other contexts it is to be a man. The norm is powerful, people who break the norm are subordinated. A person who deviates from the norm is usually not given the same scope for action and power in a certain context and faces a higher risk of disregard than people who do not deviate from the norm. For example, ageism is more often aimed at women than men [199]. However, among people subjected to ageism, it results in earlier retirement for men more often than for women. Using intersectionality as a tool to display power structures and risk of disregard in different contexts is an approach of norm criticism, with the purpose of analysing situations, contexts, organisations, companies or societies.

When individuals are categorised in the social age "senior in working life", it appears that they are attributed to certain stereotypical characteristics [2,7,9,11,106,107], for example that seniors are less inclined to change, unknowing of the latest technology and knowledge, and generally slower than younger individuals. Seniors in general are expected to have less time left in the workplace compared to younger individuals. The socio-cultural categorisation can easily result in a generalisation of senior employees as "has beens". Attitudes from managers, the organisation or company or from society often state that the time has come to leave working life. Finding oneself at the end of working life means that as a senior employee, one is expected to leave the social age of an employee and enter the social age of a pensioner, in other words, to change life phases from the social age of working life to the social age of retirement. Individuals who approach the time to leave working life often lose the status that labour force participation entails, they are stripped of their power since they will leave working life soon and the expectations on them from their surroundings change. Individuals who are involuntarily forced to move on to the next social age, in other words, become a pensioner, despite them identifying themselves as younger, risk developing symptoms of mental illness. This is probably due to the decreased status and life satisfaction that retirement entails if their sense of identity is strongly linked to their profession and to labour force participation.

Being categorised based on one's age can be a very delicate issue. This has, for example, been the case in research projects and focus group interviews where researchers have asked all participants to write their name and age on a sign and place it in front of themselves in the group interview [11,95,166,172]. A recurring occurrence when forming new focus groups was that one or several participants reacted negatively to this procedure. Occasionally, individuals refused to state their age at all since they considered their age to be a private matter that they did not wish to inform others of. At the same time, they defended their position with society being fixated on age, a fixation they did not wish to participate in, that they did not experience themselves to be older in their mind or different now than previously in their life, and therefore concluded that stating their chronological age was a negative experience.

An individual being, or experiencing to be, included or excluded in a social context based on their age can cause negative consequences [80,106,198,271]. Because of this, age is included in the legislation in many countries (directive 2002/73/EC; 2000/78/EC) and the UN Universal Declaration of Human Rights [266]. However, in these directives age is defined as chronological age, i.e. the number of years passed since the individual's year of birth. The protection in these directives covers everyone since everyone has an age.

## ORGANISATIONAL PERSPECTIVES ON SOCIAL AGE

Like in the rest of society, employees in working life are attributed to different characteristics based on their social age and life phase. Middle-aged people are perceived as the norm, accordingly, middle-aged people hold power and status [184]. Based on this, the social ages in working life can be divided into three groups:

- *Becomings.* People who are categorised in the social age group as *younger employees* are attributed to being unknowing and often being careless and not taking as much responsibility. At the same time, they have the desirable qualities of being "new", "young" and in the beginning of their careers (*becomings*). Because of this, the mistakes made by younger people are often smoothed over with remarks such as "you live and you learn", since the expectations of them being productive and knowing are not yet as high. They have the future ahead of themselves and are perceived as worth investing in.
- *Beings.* People who are of the age when they are in the middle of their working life are considered to be the norm, the social middle age. They are expected to have experience of their years in working life, to be knowledgeable and at the height of their careers. Because of this, individuals in this age group have the highest status (*beings*). This age group usually holds the power, and they have high expectations to be productive. Younger and senior individuals are compared to middle-aged individuals. Their social age, middle-age, is the reference value for both younger and senior individuals. Individuals of this age group are expected to cope and have time for everything at work. At the same time, people in this age group are usually in the middle of their family life. Consequently, many individuals in this age group experience high pressure and stress.
- *Has beens.* People who have participated in working life for a long time are categorised in the social age group of senior employees. Their career is attributed to being stagnated or declining (*has beens*), and they are considered to be degenerate and approaching the end of their working life.

    They are perceived as being of the past and are no longer taken as seriously as they previously were. Even if they have had a high status position and made significant efforts through their work, this is soon forgotten in the light of the efforts of middle-aged employees (beings) and since younger employees (becomings) have not been around to know their efforts. The expectations of performance for individuals in the social age of senior employees decline. They are no longer given as much credit for their productivity, abilities and knowledge. This results in decreased power and status, which can cause experiences of exclusion, negative stress and decreased motivation to work.

Individuals go through the different stages "becomings", "beings" and "has beens" in their working life. However, some organisations and companies are what could be described as "neophiles" and have an extreme desire for renewal. Consequently, age and ageing in these organisations can be particularly complicated, since younger and recently recruited employees are much more valued than senior employees' experiences of a long working life. In these organisations and companies, ageing is perceived as something bad, senior employees are often stigmatised and have lower status. Furthermore, studies have shown tendencies towards ageism in organisations and companies [11,26,107,08,149,166,272]. Regardless, age is, as previously mentioned, a ground for discrimination, and is regulated in the Charter of Fundamental Rights, as well as in anti-discrimination and equality acts in many countries. However, age sometimes appears to be a socially legitimate way of refusing to employ or terminating employment. In studies, age appears to be the most common cause for the experience of discrimination out of all grounds of discrimination [26]. The fact that age is experienced as the main cause of discrimination in working life is probably due to the fact that attitudes towards different ages affect access to competence development, career development, and the expectations on productivity and division of work tasks. Younger employees are regarded as careless and uncommitted [11]. Senior employees are regarded as weak, with declining competence, and are given less access to important resources in the workplace compared to younger employees. Because of this, organisations and companies too often choose not to invest in employees who are categorised as seniors in working life. Age, and being categorised in the social age group of seniors in working life, affect the possibilities of competence development, and the possibility to participate in new development projects in the workplace.

## SOCIETAL PERSPECTIVES ON SOCIAL AGE

Age and being categorised in a particular age group can be much stigmatised and a delicate subject, even if we all have an age and consequently none of us are excluded from collective categorisation into social age groups [11,26,107,108,149,166,272]. That age is a highly charged term is probably due to the existence of an elite in terms of age, in other words, an age that is perceived as better or more desirable than other ages depending on the situation and context. The question "how old are you?" is sometimes in some cultures perceived as rude and taboo to ask, and many people make excuses when they have to ask it. When questions of age arise, they are often smoothed over with sayings like "age is just a number". This statement implies that ageing is relative and that chronological age is separate from how one may feel or what age the individual experiences themselves as, in other words, their social age. Social age can be defined as something that exists in an individual, based on how old they feel, however, social age also includes other people's attitudes and how other people categorise us as ageing individuals. Therefore, categorising an individual in the social age group "seniors in working life" is a result of social norms on the societal, organisational and individual levels.

In some cultures, senior people are honoured and regarded as wise, people to collectively look up to, for example, in older Chinese and Jewish societal traditions.

However, there are cultures that have a completely opposed approach, perhaps particularly in secularised cultures, where senior people are regarded as weak and undesirable, resulting in the marginalisation of senior people [267]. Western contemporary culture celebrates and covets youth and youthfulness, often even by seniors themselves. We hear statements such as "seventy is the new fifty" as if it is more desirable to be 50 years of age rather than 70 years of age. Skincare and face creams are sold with promises of an "anti-age effect". On a societal level, individuals who are approaching retirement age or have retired are sometimes regarded as a problem. This group is perceived to go from contributing to the economy through their working life and worked hours, to becoming a weight on the social security system through their pension benefits instead. However, research on volunteer work after retirement shows that there are reasons to be critical of the perception of pensioners as a burden. This group most definitely contributes to society through volunteer work and by caring for relatives, friends, ill children and grandchildren. Unfortunately, these unpaid working hours are often taken for granted and not counted in the economy.

# Cognitive Age in Working Life

Age can also be categorised based on the context of cognition. Cognitive age concerns cognition, brain development, memory, learning ability, knowledge, competence and thinking and is significant for the ability and willingness to work and for employability (Figure 19).

## INDIVIDUAL PERSPECTIVES ON COGNITIVE AGE

Mental processes such as processing speed, reasoning speed, movement speed and working memory are slower for senior adults compared to younger adults. According to neurological examinations, cognitive processing takes longer time with increasing age, which affects speed, reaction time and the ability to focus on important information and ignore irrelevant information. The slower processing makes it more difficult for seniors to react to different stimuli in the environment, which makes it more difficult to make choices and to process and remember information. Age-related decrease in cognitive abilities, such as memory and reasoning, also relates to cardiovascular diseases and to physical abilities like walking and balancing. This causes most age-related changes in cognitive functions, like memory and attention, with increased cognitive age. Many illnesses that are diagnosed later in life actually begin to develop at 55–65 years of age. This means that approximately 70 percent of individuals in this age group probably have some degree of illness causing their cognitive functions to deteriorate.

Education, competence and the possibility of developing and using new abilities and skills are factors of importance to cognitive ageing. Based on changes in the structures of the brain, it takes longer time for senior individuals to cognitively structure and organise information, if the information is presented in a new or different way than before. However, in examinations, senior employees often display better qualitative performance in their work, even if quantitative performance generally can be lower compared to younger employees. Regardless, it is not age in itself that decreases an individual's ability to learn new things and execute work tasks. The decreased cognitive capacity among seniors is often due to a lack of motivation, lifestyle factors, illness and negative expectations. Other factors that affect mental ageing are, for example, stress, decreased self-esteem and loneliness.

The mental impact and stress of losing one's social role in society, for example, through no longer participating in the labour force, can accelerate cognitive ageing. Furthermore, senior employees are in an exposed position if they have fewer opportunities to participate in competence development and education, while they still have to work until an older age. Moreover, if senior employees have fewer opportunities for cognitive development and learning new things, they also risk being regarded as less competent and employable by organisations and companies.

## ORGANISATIONAL PERSPECTIVES ON COGNITIVE AGE

Cognitive impairments and a high degree of cognitive ageing that causes functional problems and problems in working life before 70–75 years of age are not caused by normal ageing, but primarily by illness and unhealthy changes in the brain. The cognitive age of employees can impact productivity on the organisational level. At the same time, professional skills and tacit knowledge often increase, which compensates for the potential physical limitations that can occur with increasing age or illness.

Several studies show that organisations and companies hold negative attitudes towards employees' mental and cognitive ageing [107,273,274]. In interview studies regarding managers' attitudes towards their senior employees, the participating managers made statements such as: "you cannot teach an old dog new tricks" [11,166,202,216]. Managers who consider seniors to no longer have the same ability for learning and absorbing new information do not offer access to competence development for senior employees to the same extent as managers who do not consider this to be a problem [107,108,200].

Senior employees' cognitive and mental ageing is described – often unjustifiably – as an obstacle to remain working until an older age [275]. However, longitudinal studies show that mental and cognitive ageing is postponed by remaining in working life until an older age [275–277]. Nevertheless, this presupposes that senior employees do not have decreased possibilities of exercising and keeping their cognitive abilities active through decreased opportunities of gaining new knowledge and participating in new projects. If this is the case, senior employees risk having decreased possibilities of continuously performing to high standards in a cognitive sense. Furthermore, cognitive ageing risks accelerating by entering retirement and no longer exercising cognitive functions in the same way. In particular, people who have had cognitively challenging professions risk also entering cognitive retirement when they leave working life.

## SOCIETAL PERSPECTIVES ON COGNITIVE AGE

Individuals' mental development through life, with cognitive aspects, learning and memory, is categorised as mental or cognitive ageing. Psychological or mental age describes how well an individual in a certain developmental stage and age in life adapts and reacts to demands from the surroundings, compared to individuals in other ages and developmental stages. However, the paradox in this is that in the ages of childhood and young adulthood, cognitive changes are mainly described as development and maturity, while cognitive changes after middle age are described as ageing. The word development is associated with hopes and expectations, while the word ageing is a term often associated with decline, deterioration, degeneration or something used. The development of children and young adults to "become someone" is regarded as positive and desirable, while the cognitive abilities among seniors, such as tacit knowledge and wisdom, lack the same status in Western societies. However, from a socioeconomic perspective, it is negative not to make use of seniors as a resource.

Stress affects the brain and accelerates cognitive ageing. The amount of people absent from working life, due to fatigue syndrome and stress caused by high demands, low control, bullying and mental overload, increases in society [66]. A study showed that almost one in ten employees had been diagnosed with stress-related illness caused by their working life [26]. That is, diagnoses which risk accelerating cognitive ageing.

# All Ages in Working Life

Senectus ipsa est morbus! – old age in itself is a disease – is a quote attributed to the Roman playwright Terence who lived in 195–159 B.C. This is a controversial thought, which partly appears to endure in working life. However, that ageing in itself is a disease cannot be regarded as true. What we call ageing and old age are rather a collective impact of lifestyle, social conditions, injuries, illness and age-related genetic changes, which vary between different individuals. Because of this, ageing and old age are very individual.

Some cultures and religious communities honour senior individuals and traditions, for example in China, and in Jewish and Christian communities in the past, senior people were elevated as wise and sensible. In ancient and indigenous cultures, for example, among the Inuit and the Saami, senior individuals sometimes committed altruistic suicides when they no longer were able to work and contribute to the welfare of the group. However, no scientific evidence suggests that senior people systematically were made to jump from or pushed from a precipice when they, due to old age, were unable to contribute to the welfare of the group.

Chronological, social, cognitive and biological age have been described above. The changes in demography cause a shift in chronological age. In order to make working life sustainable for all ages, opportunities must be created to shift the biological, cognitive and social age of the population as well. Several factors work together when describing an individual's age, and we must interweave different age definitions in order to grasp and understand the complexities of age and ageing. Age can vary based on the individual's categorisation in chronological, social, biological and cognitive age groups. We must assume an understanding which combines all the dimensions of age described above [1,2,7,10,60,80,106,108]. Furthermore, we must weigh in, relate and consider the different perspectives on the individual's unique, functional age category. In order to understand how all different ages relate to working life and to make working life sustainable for all ages, we must take the different ages into account separately, and consider which factors in working life primarily relate to the different age categories and ways of defining age (Figure 1).

## INDIVIDUAL LEVEL – MICRO LEVEL

Individuals' possibilities of participating in working life depend on their functional age, a combination of the individual's biological, cognitive and social age that is (Figure 20). Chronological age is in itself not significant for the individual's possibility of working. However, chronological age is a discursive resource regulated by rules, norms and laws regarding when and with which functional diversity, it is allowed to participate in working life. Furthermore, chronological age regulates the individual's personal finances, for example through chronological age regulations in social insurance systems. Age relates to the individual's employability, and to their ability and willingness to work, based on the nine determinant areas of the SwAge™ model [1,2].

Individuals make, more or less consciously, decisions in relation to their ability and willingness to participate in working life based on the four spheres for action in the SwAge™ model (read more in the Section *To Stay or to Leave?*):

- Health effects of the work environment – the individual's health in relation to their physical work environment, mental work environment, working hours, work pace and time for rest and recuperation.
- Financial incentives – the individual's personal finance.
- Social inclusion, support and sense of community – leadership, social participation and sense of community in the social work environment, in relation to social life with family and leisure activities in the personal social environment.
- The execution of work tasks and activities – the possibility of experiencing motivation, stimulation, meaningfulness and self-actualization in activities, as well as how individuals cope with their work situation based on their abilities, knowledge and competence development.

## ORGANISATIONAL LEVEL – MESO LEVEL

Age is a significant instrument of power since it categorises individuals into different groups, there are different age discourses in relation to working life and the organisation of work. The power of age in relation to work and labour force participation is multidimensional. No single individual holds the power to define age in different ways, it is rather used and exerted by everyone, both by people who make decisions and people who must abide by these decisions that is, regardless of who benefits and disadvantaged by the decision.

The main focus of organisations and companies is the possibility of production. Therefore, organisations and companies mainly assume biological and cognitive age when considering the age of an employee. The middle-aged employee is often considered to be the norm. They are in the prime of their life and are neither too young nor too old. However, younger and senior employees are often regarded as deviant from this norm and risk being perceived as inexperienced and of the past, respectively. Biological and cognitive ageing is perceived as a threat to productivity and therefore often emphasized when assessing the senior employee's work ability. Social age is a social construction caused by constructed normative roles based on expectations of an imagined lifespan in the prevailing culture and attitude in society. The social age norm in the organisational culture affects managers' attitudes towards their senior employees, and organisational measures in the work environment specifically aimed at senior employees [1,2,107]. Chronological age is not as significant to production but comes second in the organisational perspective on employees' age. At the same time, organisations and companies must abide by the laws and regulations that use chronological age as an instrument of power. In countries with an upper limit of labour force participation, organisations and companies who wish to terminate the employment of an employee who is on the verge of retirement, chronologic upper age limits can be used as a discursive resource.

## SOCIETAL LEVEL – MACRO LEVEL

On the societal level, chronological age is a discursive resource used to control the economy. In society there are expectations on individuals to participate in working life in a certain chronological age span: the socially constructed age called "working age". Social age can be used as a discursive resource to influence the culture in a society regarding when it is considered normal to enter, participate in and exit working life.

Younger people should not, according to the social norm in the Western world today, participate in the labour force. The basis for this is that the biological and mental age of children and adolescents have not yet reached sufficient maturity to cope with work. In order to reach one's full biological and functional level, as well as the ability to develop cognitive functions, gain knowledge and mature mentally, the social norm states that children should participate in education and develop cognitively without having to make a living or risk being worn out and injuring their bodies through work tasks they are not yet mature enough to execute. However, a hundred years ago children in most countries were included in the labour force. In some parts of the world, the social norm still states that children can work.

The expectations on middle-aged people are that they should work and not be left out of working life. In middle age, it is often considered and experienced as stigmatising and deviant not to participate in working life due to illness, functional diversity, caring for an ill child, unemployment or other reasons that make the individual unable to participate in working life. Moreover, seniors are distinguished from the other social ages in working life. On the societal level, there are normative expectations that senior people should retire and stop providing for themselves at a certain age. However, these expectations are mainly regulated by chronological age [1,2,18,106], for example limits concerning chronological age in social security systems and pension systems in many countries, i.e. at what chronological age an individual is allowed to retire from working life. In this, cognitive age and biological age are made invisible and subordinated. Senior people in working life are attributed to a certain age which carries implications of senior people's deviations from the biological age norm in working life, caused by biological degeneration, which partly depends on cognitive ageing which can entail risks or issues in some professions. In this, cognitive and biological age are made invisible and considered to be subordinate. There is a focus on senior people in working life regarding the issue of whether it is possible to postpone the retirement age. In this discussion, biological and cognitive age should be taken into account to a greater extent than chronological and social age [1–3,18,106].

## CONSIDER AND REFLECT

- How do you define your age?
- What significance does your chronological age have for your work ability?

# All Ages in Working Life 239

- How do you experience attitudes towards yourself based on your chronological age, both from other individuals and in different situations?
- What is your biological age, in relation to your health, physical work environment, mental work environment, working hours, work pace and need for recuperation?
- What significance does your biological age have for your work ability?
- How do you experience attitudes towards yourself based on your biological age, both from other individuals and in different situations?
- What is your social age, in relation to your social work environment and personal social environment?
- What significance does your social age have for your work ability?
- How do you experience attitudes towards yourself based on your social age, both from other individuals and in different situations?
- How do you experience your cognitive age, in relation to executing your work tasks and activities, and to your knowledge, competence, abilities and professional competence development?
- What significance does your cognitive age have for your work ability?
- How do you experience attitudes towards yourself based on your cognitive age, both from other individuals and in different situations?
- Reflect on what significance the biological, chronological, social and cognitive age of an individual have for the organisation or company they are employed in.
- Reflect on whether an individual's biological, chronological, social and cognitive age are equally or unequally important for the organisation or company, depending on the individual's profession or position.
- What significance do the biological, chronological, social and cognitive age of citizens have for society?

# Part III

# Practical Application Based on the SwAge™ Model

This part of the book describes different tools for practical application based on the SwAge™ model. For more information, please visit www.swage.org/en

### PRACTICE APPLICATION OF THE SWAGE™ MODEL

A healthy and sustainable working life includes a well-functioning systematic work environment management integrated into everyday operations, based on the nine determinant areas of the SwAge™ model. Furthermore, use all nine determinant areas for a sustainable working life and for employability in career development discussions with employees, and in the systematic work environment (see tools below). Reflect on different age perspectives and particular needs of groups and individual employees.

# Case for Reflection Supported by the SwAge™ Model

**Case 1**. Pi, 56 years old

Pi has 23 years' experience of a blue-collar profession in the construction industry. The work involves repetitive tasks and sometimes heavy lifting; however, technical solutions cover most of the heavy lifting.

Many colleagues that Pi enjoyed working with have recently quit their jobs. Her best friend at work has retired, while one of the younger employees Pi enjoyed working with has left work to study and another has started working in another industry since she didn't think the tasks were very enjoyable. Pi considered studying many years ago, but it never happened and now she feels too old to be taken seriously if she were to say that she too wants to develop her skills. For the same reason, Pi refrains when the manager asks who wants to be part of new projects in the workplace, even though she really wants to and feels that her experience in the industry could be valuable to the new projects.

Pi's workplace is male-dominated. Some younger men who have recently joined the workplace want to show their strength and capability and therefore do not use the technical solutions available, which also take longer to use. They joke, sometimes a little mockingly, with Pi because she always uses the technical lifting solutions, because she is the only woman in the work unit, because she is older and because she is religious.

Recently, Pi has experienced problems with her back and shoulders. Furthermore, Pi has high blood pressure and the medication sometimes makes her dizzy, which can be dangerous in certain work situations. She has started to think about how and what to do about her work situation. If it weren't for the fact that she needs a salary, she would have chosen to quit and devote her time to leisure interests, she is the chairman of the local football association and also the leader of two youth groups. Pi enjoys and is appreciated for her leadership.

At the same time, Pi enjoys the company and the routines that work provides in her life. She rides her bicycle to work, and not having to commute by train or car means she has sufficient time for recuperation and to pursue her leisure interests, where she can relax and recover from the static work tasks in the industry.

The manager has set up a meeting with Pi. She doesn't know why and feels worried since she experiences the manager to be unfair and favour the young men who have recently joined the workplace. What if she is to be fired? However, the manager who finds it difficult to recruit staff wants to keep Pi, and that is why he wants to talk to her. However, the company is not doing very well financially, and he does not want Pi to know that they need her so that she won't ask for a higher salary.

**To make Pi's situation sustainable and to keep Pi in the workforce until an older age:**

- What can the manager/company do?
- What can Pi do by herself?
- What can society do?

**Case 2.** Hamid, 50 years old

Hamid currently works as a chef. When he was in his twenties, he worked at a fast-food restaurant where he worked with many of his friends and enjoyed himself. However, Hamid then had the opportunity to start working in a clothing store, which he also enjoyed very much. He liked walking around and tidying up the store and talking and helping the customers who came in to shop. However, the owner of the shop was old and one day he told Hamid that he was going to close his shop and retire. Hamid was disappointed but could do nothing more than look for a new job. He then got new employment at a sawmill a few miles from home; a job that he experienced as free. He enjoyed the work even though the working environment was messy and many times dangerous with moving objects flying through the air and there were no protective covers for some of the saw blades. He also liked the colleagues at his sawmill very much.

After working ten years at the sawmill, Hamid had a workplace accident and lost two fingers on his right hand. Hamid had by this time also married Nasuc and had five children. Because of the dangerous work environment, he quit his employment at the sawmill, got a truck and bus driver's license and started driving trucks abroad for a transport company. He had difficulty sleeping when he was away from home, and several times Hamid risked falling asleep at the wheel. Additionally, because Hamid was away from home as much, Nasuc had to take care of the five children herself for long periods of time. The children were left alone at home in the early mornings because Nasuc worked as a cleaner at a school. When Hamid came home after his long journeys, both he and Nasuc were tired and they often started arguing. The situation was untenable. Hamid looked for another job and started driving the school bus instead. The situation improved, but Hamid still found it difficult to relax and sleep at night because he was paid much less. Hamid and Nasuc found it difficult to get their salary to support the large family. Hamid was therefore constantly tired and the rowdy school children could make him very angry. He applied for a new job as a driver for a construction company. Hamid was very happy when he got the job. But soon it turned out to be a problem for the company. There were flaws in the working environment and one day Hamid was hit on the knee by five boards that fell off the load when one of the other cars came driving. Hamid's knee swelled up and he had to stay home from work. The knee injury caused him to limp. The manager was threatening and aggressive when Hamid pointed out the high risks in the work environment and clearly showed that he did not like Hamid after that. Hamid's colleagues pulled away from Hamid so as not to be exposed to the manager's bad mood. He was blamed for many mistakes made by others at work and became a scapegoat. Hamid again found it difficult to sleep and also had a stomach ache and became scared and shy. When the company was at risk of bankruptcy a year later, Hamid did not receive his salary. He therefore applied for another job but found it difficult to get something new and became unemployed.

It took two years before Hamid found a new job, and the family's finances were strained. Hamid saw that a nearby restaurant wanted to hire. He applied and started in the kitchen. Hamid enjoyed working with food. After a few years, he also attended a cooking course, in the evenings after his work at the lunch restaurant, to learn more. Hamid became a very good cook and started working at a famous restaurant. However, the work pace was very fast. The stress and the damaged knee that still caused Hamid to limp, contributed to the fact that he had to start taking breathing and pain-relieving drugs to be able to sleep and be able to work. In the end, it was clear that he could not cope with the work. Hamid's wife Nasuc then saw that a cooking school was looking for an experienced chef as a teacher. Hamid applied and got the job.

Hamid has now worked at the school for five years and enjoys teaching students who want to work with food. However, Hamid does not have any formal teaching training, and this is a requirement to continue working as a teacher. The school principal is very happy with Hamid's work at the school and is thinking about how she can support Hamid so that he gets the formal qualification to teach at the school. However, Hamid still has trouble sleeping at night and doesn't know if he can afford to go on a training course to become a teacher. It's been so long since he went to school himself and he doesn't know if he can handle it. He thinks he is getting old, but at the same time, what choice does he have if he is to continue working at the school? It's a few years until he can retire, and he needs the salary. Hamid is thinking about whether he can use his experience and skills from his previous jobs and change careers again, or what he should do.

**To make Hamid's situation sustainable and to keep Hamid in the workforce until an older age:**

- What can Hamid do by himself?
- What can the manager/organisation do?
- What can society do?

**Case 3**. Mary, 41 years old

Mary works as a nurse. She has previously worked for 15 years in an intensive care unit at a large hospital, however, three years ago Mary started working at a residential care facility closer to her home. Mary enjoyed her work at the intensive care unit, the work tasks were challenging and interesting, and she could put a lot of her knowledge and competence to use. She really enjoyed the competence development offered to hospital employees, to be continuously updated on the latest knowledge in the area and in technical care. The reason for Mary to change workplaces was her father's dementia, she needed to be available closer to her parents' home. Mary's mother could not cope with caring for Mary's father on her own. Moreover, Mary's father had previously been in charge of the couple's finances, Mary's mother did not manage how to pay bills and other expenses, causing the couple to risk eviction due to unpaid rent. Mary feels that it is her responsibility to support and care for her parents in their old age. However, Mary also has three children of her own to care for, a 5-year-old daughter and two sons aged 9 and 15 years. The children's father lives in a different city since the divorce five years ago. Mary is mostly, if not entirely, responsible to care for and provide for the children. The youngest son has started to skip classes and hang out with older children, something that greatly worries Mary.

The pandemic caused a very high strain on the residential care facility. Many of the elderly people were severely afflicted by the virus and passed away. Mary felt that it was her fault that so many were infected, that she as the nurse responsible for the unit had not prevented the spread of infection. Mary enjoys working with her colleagues in the unit, though she has a bad conscience. Mary felt that too much responsibility was put on her as the manager of the unit, and that the management did not provide sufficient support and possibilities to take the measures she felt were needed. When employees became infected with the virus, Mary had to step in and work more hours, due to a lack of substitute staff. In the end, Mary was infected with the virus as well. She was ashamed that she, as a former intensive care nurse, did not know better how to protect herself from being infected.

Now, when the pandemic has ebbed, the work situation has somewhat improved, however, there is still a great lack of staff, and Mary often has to work overtime to cover for this. She knows that the demographic changes make it more difficult to recruit trained staff and that the situation risks becoming permanent.

Mary is always tired and experiences a lack of time for recuperation. She rarely has time to go on a lunch break. Additionally, she sleeps badly due to worry concerning her parents and children, and feeling bad for the patients at the residential care facility. She does not know how to cope with working until an older age now the retirement age has been postponed. How will she, with her mental exhaustion due to excessive and prolonged stress as well as sequelae effects, be able to, cope with and find the motivation to work until an even older age?

**To make Mary's situation sustainable and to keep Mary in the workforce until an older age:**

- What can the manager/organisation do?
- What can Mary do by herself?
- What can society do?

**Case 4**. Alan, 58 years old.

Alan has a university degree and works as an economist. He has been in the same workplace for the last 25 years and has never grown tired of his work tasks. Alan finds his work tasks stimulating and enjoyable. Last week the management set up a large meeting where they told employees of future re-organizations to increase profitability and competitiveness. Alan will be forced to change work tasks, something that worries him greatly. Will his competence be sufficient for his new tasks? Will he perhaps need to educate himself further and learn new skills? Perhaps he should just resign and change work places? Or would it be possible for him to leave working life for early retirement? He does experience aches in his shoulders and neck, and his eyesight has deteriorated after sitting at his desk for so many years, staring at his computer screen. He always loathed physical exercise, so his back and muscles are not very strong any more.

Alan's wife is seven years younger than Alan and will not leave working life for many years. Because of this, leaving working life does not seem like an attractive choice, even if Alan would like to have some more leisure time and to see his grandchildren a bit more. On the other hand, his grandchildren go to school and do not have a lot of time to spend with Alan. Furthermore, he cannot afford to retire unless his employer offers to compensate him financially. Since Alan has experienced his work to be so challenging, he never had a hobby or leisure activity. Though that may not be entirely true, Alan thinks, it was probably because the stimulation he got from his work tasks was enough. Furthermore, he does not have many friends outside the workplace, though he never meets his co-workers other than in the work place.

Alan easily becomes stressed, causing him difficulties in eating and sleeping. He also has issues with his stomach and suffers if he cannot eat at certain hours. Alan is pleased with his salary. He has always liked his work, his co-workers, his manager and his work place. When it comes to social environments, tasks and activities, work is important to him. Alan has, apart from being an economist, studied engineering, though he never finished his education. Perhaps, he could finish his education that might give him new possibilities, though he does not want to quit his employment.

Alan's manager knows that Alan has got expert knowledge and that his tacit knowledge is valuable. However, at the same time, it is hard to decide what Alan should do since he is very stubborn and wants to do things his way after spending many years in the workplace. Deep down, the manager would like to get rid of Alan, but the workplace is in need of employees and it would therefore be stupid to terminate employees' contracts.

**To make Alan's situation sustainable and to keep Alan in the workforce until an older age:**

- What can the manager/company do?
- What can Alan do by himself?
- What can society do?

**Case 5**. Brad, 61 years.

Brad works as an administrator at an authority. Brad has previously been married but has been separated for a few years. The two children have grown up and Brad sees them less and less. When Brad was young, his parents emphasized the importance of education, and Brad studied and graduated with an intermediate degree. However, the parents had wanted the son to go further. They were disappointed when Brad did not invest in becoming a professor but was content to be a middle manager in a department.

In the department where Brad works, they have started to introduce new technology in their work tasks. It is the organisation and the management that thinks the work needs to be made more efficient. Brad finds the new technology and digitization interesting. However, at the same time, he is concerned about whether he will be able to handle it. He doesn't want to show his anxiety and lack of knowledge of the latest technology in front of those he supervises so he loses his acting role.

Digitization has created the opportunity to work remotely, both for himself and his employees. Brad enjoys working from home a few days a week. At the same time, he does not meet his colleagues, his own boss and those he supervises every day. Brad feels that it is difficult to be a supervisor when he does not see his employees as often. He does not know how they perform their duties and tasks. Neither does he know how they feel. Do they ask for help if they need it or do the employees feel uncomfortable within their work and with the work environment? Initially, he considers requiring everyone to be present in the office instead of working from home, but he knows that this will not be accepted by either his employees or the management in the organisation. In addition, the management has understood that they can save money on premises when the employees work some days from their homes.

Some employees have indicated that they have difficulty with the digitization of the work, and Brad thinks that maybe skills development is needed both for himself and for the employees. He considers what skills development is needed, and whether the younger and older employees need the same skills development. Perhaps it is not worthwhile to give so much skills development to the oldest because it is expensive, and they are on their way out of working life to retirement. One of Brad's tasks as supervisor is to keep the budget for the department.

Recently, it seems that one of Brad's co-workers, Gita, is not doing so well mentally. She is a faithful employee, but something is not right, and others are complaining about her. At the same time, Brad knows that the complainant, Anna, is a person who, although she is very good at her job, has tended to bully in the past. Is it a question of problems with the norms in the workplace? Is there perhaps discrimination? How should Brad manage his work group remotely? How will he cope with digitalisation? The questions are many, and Brad is also thinking about whether he should stop being a supervisor, change work or go back to finishing his degree as his parents want.

**To make Brad's own and his employees' situation sustainable and to keep them all in the workforce until an older age:**

- What can Brad do by himself?
- What can Brad's manager/organisation do?
- What can the society do?

**Case 6**. Ai is 29 years old

Ai was born prematurely and her lungs were underdeveloped, but she made it through the first years of life and grove stronger. However, she easily gets problems with her trachea and lungs when she catches a cold and has difficulty tolerating strong smells, even though she has no diagnosed illness. Ai is cheerful and happy, smart and practical but has learning difficulties at school.

When Ai was 16, she started working in production at a factory. The work was one-sided, but Ai got into the routine and figured out how she could make her tasks easier. However, she has problems with her wrists and neck and has pain from time to time. She has been given an exercise program by a physiotherapist at the occupational health service, but Ai neglects to do the exercise program. It's just a bit boring and she doesn't take the time it takes to complete the training program.

Two years ago, Ai met her current partner. They have now decided to have a child and Ai is three months pregnant. She enjoys her colleagues, even if there are some who

Case for Reflection Supported                                                     249

are informal managers who like to direct and dictate and with whom it is important to get along well. Her manager is considering retirement. He is kind and fair but is the leader of 70 employees, and Ai does not feel that she has much contact with the manager. Sometimes she doubts he even knows her name.

The factory where Ai works has recently changed its production. They have been bought out by a larger company and the part of the production that Ai worked on is to be moved and instead, they are to start another production. The new production will cause dust and strong smells, even though protective equipment is in place and a new ventilation system has been installed. Ai is worried about her lungs and her unborn child.

It has now come out that there are new positions in the office in the administrative department of the company, which is to be expanded. However, Ai is worried about whether she will be able to learn what that work entails and how she should do it. She is not used to typing on a computer, filling in forms and writing reports, which is a problem for even being able to apply for the position. However, her experience in production can be good in that work, as it increases her understanding of the production chain and where there is a need for improvement. Also, she enjoys her work team very well and she doesn't really want to leave the community that exists among her colleagues there. She feels that she does not fit in working with those who now work inside the office. At the same time, she needs to think about her health and whether she will be able to continue working. But at the same time, she doesn't know if the new production will be a problem for her health in long time.

It is difficult to know what the new work in production means for her and her situation. Could she go to the manager and ask her to put in a good word so she could start doing administrative work in the new position in the office? But would she even be able to learn to work with the administrative duties of the advertised position in the office? Would it be a shame if she couldn't handle the administrative work? Also, what if she can't handle any of the jobs and instead becomes unemployed? Now that she will soon have a child, and she has many years left until retirement.

**To make Ai's own situation sustainable and to stay in the workforce until an older age:**

- What can Ai do by herself?
- What can Ai's manager/organisation do?
- What can society do to facilitate and support Ai's opportunity to work in a healthy and sustainable working life until old age?

# Nine Quick Ways to Increase Sustainability in the Organisation's Environment

1. *Health, functional diversity and illness.* Strive to promote health factors and health for everyone in the workplace. Prevent ill-health and create a trustful environment where early signs of ill-health are attended to and where work is adjusted to the needs of a person who currently has a period of reduced work ability. Promote knowledge of and keep an overview of sickness absence, establish well-functioning routines for rehabilitation from short and long-term sickness absence. Establish routines for the closest manager to remain in contact with employees on sickness absence. Promote flexibility when returning to work after sickness absence and measures concerning, for example, the need of changed or adjusted working hours or work schedule, or changed and removed certain work tasks. Utilize occupational health care when needed.
2. *Physical work environment.* Review the physical work environment, work demands, vibrations, noise, visual ergonomics, climate, chemical health risks, etc. Promote a work climate where it is natural to update and make use of ergonomic aids and technological support. Be observant and reflective to make sure that no other kind of physical workload occurs when conditions change. The physical work environment and the design of premises and surroundings should promote well-being, comfort and calmness and reduce stress.
3. *Mental work environment.* Review the mental work environment. Promote a work climate where threats and violence are prohibited. Promote a work climate where demands are made clear in relation to the employees' control of their work situation. One thing at a time rather than multi-tasking. Acknowledge if anyone puts too high demands on themselves and takes on too much. Prioritise work tasks and activities in situations with high work demands. Acknowledge employees' need for help to prioritise work tasks. The mental work environment and the design of premises and surroundings should promote well-being, comfort, calmness and stress reduction.
4. *Working hours, work pace and time for recuperation.* Establish working hours and schedules that enable sufficient time for recuperation during and between work shifts. The work pace can be high at times, however, it must be followed by periods of a lower work pace in order to enable recuperation

and reflection. When needed, support employees' ability to prioritise work tasks and activities when the work pace is high. Enable different types of recuperation, both physical and mental recuperation that is. The design of premises and surroundings should support well-being, comfort, calmness and the possibility of rest and stress reduction.
5. *Financial incentives.* Sufficient personal finance is significant for well-being and a sense of security. A work situation that promotes and enables good employability will also contribute to the employees' personal financial conditions through their work. The health effects of the work environment, social support, relations and sense of community and the employee's possibility of executing work tasks and activities do contribute to their employability.
6. *Personal social environment.* Promote a healthy balance between working life and personal life, to prevent working life from having negative impacts on the personal life and vice versa. Employees may need extra support, regard and security to cope with their work during periods when they experience stressful situations in their personal lives. Perhaps, employees need changed working hours and work schedules or support to prioritise or change work tasks.
7. *Social work environment.* A present, secure and trustful leadership that is easily accessible in the workplace, through telephone and e-mail. Promote the possibility of spontaneous, direct contact between manager and employee in the organisation, in meetings reserved for feedback and follow-up and in workplace meetings. Promote a fair and transparent organisation. Develop routines and strategies to counteract disregard and discrimination. The workplace is perceived as fair if no one is treated differently, but if rules and values apply equally to everyone. Promote social support for all employees. Promote a work climate that promotes social inclusion and the possibility of influencing decisions. Inform employees of events in the organisation that matter to them, to an appropriate extent, and consider their need for information in order to feel secure in their work and situations. Acknowledge and celebrate small and great events together.
8. *Motivation and stimulation* through work tasks and activities. Promote the employees' possibility of executing and prioritising the core tasks in their work. If possible, give employees their own areas of responsibility. Promote the possibility of rotating between different work tasks to promote their ability to execute and cope with other work tasks and activities. Continuously acknowledge employees' work efforts to make employees feel acknowledged and needed. Give constructive criticism and feedback and follow up. Establish routines and environments for informal and formal forums where employees can reflect and share their experiences and ideas for improvement. Establish routines and systems to pass suggestions on upwards and downwards in the organisation. Promote open and trustful communication between the manager and employees in meetings and informally in spontaneous conversations.

9. *Knowledge and competence.* Make use of employees' individual skills, knowledge and competence in the organisation. Promote a creative work climate that makes the most of employees' differences and talents and lets them come into their own. Establish routines and systems to pass suggestions and viewpoints on upwards or downwards in the organisation. Promote competence development throughout an employee's entire career. Encourage employees to gain new competence and try new work tasks. Reflect on whether competence development should be adjusted to some individuals or if it fits everyone.

# The Survey Tool for Investigation and Reflection of Work Ability, Employability and Work Situation

The SwAge™ model survey is an assessment tool used in research on a sustainable working life and workplace examinations to evaluate individuals' work ability, employability and resources in relation to their work situation. The questions in the survey have been developed through several research projects [4,8,26,136,278], and the first nine represent each one of the nine determinant areas of the SwAge™ model. The results from using the survey can form the basis of discussion, reflection or different measures. The points can be summarised: very good = 4; rather good = 3; rather poor = 2; very poor = 1, to a collected score in order to assess work ability and possibilities. Furthermore, the results from each of the ten first questions are examined for individuals or for a group in order to analyse factors or areas in need of particular reflection, review and measures, alternatively to celebrate successful factors. If an individual or a group has a low score, in other words, if they answer "very poor" or "rather poor" to the questions, initiate a dialogue in order to take necessary measures. The survey also includes a final open question where the participant can write down what they experience and believe needs to be strengthened and improved in their work situation and work environment for a lasting or strengthened employability and for them to experience a sustainable (working) life into the future.

## QUESTIONNAIRE: SUSTAINABLE WORKING LIFE FOR ALL AGES

(Answer what comes closest to your own experience, based on your situation)

1. **How do you assess your overall health in relation to your own possible best experience of health?**

    ☐ Very good
    ☐ Rather good
    ☐ Rather poor
    ☐ Very poor

2. How do you assess your overall ability to execute your work tasks, based on your physical work environment (i.e. physical workload, heavy lifting, unilateral movements, vibrations, chemicals, radiation, climate, air pollution)?

    ☐ Very good
    ☐ Rather good
    ☐ Rather poor
    ☐ Very poor

3. How do you assess your overall ability to execute your work tasks, based on your mental work environment (i.e. stress, mental workload, demands, exposure, threats, violence)?

    ☐ Very good
    ☐ Rather good
    ☐ Rather poor
    ☐ Very poor

4. How do you assess your overall ability to execute your work tasks, based on your working hours, work pace and your possibilities to rest and recuperate?

    ☐ Very good
    ☐ Rather good
    ☐ Rather poor
    ☐ Very poor

5. How do you assess your overall work ability, if you were to apply for your current job today?

    ☐ Very good
    ☐ Rather good
    ☐ Rather poor
    ☐ Very poor

6. How do you assess your overall ability to cope with and have time for your personal life, based on your willingness and ability to execute your work tasks?

    ☐ Very good
    ☐ Rather good
    ☐ Rather poor
    ☐ Very poor

Survey Tool for Investigation and Reflection

7. **How do you assess the overall support, sense of community and encouragement you receive in your work situation, from co-workers, managers, management, co-operative partners, etc.?**

   ☐ Very good
   ☐ Rather good
   ☐ Rather poor
   ☐ Very poor

8. **How do you assess your overall satisfaction with and stimulation through work tasks and activities?**

   ☐ Very good
   ☐ Rather good
   ☐ Rather poor
   ☐ Very poor

9. **How do you assess your overall ability to execute your work tasks based on your knowledge and competence?**

   ☐ Very good
   ☐ Rather good
   ☐ Rather poor
   ☐ Very poor

10. **How do you assess your overall ability to execute your work tasks based on your current age?**

    ☐ Very good
    ☐ Rather good
    ☐ Rather poor
    ☐ Very poor

11. **What do you experience and believe needs to be strengthened and improved in your work situation and work environment for a lasting or strengthened employability and for you to experience a sustainable (working) life into the future?**

    _____
    _____
    _____
    _____

# The Tool for Employer/Manager-Employee Work≈Situation and Career Development Conversations

Communication is crucial to make working life sustainable. Regular discussions and close dialogues are prerequisites for a healthy work situation and work environment and for the ability to direct employees and develop the organisation (see the dialogue tool for career development conversations with employees).

Dialogue facilitates reflection, insights into the needs of the organisations and sheds light on potential problems and risks of work-related ill-health. Furthermore, acknowledgement and inclusion in conversations create a sense of security and participation. If the manager takes time to communicate with their employees and acknowledge their employees it is often perceived as a reward. Greeting everyone in the morning and asking how they are make employees feel acknowledged. However, setting aside time for regular, everyday talks and for individual conversations with employees of approximately 10–30 minutes every other week is time well spent. This makes it easier to catch up on valuable ideas, follow up on individual goals, and when needed, adjust goals based on current conditions, but also to give feedback on successfully and less successfully executed tasks, initiatives and events, and to consider what changes are needed. However, structured career development conversations and salary negotiations, aside from everyday conversations, are required. In these conversations, the picture of the SwAge™ model and the nine determinant areas for a healthy and sustainable working life are useful to ensure that all aspects of the employee's work situation are covered and that none are missed.

Below you will find the SwAge™ model dialogue tool, developed to be used in career development conversations and in dialogue with employees. The dialogue tool is beneficial to use in everyday conversations and in preparation for salary negotiations.

# THE DIALOGUE TOOL FOR CAREER DEVELOPMENT CONVERSATIONS WITH EMPLOYEESS

Date:
Employee:
Manager:

## BEFORE CONDUCTING CAREER DEVELOPMENT CONVERSATIONS

The conversation builds on mutual trust, where both parties have a mutual responsibility for the conversation. Ensure that you have sufficient time to cover every step.

In the career development discussion, the employee's work situation is discussed, individual goals are set and a plan for competence development is established (Figure 21).

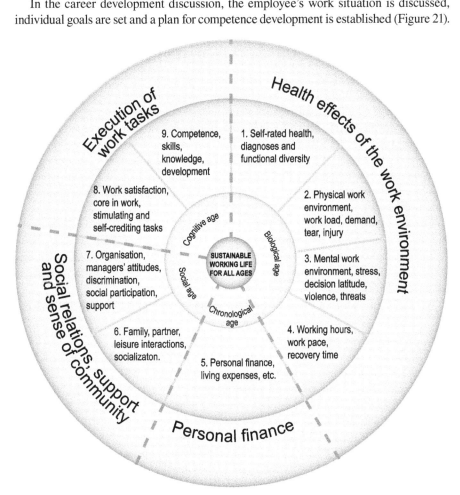

**FIGURE 21** The reflection figure in the dialogue: The four spheres of employment, and the nine influence and determinant areas that affect a sustainable working life in the SwAge™-model.

## CONDUCT YOUR CAREER DEVELOPMENT CONVERSATIONS BASED ON DIFFERENT PERSPECTIVES

- *Looking back* – follow up and give feedback on what has happened since the last career development discussion, review the goals and plan for competence development from last year together.
- *Current situation* – check in on what the work situation is currently like and acknowledge wishes for change.
- *Looking forward* – create a mutual perception together with your employee, regarding the employee's work tasks and how these should be prioritised, as well as goals and areas of development.
- *Age* – make reflections based on the employee's biological age, chronological age, social age and gender.

## DOCUMENTING THE CAREER DEVELOPMENT CONVERSATION

The discussion should be documented to facilitate follow-up on what the manager and employee have agreed upon. In the potential situation that a manager is replaced, documentation is crucial support to avoid uncertainties regarding the employee's goals and development. Sensitive information should not be documented in these templates, since the documentation of career development conversations is regarded as public documents. Personal issues, rehabilitation, misconduct, etc. are appropriate to discuss in a conversation with the employee, though the documentation should be kept in a different and appropriate manner, for example agreements on testing, written reprimands.

The discussion should be documented and signed or confirmed by both the manager and the employee, to ensure that there is consensus regarding what has been documented and to confirm that both parties agree upon the discussion and decisions made in the career development conversation. This documentation forms the basis of the next career development conversation.

**Keep the following in mind during the career development conversation:**

- Keep telephones turned off.
- Treat the career development conversation as a conversation in confidence, and agree upon when something must be passed on.
- Listen actively and ask counter-questions to make certain that both parties understand each other.
- Respect each other's perceptions or experiences even if they differ.
- Do not make any promises that cannot be kept.
- Do not forget to discuss potential specific wishes, or questions that have already been agreed to discuss in the career development conversation.
- Issues that do not belong in the career development conversation should be addressed in another context.
- Close the career development conversation with a mutual evaluation of the discussion; what has been good and what can be improved until next time.

Tool for Employer/Manager-Employee Work Situation

The employee's experience of their work situation is based on the nine determinant areas for a healthy and sustainable working life.

How do you experience your work situation?

1. Health: well-being, functional diversity, diagnoses, etc. in relation to the work situation?

2. Physical work environment with physical workload, vibrations, unilateral movements, ergonomic and technological aids, climate, chemical health risks, etc.

3. Mental work environment with stress, demands/control, threats, violence, etc.?

4. Working hours, work pace, breaks, work schedule, time for recuperation, etc.?

5. Financial incentives, scope of employment, employment regulations and employment security (salary and employee benefits are primarily discussed in the salary negotiation).

6. Personal social environment with work-life balance?

7. Social work environment with social inclusion, sense of community, security in the work unit, social support, leadership, informal leaders, co-workers, co-operation, disregard, discrimination, etc.?

8. The experience of core work tasks, satisfaction, motivation and stimulation with work tasks and activities, percentage and balance between different work tasks, etc.?

9. Knowledge, competence, possibility of using skills, excessively difficult/easy work tasks, challenging work tasks etc.?

## INDIVIDUAL GOALS AND COMPETENCE DEVELOPMENT

Proceed from the goals of the organisation. Break these down to form a basis for the individual's goals and the organisation's purpose with the employee's work tasks.

The goals should be concrete and feasible to implement and follow-up.

| Individual goals | Comments | When | Responsible | Follow-up |
|---|---|---|---|---|
| | | | | |
| | | | | |
| | | | | |

| What can and should be developed? | How should it happen? | | | |
|---|---|---|---|---|
| Competence development Short-term development goals (approx. one year) | Comments | When | Responsible | Follow up |
| | | | | |
| | | | | |

| Competence development Long-term development goals (approx. three years) | Comments | When | Responsible | Follow-up |
|---|---|---|---|---|
| | | | | |
| | | | | |

**Notes:**

# A Tool for Systematic Workplace Management and Action Plan for a Sustainable Working Life for All Ages

The different areas of the SwAge™ model can be used as a checklist to manage and facilitate the examination of risks and issues in the work environment, to take measures to amend these and to follow up on the systematic work environment management.

To conclude, the four different spheres for action in the SwAge™ model contain nine different determinant areas (Figure 1) for the ability and willingness to work and for a sustainable working life.

The first sphere for action, *the health effects of the work environment*, includes the following determinant areas:

1. Diagnosis, illness and functional diversity
2. Physical work environment
3. Mental work environment
4. Possibility to rest and recuperate in relation to work pace, work schedule and breaks

The second sphere for action, *financial incentives*, includes the determinant area:

5. Personal finance

The third sphere for action, *social inclusion, support and sense of community*, includes the determinant areas:

6. Personal social environment
7. Social work environment

The fourth sphere for action, *execution of work tasks*, includes the determinant areas:

8. Work satisfaction, motivation and stimulation
9. Knowledge, skills and competence

Make a workplace analysis, proceed from the areas of the SwAge™ model and fill in the matrix.

Below you will find a picture of the matrix *Workplace analysis and measures – the SwAge™ model* (Table 2) with an example of how a workplace has used it (Table 3).

## HOW TO CONDUCT THE ANALYSIS

Use the SwAge™ model (Figure 22) as a reflection tool, a checklist to conduct the analysis, and fill in the matrix (Table 2).

1. **Collect data and information for the workplace analysis.** It is a good idea for the manager and employee to collect information together. This can for example be achieved through discussions in workplace meetings, by walking around the workplace and examining the work environment or by using employee surveys.
2. **Examine the work environment and make a record in the matrix of all factors assessed to be risks or issues in each of the nine determinant areas** (please note the different areas in the matrix, Table 2).

   - It is better to facilitate an overview of the situation by considering data and information from several sources. Therefore, make a list of all sources for the information and data used in the risk assessment of the workplace analysis.
   - Regardless of which source is used to assess risk in the workplace, make sure not to rely on one single source of information.
   - In order to study the workplace and situation when conducting the workplace analysis, use methods such as observations, surveys, interviews, focus groups, available data and examinations (for example statistics of sickness absence) or other technology. Another way is to discuss the work environment together during regular workplace meetings and meetings with the work unit. In order to collect data for the workplace analysis by observing work situations and the work environment, the observer can for example review, observe and study a section of work and work processes in order to assess whether there are obvious aspects of the work situation that can cause problems. This approach makes clear for example how work is executed, work postures, work pace, work conditions, etc. Observations are most efficient if they are conducted while an employee in the work situation describes what happens when they execute their work task. It increases the possibility of reflecting on risks for both employees and managers, even possible solutions in a particular work situation.
   - Confirm your results. Verify, consider and reflect together with your employees and others to examine whether they agree that the results make up a probable representation of the work situation.
   - Examine whether and what the problem means to a particular place, in other words, to the different physical places in the work environment. It is important to reflect on, determine and document potential underlying factors of risks and problems. For example, consider whether risks and problems are associated with variations, loose items or other factors

in the design of the workplace, whether and how risks and problems affect employees of different ages, with different functional diversity, with different levels of inclusion, communication and abilities, as well as whether and how staffing in different situations and at different times affect whether the work task is executed in a more or less risky manner.

3. **Fill in the matrix based on the examination and risk assessment of factors associated with different parts of the SwAge™ model in the work situation and make a priority list of severe and acceptable risks in the workplace.**

- Evaluate the organisation's capacity to deal with the most important causes of risks and problems. Identify areas where the organisation appears to have less capacity.
- Acknowledge that the identified risks and problems potentially could affect someone!
- Ensure cooperation between employer and employees for lasting and efficient risk prevention.
- Clarify potential problem areas to employees and work units.
- Acknowledge areas where the organisation does well. Strengthen and pay attention to these areas.
- Open up and facilitate a discussion and communication of the results with and between employees and their representatives.
- Discuss and reflect on potential solutions, for example with help from a representative selection of the staff.
- Facilitate initiatives and suggestions to reach an agreement on practical improvements.
- Compare current problems to the organisation in general and in relation to standards.
- Identify how and in what way it is possible to take appropriate measures in order to decrease the distance between the current situation and the goal situation: a sustainable workplace and organisation where employees do not run the risk of developing illness, injury or want to leave the workplace.

4. **Evaluate acceptable and acute, severe and unacceptable risks in the workplace, as well as the order in which these should be targeted. Make a plan for measures and actions. Divide responsibility, decide who is responsible for a certain measure and at what time the measure should be taken. Do not forget to set dates for follow-up to make sure that the measures have been implemented and are not forgotten.**

- Begin by filling in the matrix and establishing a plan for measures and actions in order to counteract the risks and problems.
- Clarify and inform everyone in the workplace of the action plan, of both identified risks and measures.
- Make sure the information is clear, comprehensible, up-to-date and easily accessible for everyone. The information must be repeated for everyone to assimilate the message and learn new routines and make sure to provide continuous information.

- Divide responsibility to ensure that measures are taken. Clearly document who is responsible for the implementation of each measure in the matrix. Establish and document a schedule for implementation. Do not forget to add dates for follow-up.
- Acknowledge whether there are any threats to the implementation of measures and their underlying causes and rectify these threats and causes.
- Facilitate the everyday risk assessment of the workload in the organisation, by being observant of the most important and most frequently identified risk factors in the everyday work situation. Reflect on these risk factors in relation to the SwAge™ model. Through this approach, you can continuously control and re-evaluate the risk assessment, measures and new potential threats in the work situation.
- Discuss how to assess whether the measures have been successful and benefit the employees, organisation and productivity.
- Identify areas in the organisation with potential for improvement.
- Do not forget to include positive and successful factors in the organisation! Identify areas that appear to have good current praxis to build on, and that can be strengthened further in order to prevent risks of ill-health, occupational injuries and employees leaving the workplace prematurely.
- Develop a mutual goal of what the work situation should be like in the workplace, and for implemented measures to be lasting and sustainable.
- Establish a work environment policy that displays the overall goals of the employer and workplace, and the long-term direction of the work environment in the organisation. The workplace analysis is an important starting point for the work environment policy. Questions to reflect on and consider when deciding on the work environment policy can for example be:
  - How do employees experience their work?
  - Who and what contribute to a good atmosphere and work environment?
  - What do we want the work environment to be like?
  - Why is work environment management important to us, to employees and the employer that is?
  - When and in what situations do we address questions and discussions of the work environment?
  - Do we assess risks and work environment problems efficiently?
  - Do we do it in a way that makes us discover whether different risks affect different employees, of different ages, of different genders, with different functional diversity or with different linguistic diversity?
  - Do we consciously rectify the risks we are aware of, do we follow up and assess whether measures have had the desired effect?
  - Are any work environment regulations particularly important to our organisation?

5. **Follow up and verify the results of measures according to the schedule.**
   - Acknowledge and celebrate improvements!
   - Make a revised action plan, schedule and division of responsibility for measures that are not taken or are not satisfactory.

# Tool for Systematic Workplace Management and Action Plan

6. **Even if reflection and improvements are continuous, a more thorough review and workplace analysis based on the nine determinant areas for a sustainable working life in the SwAge™ model is needed at least once a year.** Therefore, start over with item 1 in the list above and assess the risks in the work environment and the workplace again, this is a continuous task. You can for example ask yourselves: What is the work environment currently like? Have previous risks been rectified correctly? Have new risks occurred?

7. **There is also a field in the matrix to add an assessment of whether everyone in the workplace has sufficient knowledge of and education in, for example, work environment risk factors, work injury reports, routines, aids, fire alarms, emergency exits and co-operation (Table 2).**

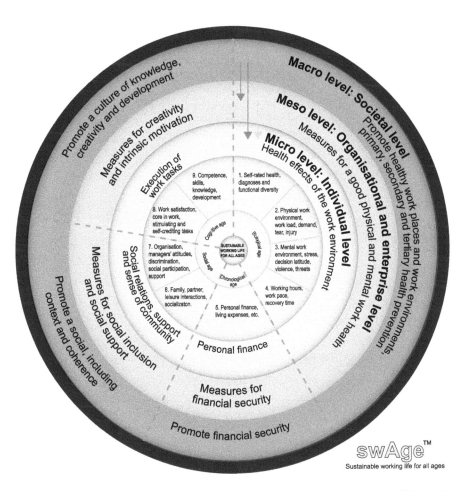

**FIGURE 22** The reflection figure, the SwAge™ model (sustainable working life for all ages). A theoretical model of determinant areas for a sustainable working life for all ages [1].

# THE MATRIX TOOL FOR SYSTEMATIC WORKPLACE MANAGEMENT AND ACTION PLAN FOR A SUSTAINABLE WORKING LIFE

## TABLE 2
The matrix for systematic workplace management and action plan for a sustainable working life for all ages – the SwAge™ model

| Date: | Name: | Results of workplace analysis and risk assessment | |
|---|---|---|---|
| Sphere for action: | Determinant area: | Check-up done | Problem, risks and risk sources in the area If/how/what is functioning well in the area |
| Health effects of the work environment (and associations with – Biological age) | Functional diversity, diagnoses, self-rated health | | |
| | Physical work environment: workload, vibration, strain, hazardous substances, climate, access to tools, etc. | | |
| | Mental work environment: stress, demands, control, threats, violence, etc. | | |
| | Working time, work pace, recovery: schedule, shifts, breaks, breaks, etc. | | |
| Finance (and associations with – Chronological age) | Personal finance, financial security, employability, insurance, etc. | | |
| Social support & participation (and associations with – Social age) | Personal social environment: personal life, family life, leisure time in relation to work | | |
| | Social work environment: social support, discrimination, participation, attitudes, leadership | | |
| Execution of work tasks (and associations with – Cognitive age) | Work tasks, activities, stimulation, motivation, job satisfaction | | |
| | Competence, knowledge ability, employability, development in relation to task | | |
| Knowledge and education in the workplace about occupational hazards, work injury notification, routines, tools, fire alarms, escape routes, collaboration, etc. | | | |

Notes of sources and methods for the execution of the workplace analysis, and other comments:

_____

_____

_____

_____

# Tool for Systematic Workplace Management and Action Plan

| | | | Action plan | | |
|---|---|---|---|---|---|
| Severe risk | Other risk | Measures and actions for prevention and promotion in the area | Responsible for the action | Should be done by | Policy/follow-up/ check-up |

## An example of a Used Systematic Workplace Management and Action Plan Matrix for a Sustainable Working Life

Below is an example of a completed matrix. The example is intended to describe how to use the matrix and what help the matrix is in workplace analysis, reflection and discussion to find out problems, measures, timetable for activities, etc. and to identify well-functioning parts of the work situation.

### TABLE 3
### The matrix for systematic workplace management and action plan for a sustainable working life for all ages – the SwAge™ model

| Date: | Name: | | Results of workplace analysis and risk assessment | | |
|---|---|---|---|---|---|
| Determination spheres: | Determination areas: | Check-up done | Problem, risks and risk sources in the area / If/how/what is functioning well in the area | Serious risk | Other risk |
| Health effects of the work environment (and associations with – Biological age) | Functional diversity, diagnoses, self-rated health | X | 1. Employees with blood pressure medication, which can cause side effects like dizziness and balance problems, execute work tasks on a roof.<br>2. One employee states that they experience symptoms of fatigue syndrome.<br>3. One employee has been diagnosed with hearing loss, which has deteriorated with increasing age. | 1. x | 2. x<br>3. x |
| | Physical work environment: workload, vibration, strain, hazardous substances, climate, access to tools, etc. | X | 1. Risk of falling when working at heights.<br>2. Working with arms above shoulder height.<br>3. Computer workspaces have old equipment and lack height-adjustable desks, which causes problems in the neck and shoulders for short and tall employees.<br>4. Several employees are getting older and their security at work is at increased risk to due to biological ageing. | 1. x | 2. x<br>3. x<br>4. x |
| | Mental work environment: stress, demands, control, threats, violence, etc. | X | 1. Several employees describe that they experience occupational stress. These employees have been interviewed regarding what work procedures are experienced as particularly stressful. Interviews showed that the fragmentation between many different work tasks everyday was experienced as particularly stressful. The fragmentation made it difficult to have sufficient time to execute the most prioritised tasks. This contributed to feelings of insufficiency, shame, guilt and disappointment.<br>2. One employee was threatened by the parent of a student through social media. The employee thought the threats were offensive but did not take the threats very seriously. | | 1. x<br>2. x |

# Tool for Systematic Workplace Management and Action Plan

## Action plan

| Measures and actions for prevention and promotion in the area | Responsible for the measure | Should be done by | Policy/follow-up/check-up |
|---|---|---|---|
| 1. Replace employee to work tasks that do not risk contributing to falls due to medication, alternatively the employee contact their physician regarding whether they can change to another medication without this side-effect.<br>2. Examine whether and what effects the work environment and work situation have on this.<br>3. Examine if hearing loss can contribute to increased risk of accidents or injuries and if possible take measures. | 1. Anne<br>2. Anne<br>3. Anne | 1. As soon as possible, 20/8<br>2. 1/9<br>3. 28/8 | 1. 20/9<br>2. 15/9<br>3. 15/9 |
| 1. Purchase updated personal fall protective equipment and review policy and safety information.<br>2. Difficult to eliminate the work procedure, but a long-term solution should be developed. Meanwhile, make sure the procedure is limited and that work tasks are rotated to decrease strain.<br>3. Purchase height-adjustable desks. Arrange for an ergonomist from occupational health care to examine adjustments on chairs, desks, lamps, screens, aids, etc. for different employees in computer workstations.<br>4. Ongoing examination of workplace due to increased risks of biological ageing. | 1. Carol<br>2. Mike<br>3. Ali<br>4. Anne | 1. 20/8<br>2. Begin immediately – should be done by 20/5 next year<br>3. 30/10<br>4. By annual review | 1. 20/9<br>2. 25/5 next year<br>3. 25/5 next year<br>4. By annual review |
| 1. New employee or trainee will be recruited to part-time position and be responsible for the function and maintenance of photocopier; organising, ordering and packing material; emptying dishwasher, making sure equipment works and is present; etc.<br>2. Measures to report threats and contact parent regarding their behaviour. Supportive group for conversation together with occupational health care regarding how we deal with threats. | 1. Anne<br>2. Anne | 1. 1/12<br>2. Contact parent immediately, supportive conversation during November | 1. 1/4 next year<br>2. 1/2 next year |

*(Continued)*

## TABLE 3 (Continued)
## The matrix for systematic workplace management and action plan for a sustainable working life for all ages – the SwAge™ model

Date:  Name:  Results of workplace analysis and risk assessment

| Determination spheres: | Determination areas: | Check-up done | Problem, risks and risk sources in the area If/how/what is functioning well in the area | Serious risk | Other risk |
|---|---|---|---|---|---|
| | Working time, working rate, recovery: schedule, shifts, breaks, breaks, etc. | X | 1. Lack of recuperation and tiredness increase the risk of injuries and illness. New work schedules with shift work decrease the possibility of sufficient recuperation, particularly for employees who commute to work.<br>2. There is a particularly increased risk of insufficient recuperation for senior employees, who in particular complain that they always work and lack energy for leisure activities. | | 1. x<br>2. x |
| Finance (and associations with – Chronological age) | Personal finance, financial security, employability, insurance, etc. | X | 1. Some employees have expressed worry that their services will no longer be needed after next year's re-organisations. Particularly employees with chronic illness, employees who approach retirement age or who have an older education and have not participated in competence development, or who have not participated in projects concerning new technology. This is described as an uncertainty and source of stress in the daily work, which may be a risk of ill-health and issues in the workplace. It is expensive for the organisation if anyone develops ill-health or injury due to these worries. Moreover, it is expensive to recruit new employees if current employees' knowledge and competence are not put to use and continuously updated. | | 1. x |
| Social support & participation (and associations to – Social age) | Personal social environment: personal life, family life, leisure time in relation to work | X | 1. Two employees, who both have children in pre-school and school, find it difficult to be on time for the Monday meeting at 7 am. Since the bus no longer stops outside the pre-school they must drop off their younger children and then ride bikes to school. Because of this, they are usually not on time to catch the first bus to work in the morning if issues arise when dropping off their children. This is very stressful for the parents, who suffer bad conscience both towards their children and their co-workers. The risk of stress reactions increases when working life clashes with employees' personal lives.<br>2. Senior employees have criticised the parents for being late to the morning meeting. Increased risk of stress reaction. | | |

# Tool for Systematic Workplace Management and Action Plan

|  | Action plan |  |  |
|---|---|---|---|
| Measures and actions for prevention and promotion in the area | Responsible for the measure | Should be done by | Policy/ follow-up/ check-up |
| 1. Examine how new schedule and planning works. If necessary change back to the old schedule or try another one.<br>2. Examine work pace and breaks to make sure no one skips them and has less time for recuperation and renewed energy through food and drink. | 1. Ali<br>2. Yasmine | 1. 12/12<br>2. 12/12 | 1. 12/3 next year<br>2. 12/3 next year |
| 1. Introduce competence development and rotation in the workplace to learn new work tasks that can be useful and increase employability after the re-organisations. Utilise senior employees' tacit knowledge from previous re-organisations. | Anne | 1. Done by 20/9 next year | 1. 12/12 next year |
| 1. Since the workplace has a flex schedule with the possibility of flexing between 7 am and 9 am, the Monday meeting will take place at 9.15 am instead.<br>2. Joint discussion at the next workplace meeting regarding the implications of being in different ages and life phases, to become more tolerant of each other's life situations and life phases. | Anne | 1. next Monday meeting<br>2. next workplace meeting | 1. 1/11<br>2. 1/11 |

(*Continued*)

## TABLE 3 (Continued)
## The matrix for systematic workplace management and action plan for a sustainable working life for all ages – the SwAge™ model

| Date: | Name: | | Results of workplace analysis and risk assessment | | |
|---|---|---|---|---|---|
| Determination spheres: | Determination areas: | Check-up done | Problem, risks and risk sources in the area If/how/what is functioning well in the area | Serious risk | Other risk |
| | Social work environment: social support, discrimination, participation, attitudes, leadership | X | 1. In the career development conversation, an employee has expressed their experience of disregard on three grounds of discrimination: age, gender and religion.<br>2. They experience the presence of an informal leader.<br>3. They experience being bullied by said informal leader. | | 1. x<br>2. x<br>3. x |
| Execution of work tasks (and associations to – Cognitive age) | Work tasks, activities, stimulation, motivation, job satisfaction | X | 1. An employee shows signs of dissatisfaction with the content of their work tasks, and that changes have eliminated work tasks that were much more stimulating. Dissatisfaction and boredom indicate an increased risk of carelessness and lack of occupational safety. Furthermore, it may cause a stress reaction due to a lack of control of one's work situation and the experience of no longer being valued in the organisation. | | 1. x |
| | Competence, knowledge abilities, employability and development in relation to work tasks | X | 1. Some employees have experienced less areas of responsibility and possibility of using their competence since the last re-organisation. This is stressful and increases the risk that competence is lost in the organisation.<br>2. The changed learning patterns of senior individuals have not been considered in the latest technology course for open-plan offices, because of this, some people have been unable to assimilate the new way of working. Insufficient knowledge results in risks of injury. | | 1. x<br>2. x |
| Knowledge and education in the workplace about occupational hazards, work injury notification, routines, tools, fire alarms, escape routes, collaboration, etc. | | X | | | |

Notes of sources and methods for the execution of the workplace analysis and other comments:
- Discussions of the work environment in workplace meetings, observations in the work environment, interviews with employees who are at particular risk and the nine determinant areas of the SwAge™ model for a sustainable working life, were used as sources to gather information and data regarding the work situation and work environment for the execution of the workplace analysis.
- Sources to find measures and plans to take measures were the ILO webpage with information on different work environment risks. Discussions took place with employees, particularly risk groups, labour unions, union representatives and management in order to find out how our workplace should manage the findings of the workplace analysis.

# Tool for Systematic Workplace Management and Action Plan

| | Action plan | | |
|---|---|---|---|
| Measures and actions for prevention and promotion in the area | Responsible for the measure | Should be done by | Policy/ follow-up/ check-up |
| 1. We will invite an expert to the next workplace meeting to discuss the grounds of discrimination and what they entail. The implementation of the equality policy will be made clear.<br>2+3. Communication and delegation of authority will be made clear to eliminate the experience of informal leaders in the workplace. Occupational health care will have supportive conversations with the manager, the person who has been described as an informal leader and the affected person to find a solution and improve communication. The safety representative will also be present. In the next workplace meeting, we will discuss measures to increase well-being, the sense of security and community and support in the work unit. | Anne<br>Ali | 1. next workplace meeting, 20/1 next year<br>2+3. 1/11 | 1+2+3. 20/2 next year |
| 1. Establish rotation between different work tasks to make more employees experience stimulation and development in the work situation. Increase motivation by showing the significance of all work tasks for the profit and success of the organisation. | 1. Anne, Yasmine, Mike | 1. 12/1 next year | 1. 6/4 next year |
| 1. Consider the competencies of these employees that could be utilised better to enable them to use their skills and knowledge. Review the requirements of new occupational roles.<br>2. Review competence development and information and make sure it is appropriate for different cognitive ages. Furthermore, make sure informational signs make sense and are comprehensible. | 1. Anne, Carol | 1. 1/3 next year<br>2. 1/12 | 1. 1/9 next year<br>2. 1/4 next year |
| 1. The course in cardiopulmonary resuscitation will once again be carried out in the spring<br>2. A fire drill will, as usual, be carried out on a day unknown to us in advance during the autumn (probably in November) | 1. Occupational health care<br>2. Maryam | 1. 1/3<br>2. 30/ | 1. 1/9 next year<br>1/9 next year |

- A solution to work tasks executed with arms and hands above shoulder height must be found eventually.
- Job design and recruitment of employees for service tasks in the office will be developed together with the HR department.

# Examples of Measures Taken in Different Workplaces

Below you will find measures related to the four spheres for action in the SwAge™ model for a healthy and sustainable working life that have been developed together with managers, HR, employees, trade unions and occupational health care.

*Health effects of the work environment*:

- Maintain a well-functioning systematic work environment management and use the SwAge™ model as a checklist (read more in Chapter *A Tool for Systematic Workplace Analysis and Action Plan for a Sustainable Working Life for All Ages*)
- Make sure work tasks and the work environment are adjusted to and take the employees' functional diversity, age effects, diagnoses and health status into account
- Create an organisational culture that promotes the use of ergonomic aids to prevent occupational injuries
- Comply with limit values concerning chemicals, radiation, ergonomic factors and climate in the work environment
- Use and comply with country-specific work environment recommendations and directives in order to minimise health effects and risks of occupational injuries
- Implement rotation, variation and changes in work tasks to decrease the risk of unilateral movements and workload
- Counteract negative stress reactions due to excessively high work pace
- Counteract negative stress reactions due to fragmentation between too many work tasks and fragmented work efforts during the work shift
- Counteract ethical stress by making sure that employees have sufficient possibilities and resources, and sufficient time to execute their work tasks to qualitative standards, to make them feel satisfied at the end of the work shift
- Promote a balance between the demands employees face in their work situation and the control they have over it
- Minimise the risk of threats and violence in the work environment
- Create well-functioning work schedules and working hours that facilitate sufficient recuperation
- Make sure that the work pace is not unreasonably demanding in a physical and mental sense

# Examples of Measures Taken in Different Workplaces 275

- Provide possibilities to take pauses and (snack) breaks during a work shift, adjusted to the demands in the work situation, individual needs (for example, health and age) and in proportion to the duration of the work shift
- Provide possibilities for or facilitate physical exercise and activities to maintain and increase mental and physical well-being and health
- Provide possibilities for healthy dietary choices and nutrition during a work shift, it is important for the creation and maintenance of good (occupational) health
- Provide access to occupational health care that supports the prevention of ill-health, manages age-related risks and risks in the work environment and implement measures in the workplace to promote health and prevent ill-health

*Financial incentives*:

- Provide salary and different financial advantages in relation to work efforts
- Arrange activities and take measures for employees' maintained employment and to prevent them from being pushed out of working life
- Apply occupational health and safety and risk assessments to decrease the occurrence of health issues, sickness absence, sick pay and unemployment
- Offer ongoing competence development and knowledge development for employees to have maintained employability in the event of changes and reorganisations in the workplace and in working life with new types of work tasks, work situations and occupations throughout life

*Social inclusion, support and sense of community*:

- Apply situational and age-conscious leadership based on the different parts of the SwAge™ model, to make sure employees are acknowledged and supported based on their situation, age, ability, opportunities and needs
- Facilitate employees' social integration and inclusion in the work unit, and try to make employees get to know each other in (and outside) the workplace
- Promote a sense of community and security in work units where employees dare to admit flaws and errors, but also to knowledge and skills
- Provide possibilities for joint pauses and breaks
- Acknowledge good initiatives and celebrate successes to increase the sense of community
- Discuss problems in the (social) work environment and approaches in the work situation and the work unit using the case method, make sure everyone can contribute, join in and be acknowledged for their experience and knowledge
- Establish zero tolerance policies against discrimination and disregard, for example discrimination on the ground of age
- Communicate information to make sure that everyone is informed and feel included in what happens in and around the organisation (however, make

sure the amount of information is appropriate to avoid employees being overwhelmed by excessive information that is not important to them)
- Eliminate risks of informal leaders and lack of structure in the organisation
- Provide social support to increase senior employees' self-esteem in the organisation
- Utilize senior employees in mentorship with younger and recently recruited employees to increase the sense of community and inclusion
- Implement work schedules that work well in relation to the employees' personal social environment and needs, to make sure that they have sufficient energy to tend to personal social relationships, to be active and engage in activities in their leisure time

*Execution of work tasks*:

- Contribute to professional pride and pride in work tasks and the workplace
- Contribute to employees' understanding and knowing of each other's work tasks, regardless of profession and position in the hierarchy making sure that employees understand that everyone is equally important to daily operations, productivity and achieving the goals of the organisation
- Provide possibilities for everyone to use their knowledge, skills and competence in their work tasks, profession and workplace
- Promote employees' possibility of primarily working with the core of their particular work tasks and make sure this is not de-prioritised through fragmentation between too many work tasks, work procedures and a flattened organisation
- Contribute to employees' experience of balance between their work efforts and the "rewards" and appreciation they receive for their efforts in the work situation
- Increase motivation and work satisfaction by acknowledging and making use of good initiatives, work efforts, problem-solving, etc.
- Provide employees with responsibility for different areas in the organisation, in order to increase motivation, stimulation, responsibility and pride in the work situation and work tasks
- Provide possibilities for some type of career ladder over the lifespan for all occupational groups
- Improve employees' employability through rotation between different work tasks, to make employees master more than one task if they should need to be relocated in reorganisations, changes, or if they need to change careers
- Rotate and change work tasks to decrease monotony and increase the possibility of stimulation through various work tasks
- Provide access to competence development and new knowledge, regardless of age, adjusted to general cognitive functions in different ages
- Promote an organisational culture where senior employees are also included in new projects, and where senior employees' experiences and knowledge are utilized

- Acknowledge experienced and senior employees and give them the opportunity to transfer their knowledge to others
- Provide possibilities to acknowledge and potentially utilize knowledge, skills and competence that the employees have gained in their leisure time, in the organisation
- Promote the transfer and exchange of tacit knowledge from senior employees, and "new" knowledge from recently recruited employees, from employees who have recently participated in courses or competence development. This can also be done by using for example the case method, joint discussions and reflections for problem-solving and practical learning
- Provide the possibility of changing work tasks, area of competence and career, regardless of age

# Tools for Evaluations and Follow-Up of Actions and Measures

In order to make working life sustainable for all ages, the organisation should implement continuous development and review whether the nine determinant areas in the four spheres for action in the SwAge™ model are optimal for individuals' willingness and ability to work. Needs for different measures will probably arise at some point when conducting workplace analyses. Some type of evaluation can be useful in order to find out whether these measures have had an impact, and whether the impact has been expected or desired. It is important to consider and discuss the type of evaluation to be used prior to implementing the measure. Different types of evaluations require different types of collected data and material, which can be difficult to reconstruct after the measure has been implemented. There are several different types of evaluations. In the section below you will find an introduction to process evaluations, impact evaluations and economic evaluations.

## PROCESS EVALUATIONS

Results of measures and actions are associated with both the objectives sought to be achieved through the efforts of the measure and the implementation process itself. It is not enough to set objectives for the expected results of measure, there are almost always obstacles to achieving objectives. In order to understand what obstacles have occurred in the implementation process of a measure and when trying to achieve the expected results, the implementation process itself should be followed by a process evaluation. A process evaluation carefully describes what happens during the process through close observation and follow-up on the process, and through keeping a strategic overview of objectives set for the implementation of measures, and what contributed to the results of the measure. Factors that can be affected should be chosen and examined in order to keep the evaluation simple, manageable and uncomplicated. When there is a clear picture of what the process of implementing measures should look like, it is possible to proceed and implement the measure while simultaneously collecting material and information to follow the implementation process of the measure and to evaluate what factors affect the process and results. The process evaluation identifies internal and external weaknesses, risks, threats, opportunities and strengths to the implementation, and how results were achieved because of or despite these factors. For what reason were objectives not achieved? For what reason were objectives achieved better than expected? Were participants positive, negative or did they feel forced to participate in the implementation of the measure? Did disturbances occur in the process? For example re-organisations or mandatory education for all managers in the workplace during the implementation of the measure. Did anything occur internally or externally, contributing to excessive stress and worry among employees? Did society face a recession, with budget cuts that affected how the results were achieved? Were changes made in the social insurance system or pension system during the time of the measure? Did a new employee, who was not introduced to, did not prioritise or was not interested in the implementation of the measure, join the organisation during implementation? Was there a lack of knowledge regarding how to implement the measure? Was the knowledge so great that results exceeded expectations? Did attitudes in society change, or were the needs of the measure brought up to date by the media, which made it particularly easy and successful to implement the measures with good results?

Furthermore, people may have different opinions regarding measures and achievement of objectives. For example, if everyone has not received the same information, or if people have different understandings of the information, this could cause different perceptions of the process and objectives of the measure.

Observations, surveys, interviews or discussions with managers and employees can be used to describe the process [279]. In a greater mapping of the process of the result or measures or a more systematic evaluation of effects, it is possible to observe, send out surveys, interview and/or discuss with authorities, politicians, researchers or other parties who can supply information. After a while, a perception of what factors the measure has influenced will emerge. If they have changed in the direction of the objectives, the intervention can preliminarily be said to have had an effect. It often takes a long time before an implemented measure and intervention can be said to have had a lasting long-term effect and increased the long-term possibility of a sustainable working life for all ages in the organisation or in the workplace. This often requires that the implemented measure and intervention is based on a sufficient amount of participants, that the intervention is implemented in roughly the same way and that it lasts for a long period of time. Furthermore, the possibility of comparing whether the participants' attitudes, norms, knowledge and characteristics have changed after the measures have been implemented is important, likewise whether policy documents and written information in the organisation or company have changed. This is described as reflexive monitoring when the examined group monitor themselves [280,281].

The amount of time the evaluation and follow-up of effects take can affect the result. At times, the concrete results of a measure evaluation conducted a short time after the implementation may seem non-existent or insignificant. However, the measure may have changed attitudes and instilled new insights in individuals, groups, companies, organisations, authorities, decision-makers, etc. These changes can contribute to long-term results that cannot be determined at this point. At the same time, some objectives may be achieved in the short-term perspective, though new re-organisations or budget cuts in the organisation can contribute to a loss of effects in the long-term perspective. Furthermore, participants may feel acknowledged through the measures, and this may cause successful results in the short-term perspective, however, this may decrease and stabilise with a lower effect after some time when the novelty has worn off. Therefore, it can be difficult to prove how successful a measure has been in the short-term perspective. Common challenges when conducting evaluations are lack of time to follow-up, lack of time to conduct evaluations, difficulties in finding and collecting appropriate data and difficulties in establishing an overview if assessments are considered to be too subjective. In the evaluation, it is also important to assess where it is most profitable to implement a measure, i.e. at the Individual level, Organisational/Enterprise level, Community level and Country level.

Despite different problems when implementing evaluations, it must be possible to show what works and what does not work. Learning in the organisation increases through reflecting on the results. Evaluations can provide tools for development and improvement, and chances to discuss the purpose, resources, results, obstacles and possibilities in the organisations, as well as the potential for development, based on all nine determinant areas of the SwAge™ model, for a future healthy and sustainable working life.

## IMPACT EVALUATIONS

The purpose of an impact evaluation is to partly answer whether the results of the measures correspond to the set and expected objectives, and partly whether it really was the measure itself that caused the results. An impact evaluation can be conducted with different ambition levels, for example it can only contain a summary of whether the objectives were achieved, or it can be more advanced and also include a process evaluation or an economic evaluation. The starting point of an impact evaluation is to set and specify the expected objectives and results. Have an open discussion of what is possible to strive for and what is unrealistic. What kind of evidence is possible to find through the evaluation? It is important to keep in mind that the objectives should not be too vague or inconsistent. An impact evaluation can be regarded as a tool to develop and improve the organisation. Improvement, not radical change, is a realistic expectation to have on the objectives. It is impossible to discuss the impacts and effects of the evaluation without having clear and realistic objectives. Is the purpose that employees should work until an older age? That a larger amount of employees should return to work after long-term sickness absence? To make a workplace analysis once or twice a year, using the SwAge™ model as a checklist? To include age in the equality plan? To decrease the number of incidents in the work environment of nurse assistants? To increase employees' employability or to increase employees' productivity through providing access to relevant competence development for a larger amount of employees? How do we measure the objectives we want to achieve? What values and effects should be achieved for the measure to be considered successful? In the evaluation, it is also important to assess where it is most profitable to implement a measure, i.e. at the Individual level, Organisational/Enterprise level, Community level and Country level.

## ECONOMIC EVALUATIONS

It can be significant to evaluate the utility in relation to the cost when implementing different measures to increase the possibility of a sustainable working life for all ages. The purpose of an economic evaluation is to assess effects and costs in relation to resources used. By calculating the cost efficiency of a measure, the evaluation can estimate whether the cost of a measure is reasonable in relation to the effects or impacts of the same measure. However, cost efficiency is a relative term, a measure is not cost efficient in itself, since the implementation of measures always comes with costs, instead, cost efficiency is always measured in relation to a compared option. The compared option can be another type of measure, a measure for another group of individuals or the option of not implementing any measure at all. However, preventive measures can be difficult to measure, since the intention is to prevent that something occurs. Naturally, it is difficult to measure something that has not occurred. A way to work around this is to have a comparative group in which no measure is implemented, in other words, "treatment as usual" (TAU). Four areas must be considered prior to an economic evaluation of actions and implemented measures, in order to measure what effects and impacts the measures entailed compared to other measures, and to "TAU" [282–284]. These four areas are:

1. *Level of evaluation.* Is it of the highest interest to focus on individuals, on the organisation and company or on society? Or is it significant to include all three levels in the evaluation?
2. *Time.* Should the evaluation only concern what happens during the time of the evaluation itself, or should it also include follow-up or an estimate of what future effects the measure will have?
3. *Affected costs.* Which ones are affected?
4. *Most relevant costs.* Which ones are the most relevant to include in the evaluation?

An economic evaluation measures costs in monetary terms, but differs in the measure of health effects. Which analysis to choose depends on what objectives the organisation wishes to achieve, but also on the possibility and access to different data required to conduct the economic evaluation. A few different types of economic evaluations and economic analyses for different actions and measures are cost-effectiveness analysis, cost minimisation analysis, cost-benefit analysis and cost-utility analysis.

*Cost-effectiveness analysis* evaluates the effect of a measure. For example, a particular measure in relation to sickness absence in the workplace or the number of additional life years. However, it does not provide much information regarding whether the individual experiences well-being in the workplace, or of potential sickness absence in the workplace, i.e. that individuals attend work despite being ill, for example due to financial reasons, or because the workplace is understaffed and they do not want to burden their co-workers further. Working life is complex and several of the nine determinant areas for a sustainable working life usually have different simultaneous effects. The number of days of sickness absence is a quantitative measure that provides insufficient information on the individual's experience of their ability and willingness to work and of their quality of life. Cost-efficiency analysis cannot be used to compare different determinant areas with each other since it is impossible to value measures for preventing sickness absence through better physical work environment against measures for better competence development to increase employees' employability.

*Cost minimisation analysis* is a type of cost-effectiveness analysis. This approach evaluates whether two measures with different costs have the same effect. The measure with the lowest cost is considered to be the most cost-efficient.

*Cost-benefit analysis* evaluates both cost and effect in monetary terms. The analysis evaluates the "benefit" of a measure in relation to the "profitability" of the same measure. However, there are partly ethical considerations to make regarding whether this analysis is appropriate to apply to measures with the primary purpose of making working life sustainable for all ages.

*Cost-utility analysis* is the most frequently used economic evaluation of measures and actions. A cost-utility analysis can use more than one measure of effect, for example, it is possible to use a measure of effect that combines the two dimensions of work satisfaction and duration of working life. Cost-utility analyses in health care often use a combined, two-dimensional measure of effect called *quality-adjusted life years* (QALY). QALY is constructed so that one life year is multiplied by the life quality during the life year. This cost-utility analysis facilitates comparisons between different areas of medical interventions and provides a more comprehensive overview of the health effects of one or several medical interventions. QALY is weighted and stated as a number between 0 and 1, where 1 corresponds to full health and 0 corresponds to death. A person who lives two years with the life quality weighted as full health has 2 QALY ($1 \times 2 = 2$). A person who lives two years with a life quality of 50 percent has 1 QALY ($0.5 \times 2 = 1$). The results of this cost-utility analysis are presented as an incremental cost-effectiveness ratio, also known as ICER. ICER is the ratio (cost per additional measure of effect) calculated based on the differences in costs for two different actions and measures, divided by the difference in the measure of effect for the two actions and measures. The cost per additional measure of effect can be interpreted as the costs for society, the organisation or the company to "buy" an additional entire individual year in working life (i.e. if the employee stays in working life) through a measure, compared to not taking any measure or compared to a different measure in the workplace. ICER is calculated using the following formula:

$$\text{ICER} = \frac{\text{Cost 1} - \text{Cost 2}}{\text{Measure of effect 1} - \text{Measure of effect 2}}$$

The interpretation and reflection of different costs for two measures, in relation to the difference between the measure of effects for both measures, in other words, the ICER can be facilitated through the use of a cost efficiency plan.

In the diagram, four dimensions are used to visualise and clarify the interpretation of ICER. If the cost-utility analysis shows that the implemented measure has a lower cost and better effect, you will find this measure in the bottom right corner of the diagram (Figure 23). This part of the diagram shows that the measure was cost-effective and dominant. However, if the measure has a higher cost and is not very effective, the implemented measure is dominated and not as cost-effective. The

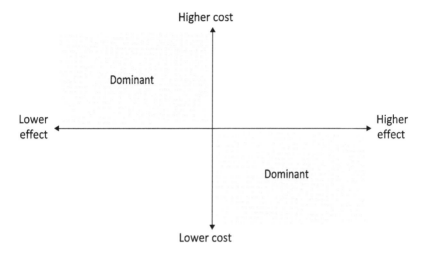

**FIGURE 23** Visualised and clarified interpretation of incremental cost-effectiveness ratio (ICER).

interpretation of the results is not as clear-cut if ICER is calculated to be in any of the other dimensions.

The willingness to pay in organisations, companies or societies affects whether a measure is implemented or not. In the top right corner, new measures have a higher cost and effect simultaneously, while the measures have a lower cost and effect simultaneously in the bottom left corner. However, the chosen and prioritised measure should not only be based on a cost-efficiency analysis and the cost per additional measure of effect, other aspects must be considered, for example:

- The principle of human dignity, that all human beings have equal value, regardless of personal characteristics or functions in society
- The principle of need and solidarity, that resources first and foremost should be provided to people with the greatest actual or expected need
- The principle of cost-effectiveness, that the relation between cost and effect should be reasonable

The willingness to pay for different measures is set when reflecting on ICER and weighing in the principles mentioned above. A higher cost per additional measure of effect is generally accepted when extensive issues or hardships occur, or when few alternative measures and possibilities for action are present. In the evaluation, it is also important to assess where it is most profitable to implement a measure, i.e. at the Individual level, Organisational/Enterprise level, Community level and Country level.

Table 4 is an example matrix that can be used to conduct economic evaluations of implemented measures.

**Economic evaluation matrix of measures based on the SwAge™-model**

## TABLE 4
## Example of economic evaluation matrix of measures based on the SwAge™ model

|  |  | No measure 0 = Treatment as usual (TAU) || Measure group 1 ||
|---|---|---|---|---|---|
|  |  | Mean value | SD | Mean value | SD |
| Resource cost |  |  |  |  |  |
| Measure area at the level: | 1. Health |  |  |  |  |
|  | 2. Physical environment |  |  |  |  |
|  | 3. Mental environment |  |  |  |  |
| – Individual level | 4. Working hours, work pace and time for recuperation |  |  |  |  |
| – Organisational/ Enterprise level | 5. Financial incentives |  |  |  |  |
| – Community level | 6. Personal social environment |  |  |  |  |
| – Country level | 7. Social work environment |  |  |  |  |
|  | 8. Stimulation and motivation through work tasks |  |  |  |  |
|  | 9. Knowledge, competence, development |  |  |  |  |
| Sickness absence |  |  |  |  |  |
| Sickness benefit |  |  |  |  |  |
| Unemployment/ termination |  |  |  |  |  |
| Medication |  |  |  |  |  |
| Health care | Primary care |  |  |  |  |
|  | Inpatient care |  |  |  |  |
|  | Specialist consultation |  |  |  |  |
|  | Municipal care service |  |  |  |  |
| Societal service | Social service |  |  |  |  |
|  | Adjustments for housing and living |  |  |  |  |
|  | Education/retraining |  |  |  |  |
|  | Social security system services |  |  |  |  |
|  | Police, penal system services |  |  |  |  |
| Voluntary efforts |  |  |  |  |  |
| Total resource costs |  |  |  |  |  |
| Productivity costs |  |  |  |  |  |
| Total costs |  |  |  |  |  |
| Total costs per week |  |  |  |  |  |

| Potential measure group 2 | | Mean difference between groups TAU – Measure (in case of several measures, include comparison between Measure 1- Measure 2 and TAU – Measure 2) | | Adjusted difference between groups (age, gender, education, income, etc.) |
|---|---|---|---|---|
| Mean value | SD | TUA-Measure | (95% confidence interval) | |

# References

1. Nilsson, K. (2020). A sustainable working life for all ages – the SwAge-model. *Applied Ergonomics* 86:1–13.
2. Nilsson, K. (2016). Conceptualization of ageing in relation to factors of importance for extending working life – a review. *Scandinavian Journal of Public Health* 44(5):490–505.
3. Nilsson, K., Rignell-Hydbom, A. & Rylander, L. (2011). Factors influencing the decision to extend working life or to retire. *Scandinavian Journal of Work Environment & Health* 37(6):473–480.
4. Nilsson, K. (2005). Who can and want to work until 65 years or beyond? A study with employees in health and medical care. *Arbete och hälsa* 14:1–35.
5. Nilsson, K. (2006). Older workers attitude to an extended working life. Differences between occupations in health and medical care. *Arbetsliv i omvandling* 10:1–69.
6. Nilsson, K. (2012). Why work beyond 65? Discourse on the decision to continue working or retire early. *Nordic Journal of Working Life Studies* 2(3):7–28.
7. Nilsson, K. (2013). *To work or not to work in an extended working life? Factors in working and retirement decision.* Lund: Lund University.
8. Nilsson, K. (2017). The influence of work environmental and motivation factors on seniors' attitudes to an extended working life or to retire. A cross sectional study with employees 55–74 years of age. *Open Journal of Social Sciences* 5(7):30–41.
9. Nilsson, K. (2015). *Gender and old-age pension – a cross-sectional study on differences between men and women in being able and willing to work until age 65 or longer.* Report 2015:4. Occupational and Environmental Medicine. Lund: Lund University, pp. 1–26.
10. Nilsson, K. (2020). When is work a cause of early retirement and are there any effective organizational measures to combat this? A population-based study of perceived work environment and work-related disorders among employees in Sweden. *BMC Public Health* 20(1):1–15.
11. Nilsson, K. (2017). Active and healthy ageing at work – a qualitative study with employees 55–63 years and their managers. *Open Journal of Social Sciences* 5(7):13–29.
12. Nilsson, K. & Nilsson, E. (2021). Management, measures and maintenance: success and setbacks in interventions promoting a healthy and sustainable employability and working life for all ages. *Springer Nature* 220:17–24. https://doi.org/10.1007/978-3-030-746 05-6_3.
13. Nilsson, K. & Nilsson, E. (2021). Organisational measures and strategies for a healthy and sustainable extended working life and employability—a deductive content analysis with data including employees, First Line Managers, Trade Union Representatives and HR-Practitioners. *International Journal of Environmental Research and Public Health* 18:5626.
14. Nilsson, K., Rignell-Hydbom, A. & Rylander, L. (2016). How is self-rated health and diagnosed disease associated with early or deferred retirement: a cross sectional study with employees aged 55–64. *BMC Public Health* 16(1):1–9.
15. Trist, E.L. & Bamforth, K.W. (1951). Some social and psychological consequences of the Longwall method of coal-getting. *Human Relations* 4(1):3–38–178.
16. Trist, E.L. & Murray, H. (1993). *The social engagement of social science: a Tavistock anthology.* Volym II. The socio-technical perspective. Philadelphia: University of Pennsylvania Press.

17 Månsson, N.O. & Råstam, L. (2001). Self-rated health as a predictor of disability pension and death – a prospective study of middle-aged men. *Scandinavian Journal of Public Health* 29(2):51–158.
18 Nilsson, K., Östergren, P.-O., Kadefors, R. & Albin, M. (2016). Has the participation of older employees in the workforce increased? Study of the total Swedish population regarding exit working life. *Scandinavian Journal of Public Health* 44(5):506–516.
19 Nordenfelt, L. (2003). An the evolutionary concept of health: health as natural function. In: Nordenfelt, L. & Liss, P.-E. (red.). *Dimensions of health and health promotion*. Amsterdam: Rodopi Press, pp. 37–53.
20 Whitbeck, C. (1981). A theory of health. In: Caplan, A.L., Engelhardt, H.T. & McCartney, J.J. (red.). *Concepts of health and disease: interdisciplinary perspectives*. Boston: Addison-Wesley, pp. 611–626.
21 Pörn, I. (1984). An equilibrium model of health. In: Lindahl, I. & Nordenfelt, L. (red.). *Health, disease and causal explanations in medicine*. Dordrecht: Reidel, pp. 3–9.
22 Pörn, I. (2000). What is health? In: Klockars, K. & Österman, B. (eds.). *Concepts of health: philosophical and ethical perspectives on quality of life, health and care*. Falköping: Elanders Gummesson, pp. 14–28.
23 Nordenfelt, L. (1991). *Quality of life and health. Theory and criticism*. Stockholm: Almqvist & Wiksell.
24 Mather, M. (2010). Aging and cognition. *Cognitive Science* 1(3):346–362.
25 Eriksson, K. (1989). *The idea of health*. Stockholm: Almqvist & Wiksell.
26 Nilsson, K. (2017). *Sustainable working life in healthcare - study on how 11,902 employees experience their working situation and the possibility of working in an extended working life*. Report 2017:13. Occupational and Environmental Medicine. Lund: Lund University, pp. 1–84.
27 Börsch-Supan, A., Brugiavini, A. & Croda, E. (2009). The role of institutions and health in European patterns of work and retirement. *Journal of European Social Policy* 19(4):341–358.
28 International Labour Organisation (ILO). Occupational safety and health. *World Statistics*. https://www.ilo.org/moscow/areas-of-work/occupational-safety-and-health/WCMS_249278/lang--en/index.htm.
29 Mein, G., Martikainen, P., Stansfeld, S. & Marmot, M. (2003). Is retirement good or bad for mental and physical health functioning? Whitehall II longitudinal study of civil servants. *Journal of Epidemiology and Community Health* 57(1):46–49.
30 Westerlund, H. et al. (2009). Self-rated health before and after retirement in France (GAZEL): a cohort study. *Lancet* 374:1889–1896.
31 Westerlund, H. et al. (2010). Effect of retirement on major chronic conditions and fatigue: French GAZEL occupational cohort study. *BMJ* 341:c6149.
32 Quaade, T., Engholm, G., Johansen, A.M. & Möller, H. (2013). Mortality in relation to early retirement in Denmark: a population-based study. *Scandinavian Journal of Public Health* 30:216–222.
33 Ahola, K. et al. (2012). Work-related exhaustion and telomere length: a population-based study. *PLoS One* 7(7):e40186.
34 Seitsamo, J. & Klockars, M. (1997). Aging and changes in health. *Scandinavian Journal of Work Environment & Health* 23(suppl. 1):27–35.
35 Karlsson, N.E., Carstens, J.M., Gjesdal, S. & Alexandersson, K.A.E. (2008). Work and health. Risk factors for disability pension in a population-based cohort of men and women on long-term sick leave in Sweden. *European Journal of Public Health* 18(3):224–231.
36 Karpansalo, M., Manninen, P., Kauhanen, J., Lakka, T. & Salonen, J.T. (2004). Perceived health as a predictor of early retirement. *Scandinavian Journal of Work Environment & Health* 30(4):287–292.

# References

37 Pietiläinen, O., Laaksonen, M., Rahkonen, O. & Lahelma, E. (2011). Self-rated health as a predictor of disablility retirement – the contribution of ill-health and working conditions. *PLoS One* 6(9):e25004. https://doi.org/10.1371/journal.pone.0025004.
38 Idler, E.L. & Benyamini, Y. (1997). Self-rated health and mortality: a review of twenty-seven community studies. *Journal of Health and Social Behavior* 38(1):21–37.
39 Singh-Manoux, A. et al. (2007). Self-rated health and mortality: short- and long-term associations in the Whitehall II study. *Psychosomatic Medicine* 69:138–143.
40 Strawbridge, W.J. & Wallhagen, M.I. (1999). Self-rated health and mortality over three decades. *Research on Aging* 21(3):402–416.
41 Nilsson, K. & Nilsson, E. (2022). Can they stay or will they go? A cross sectional study of managers' attitudes towards their senior employees. *International Journal of Environmental Research and Public Health* 19:1057.
42 Caffaro, F. et al. (2018). Being a farmer at old age: an ergonomic analysis of work-related risks in a group of Swedish farmers aged 65 and over. *Journal of Agromedicine* 23:78–91.
43 Bonde, J.P.E. et al. (2022). Occupational risk of COVID-19 across pandemic waves: a two-year national follow-up study of hospital admissions. *Scandinavian Journal of Work, Environment and Health* 48(8):672–677.
44 Nilsson, K., Pinzke, S. & Lundqvist, P. (2010). Occupational injuries to senior farmers in Sweden. *Journal of Agricultural Safety and Health* 16(1):19–29.
45 Nagel, C. et al. (2022). Nurses' work environment during the COVID-19 pandemic in a person-centred practice: a systematic review. *Sustainability* 14(10):5785.
46 Nilsson, K. & Pinzke, S. (2011). Occupational accidents among elderly farmers in Sweden. *Work* 41:5324–5326.
47 von Bonsdorff, M.E. et al. (2011). Work strain in midlife and 28-year work ability trajectories. *Scandinavian Journal of Work Environment & Health* 37(6):455–463.
48 Sauré, P. & Zoabi, H. (2011). *Retirement age across countries: the role of occupations.* Zürich: Swiss National Bank working papers 6/2012. [citerad 2020-09-27]. Hämtad från: http://papers.ssrn.com/sol3/papers.cfm?abstract_id=1940452.
49 Pohjonen, T. (2001). Perceived work ability of home care workers in relation to individual and work-related factors in different age groups. *Occupational Medicine* 51(3).209–217.
50 Pálsdóttir, A.M., Wissler, S.K., Nilsson, K., Petersson, I.F. & Grahn, P. (2015). Nature-based rehabilitation in peri-urban areas for people with stress-related illness – a controlled prospective study. *Acta Horticulturae* 8. https://doi.org/10.17660/ActaHortic.2015.1093.2.
51 Hagberg, M. & Wahlström, J. (2019). Ergonomic factors and diseases of the locomotor organs and especially the ergonomics of the computer workstation. In: Edling, C., Nordberg, G., Albin, M. & Nordberg, M. (eds.). *Occupational and environmental medicine – a textbook on health and the environment.* Lund: Studentlitteratur, pp. 117–126.
52 Wolf, J. et al. (2018). *Preventing disease through a healthier and safer workplace.* Geneva: World Health Organization.
53 Nordander, C. et al. (2016). Exposure-response relationships for work-related neck and shoulder musculoskeletal disorders – analyses of pooled uniform data sets. *Applied Ergonomics* 55:70–84.
54 Nylén, P. (2016). Sensory aging – visual ergonomics and lighting. In: Vingård, E. (ed.). *Healthy workplaces for women and men of all ages.* Report 2016:8. Stockholm: The Swedish Work Environment Authority, pp. 86–95.
55 Ringen, K. et al. (2022). Hearing impairment and tinnitus among older construction workers employed at DOE facilities. *American Journal Industrial Medicine* 65:644–651. https://doi.org/10.1002/ajim.23406.

56 The National Institute for Occupational Safety and Health (NIOSH). https://www.cdc.gov/niosh/topics/productiveaging/safetyandhealth.html.
57 Bohgard, M. & Albin, M. (2019). Chemical health risks in the work environment. In: Bohgard, M. et al. (eds.). *Work and technology on human terms*. Stockholm: Prevent.
58 Kjellstrom, T., Lemke, B. & Otto, M. (2017). Climate conditions, workplace heat and occupational health in South-East Asia in the context of climate change. *WHO South-East Asia Journal of Public Health* 6(2):15–21.
59 Kadefors, R., Nilsson, K., Östergren, P-O., Rylander, L. & Albin, M. (2018). Social inequality in working life expectancy in Sweden. *Zeitschrift für Gerontologie und Geriatrie* 52(1):52–61.
60 Nilsson, K. (2016). The ability and desire to extend working life. In: Vingård, E. (ed.). *Healthy workplaces for men and women of all ages*. Report 2016:8. Stockholm: The Swedish Work Environment Authority, pp. 31–49.
61 Nilsson, K. & Lundqvist, P. (2018). Healthy and sustainable workplaces for older workers in agriculture – a Swedish perspective. In: *Abstract from 20th congress of international association of rural health medicine (IARM)*. Tokyo, pp. 48–51.
62 Pinzke, S., Nilsson, K. & Lundqvist, P. (2014). Farm tractors on Swedish public roads – age-related perspectives on police reported incidents and injuries. *Work* 49(1):39–49.
63 Hovbrandt, P. et al. (30 August 2021). Psychosocial working conditions and social participation. A 10-year follow-up of senior workers. *International Journal of Environmental Research and Public Health* 18(17):9154.
64 Jahoda, M. (1981). Work, employment, and unemployment: values, theories, and approaches in social research. *American Psychology* 36(2):184–191.
65 Erikson, E. & Erikson, J. (1981). On generativity and identity: from a conversation with Erik and Joan Erikson. *Harvard Educational Review* 51(2):249–269.
66 Burke, H. (2023). *Living and working in Europe*. Luxembourg: Publications Office of the European Union.
67 Nagel, C. & Nilsson, K. (2022). Nurses' work-related mental health in 2017 and 2020—a comparative follow-up study before and during the COVID-19 pandemic. *International Journal of Environmental Research and Public Health* 19:15569.
68 Restrepo, J. & Lemos, M. (2021). Addressing psychosocial work-related stress interventions: a systematic review. *Work* 70(1):53–62.
69 Belloni, M., Carrino, L. & Meschi, E. (2022). The impact of working conditions on mental health: novel evidence from the UK. *Labour Economics* 76:102176.
70 WHO. (2019). *Mental disorders*. Geneva: Word health organisation. https://www.who.int/mental_health/ management/en/.
71 WHO. (2019). *Protecting workers' health*. Geneva: Word health organisation. https://www.who.int/en/news-room/ fact-sheets/detail/protecting-workers'-health.
72 Siegrist, J., Wahrendorf, M., von dem Knesebeck, O., Jürges, H. & Bösch-Supan, A. (2007). Quality of work, well-being and intended early retirement of older employees – baseline results from the SHARE. Study. *European Journal of Public Health* 17(1):62–68.
73 Selye, H.A. (1936). A syndrome produced by diverse nocuous agents. *Nature* 138:32.
74 Cullberg, J. (1984). *Dynamic psychiatry*. Stockholm: Natur och Kultur.
75 Cullberg, J. (2001). *Crisis and development*. Stockholm: Natur och Kultur.
76 Lazarus, R.S. (1993). Coping theory and research: past, present and future. *Psychosomatic Medicine* 55(3):234–247.
77 Folkman, S. (1984). Personal control and stress and coping processes: a theoretical analysis. *Journal of Personality and Social Psychology* 46(4):839–852.
78 Folkman, S. & Lazarus, R.S. (1980). An analysis of coping in a middle-aged community sample. *Journal of Health and Social Behavior* 21(3):219–239.

# References

79 Rotter, J.B. (1966). Generalized expectancies for internal versus external control of reinforcement. *Psychological Monographs – General and Applied* 80(1):1–28.
80 Nilsson, K. (2019). *Stay or go? The SwAge model = sustainable working life for all ages.* Report 5, S2018:10. Delegation for Senior Workforce. Stockholm: Government Offices of Sweden, pp. 1–67.
81 Bravo, G. et al. (March 2022). Do older workers suffer more workplace injuries? A systematic review. *International Journal of Occupational Safety and Ergonomics* 28(1):398–427.
82 International Labour Organization. Health and safety at the workplace. https://www.ilo.org/global/topics/dw4sd/themes/osh/lang--en/index.htm.
83 Conway, S.F., Farrell, M., McDonagh, J. & Kinsella, A. (2022). Farmers don't retire': re-evaluating how we engage with and understand the 'older' farmer's perspective. *Sustainability* 14:2533.
84 Fix, J. et al. (2021). Gender differences in respiratory health outcomes among farming cohorts around the globe: findings from the AGRICOH consortium. *Journal of Agromedicine* 26(2):97–108.
85 Reed, D.B., Rayens, M.K., Winter, K. & Zhang, M. (2008). Health care delay of farmers 50 years and older in Kentucky and South Carolina. *Journal of Agromedicine* 13(2):71–79.
86 van der Meer, M.J. (2008). The sociospatial diversity in the leisure activities of older people in the Netherlands. *Journal of Aging Studies* 22(1):1–12.
87 Grahn, P., Pálsdóttir, A.-M. & Nilsson, K. (2015). *Green rehabilitation in the countryside.* Report 2015:48/09. Alnarp: Swedish University of Agricultural Sciences.
88 Lavesson, L. & Nilsson, K. (2011). The importance of nature for health and lifestyle. In: Ottosson, J., Lundqvist, S. & Johnsson, L. (eds.). *Green entrepreneur. Nature experience and health.* Alnarp: Swedish University of Agricultural Sciences, pp. 90–107.
89 Karasek, R. & Theorell, T. (1990). *Healthy work: stress, productivity, and the reconstruction of working life.* New York: Basic books.
90 Hovbrandt, P., Håkansson, C., Karlsson, G., Albin, M. & Nilsson, K. (2017). Prerequisites and driving forces behind an extended working life among older workers. *Scandinavian Journal of Occupational Therapy* 26(3):1–13.
91 Boedeker, W. & Berkels, H. (2005). Trends in sickness absence and early retirement due to mental disorders in Germany. In: Järvisalo, J., Andersson, B., Boedeker, W. & Houtman, I. (red.). *Mental disorders as a major challenge in prevention of work disability.* Experiences in Finland, Germany, the Netherlands and Sweden. Social security and health reports Kela:2005:66. Helsinki: The Social Insurance Institution.
92 Cowen Forssell, R. (2019). *Cyberbulling. Transformation of working life and its boundaries.* Doktorsavhandling, fakulteten för kultur och samhälle. Malmö: Malmö universitet.
93 SCB. (2018). *Threats, violence and anxiety 1980–2017.* Solna: Statistic Sweden.
94 Kadefors, R. Nilsson, K. Rylander, L. Östergren, P-O. & Albin, M. (2018). Occupation, gender and work-life exits: a Swedish population study. *Ageing & Society* 38(7):1332–1349.
95 Nilsson, E. & Nilsson, K. (2017). Time for caring? Elderly care employees' occupational activities in the cross draft between their work priorities, "must-do's" and meaningfulness. *International Journal of Care Coordination* 20(1–2):8–16.
96 Lipsky, M. (1980). *Street level bureaucracy – dilemmas of the individual in public services.* New York: Russel Sage Foundation.
97 Johnston, D.W. & Wang-Sheng, L. (2009). Retiring to the good life? The short-term effects of retirement on health. *Economic Letters* 103(1):8–11.
98 Sjösten, N. et al. (2010). Influence of retirement and work stress on headache prevalence: a longitudinal modelling study from GAZEL Cohort Study. *Cephal* 31(6):696–705.

99 Muto, T. et al. (2003). Health status and lifestyles of elderly Japanese workers. In: Kumashiro, M. (red.). *Aging and work*. London: Taylor & Francis, pp. 72–84.
100 Ejlertsson, L., Heijbel, B., Ejlertsson, G. & Andersson, I. (2018). Recovery, work-life balance and work experiences important to self-rated health: a questionnaire study on salutogenic work factors among Swedish primary health care employees. *Work* 59(1):155–163.
101 Europeiska unionen. (2022). *Employment and social developments in Europe 2022*. Luxembourg: Publications office of the European Union.
102 OECD. (2019). *Pensions at a glance 2021. OECD and G20 indicators*. Paris: OECD Publishing. https://stat.link/7bkwjc.
103 OECD. (2019). *Health at a glance 2019: OECD indicators*. Paris: OECD Publishing. https://doi.org/10.1787/4dd50c09-en.
104 WHO & ILO. (2022). *Caring for those who care: guide for the development and implementation of occupational health and safety programmes for health workers*. Geneva: World Health Organization and the International Labour Organization.
105 Mykletun, R. & Furunes, T. (2011). The ageing workforce management programme in Vattenfall AB Nordic, Sweden. In: Ennals, R. & Salomon, R.H. (eds.). *Older workers in a sustainable society. Labor, education & society*. Frankfurt: Peter Lang Verlag, pp. 93–106.
106 Nilsson, K. (2017). Best before date on the workforce? The importance of different age concepts for older people in working life. In: Krekula, C. & Johansson, B. (eds.). *Age, power and organization – theory and empiricism*. Lund: Studentlitteratur, pp. 169–182.
107 Nilsson, K. (2018). Managers' attitudes to their older employees – a cross- sectional study. *Work* 59(1):49–58.
108 Nilsson, K. (2019). *Managers' attitudes towards their older employees*. Report 15, S2018:10. Delegation for Senior Workforce. Stockholm: Government Offices of Sweden, pp. 1–47.
109 Nilsson, K. (2017). Interventions to reduce injuries among older workers: a review of evaluated intervention projects. *Work* 55(2):471–480.
110 Eurostat. *European health interview survey (EHIS)*. https://ec.europa.eu/eurostat/web/microdata/european-health-interview-survey.
111 Åkerstedt, T. et al. (2013). Having to stop driving at night because of dangerous sleepiness – awareness, physiology and behaviour. *Journal of Sleep Research* 22(4):380–388.
112 Myllyntausta, S. et al. (2023). Association of working hours with accelerometer-based sleep duration and sleep quality on the following night among older employees. *Sleep Epidemiology*. https://doi.org/10.1016/j.sleepe.2023.100060.
113 Härmä, M. (2012). Is retirement beneficial or harmful to mental health? *Scandinavian Journal of Work Environment & Health* 38(5):391–392.
114 Laaksonen, M. et al. (2012). Trajectories of mental health before and after old-age and disability retirement: a register-based study on purchases of psychotropic drugs. *Scandinavian Journal of Work Environment & Health* 38(5):409–417. https://doi.org/10.5271/sjweh.3290.
115 Åkerstedt, T. et al. (2021). Acute and cumulative effects of scheduling on aircrew fatigue in ultra-short-haul operations. *Journal of Sleep Research* 30:e13305.
116 Nilsson, K. (2019). A Sustainable Working Life for All Ages – The swAge-Model. In: Bagnara, S., Tartaglia, R., Albolino, S., Alexander, T. & Fujita, Y. (eds.). *Proceedings of the 20th Congress of the International Ergonomics Association (IEA 2018). IEA 2018. Advances in Intelligent Systems and Computing*, vol 826. Springer: Cham, https://doi.org/10.1007/978-3-319-96065-4_25. https://link.springer.com/chapter/10.1007/978-3-3 19-96065-4_25#author-information.
117 Mulders, J.O. (2019). Attitudes about working beyond normal retirement age: the role of mandatory retirement. *Journal of Aging & Social Policy* 31:106–122.

# References

118. Marcum, J.L., Browning, S.R., Reed, D.B. & Charing, R.J. (2011). Determinants of work hours among a cohort of male and female farmers 50 years and older in Kentucky and South Carolina (2002–2005). *Journal of Agromedicine 16*(3):163–173.
119. Boorse, C. (1977). Health as a theoretical concept. *Philosophy of Science 44*(4):542–573.
120. Rongen, A., Robroek, S.J.W., van Lenthe, F.J. & Burdorf, A. (2013). Workplace health promotion: a meta-analysis of effectiveness. *American Journal of Preventive Medicine 44*(4):406–415.
121. WHO. (1986). *The Ottawa charter for health promotion.* Geneva: World health organization. [citerad 2020-09-27]. Hämtad från. https://www.who.int/healthpromotion/conferences/previous/ottawa/en/.
122. WHO. (2020). *Guidelines on physical activity and sedentary behaviour.* Geneva: World Health Organization.
123. Lupton, D. (1995). *The imperative of health: public health and the regulated body.* London: Sage.
124. Buchanan, D.R. (2000). *An ethic for health promotion. Rethinking the sources of human well-being.* Oxford: Oxford University Press.
125. Weber, M. (2013). *Economy and society.* Berkeley: University of California Press.
126. Taylor, F.W. (2007). *The principles of scientific management.* Sioux Falls: NuVision Publications LLC.
127. Mayo, E. (1939). *The social problems of industrial civilization.* New York: Ayer.
128. McGregor, D. (1960). *The human side of enterprise.* New York: McGraw-Hill.
129. Huzzard, T. (2003). *The convergence of the quality of working life and competiveness.* Stockholm: The Swedish National Institute of Working Life.
130. Du Gay, P. (2000). *In praise of bureaucracy.* London: Sage.
131. Bonde, J.P.E. et al. (2023). Occupational risk of SARS-CoV-2 infection: a nationwide register-based study of the Danish workforce during the COVID-19 pandemic, 2020–2021. *Occupational and Environmental Medicine 80*:202–208.
132. WHO. (1948). *Constitution.* Geneva: World health organization.
133. Thaler, R.H. & Sunstein, C.S. (2009). *Nudge. Improving decisions about health, wealth and happiness.* London: Penguin Books Ltd.
134. Takashi, M. et al. (2003). Health status and lifestyles of elderly Japanese workers. In: Kumashiro, M. (red.). *Aging and work.* London: Taylor & Francis, pp. 72–85.
135. Kunze, F. & Reas, A.M.L. (2015). It matters how old you feel: antecedents and performance consequences of average relative subjective age in organizations. *Journal of Applied Psychology 100*(5):1511–1526.
136. Löfqvist, L. & Nilsson, K. (2019). *Employees within Sustainable working life for all ages in the city of Helsingborg.* Report No. 12/2019. Lund: Occupational and Environmental Medicine, Lund University, pp. 1–102.
137. Nilsson, K. (2005). *Pension or working life?* Malmö: The Swedish National Institute of Working life, pp. 1–280.
138. Nilsson, K. & Löfqvist, L. (2023). Older employees' experience of their conditions, work situation and work ability and need for measures for a healthy and sustainable working life for all ages - follow-up survey in 2022 in a Swedish municipality. Report No. 4/2023. Occupational and environmental medicine Lund University, pp. 1–110.
139. Gratton, L. & Scott, A. (2017). The corporate implications of longer lives. Massachusetts Institute of Technology (MIT). *Sloan Management Review 58*(3):63–70.
140. di Fabio, A. (2017). A review of empirical studies on employability and measures of employability. In: Maree, K. (red.). *Psychology of career adaptability, employability and resilience.* Cham: Springer, pp. 107–123.
141. McDowell, M., Thom, R., Frank, R.H. & Bernanke, B. (2012). *Principles of economics.* 3 uppl. New York: McGraw-Hill Europe.

142 Kahneman, D. & Tversky, A. (2000). *Choices, values, and frames*. Cambridge: Cambridge University Press.
143 Larrick, R.P., Nisbett, R.E. & Morgan, J.N. (1993). Who uses the cost-benefit rules of choice? Implications for the normative status of microeconomic theory. *Organizational Behavior and Human Decision Processes 56*(3):331–347.
144 Arkes, H.R. & Blumer, C. (1985). The psychology of sunk cost. *Organizational Behavior and Human Decision Processes 35*(1):124–140.
145 Midtsundstad, T. Hilsen, A-I. & Nilsson, K. (2022). *Industry-specific senior policy. Challenges, experiences and proposed solutions*. Report 2022:21. Oslo: Fafo, pp. 1–142.
146 Bouziri, H., Descatha, A., Roquelaure, Y., Dab, W. & Jean, K. (2022). Can we distinguish the roles of demographic and temporal changes in the incidence and prevalence of musculoskeletal disorders? A systematic review. *Scandinavian Journal of Work, Environment & Health 48*(4):253–263.
147 Herzberg, F. (1966). *Work and the nature of man*. New York: World Publishing Company.
148 Herzberg, F., Mausner, B. & Bloch Snyderman, B. (1959). *The motivation to work*. New York: Wiley & Sons Inc.
149 Nilsson, K. (2011). Attitudes of managers and older employees to each other and the effects on the decision to extended working life. In: Ennals, R. & Salomon, R.H. (red.). *Older workers in a sustainable society. Labor, education & society*. Frankfurt: Peter Lang Verlag, pp. 146–157.
150 Löfqvist, L. & Nilsson, K. (2019). Managers - sustainable working life in the city of Helsingborg. Report No. 17/2019. Lund: Occupational and Environmental Medicine Lund University, pp. 1–72.
151 Nilsson, K. & Löfqvist, L. (2023). Managers' attitudes towards their older employees' conditions, work situation and work ability and the need for measures for a healthy and sustainable working life for all ages - follow-up survey in 2022 in a Swedish municipality. Report No. 3/2023. Occupational and environmental medicine Lund University, pp. 1–82.
152 van Ours, J.C. (2009). *Will you still need me: when I'm 64?* Bonn: Institute for the Study of Labour.
153 Johnson, H.M. (1962). *Sociology: a systematic introduction*. London: Latimer Trend & Co Ltd.
154 Donehower, G. (2023). Gender and the total work of older workers in Asia. Economics Working Paper Series 687, *Asian Development Bank*.
155 d'Albis, H. (2023). The employment of older workers. In: Bloom, D.E., Sousa-Poza, A. & Sunde, U. (eds.). *The Routledge handbook of the economics of ageing*. London: Routledge, pp. 363–380.
156 Farrants, K., Dervish, J., Marklund, S. & Alexanderson, K. (2023). Health and morbidity among people in paid work after 64 years of age: a systematic review. *Social Sciences & Humanities Open 8*(1):100571. https://www.sciencedirect.com/science/article/pii/S2590291123001766.
157 Repetti, M. & Calasanti, T. (2023). *Retirement migration and Precarity in later life*. Bristol: Policy Press.
158 Rothstein, B. (2008). Trust and social capital. In: Lindvert, J. & Schierenbeck, I. (eds.). *Comparative politics*. Malmö: Liber, pp. 218–244.
159 Krekula, C. (2019). Time, precarisation and age normality: on internal job mobility among men in manual work. *Ageing & Society 39*(10):2290–2307.
160 Zhan, Y., Wang, M. & Daniel, V. (2019). Lifespan perspectives on the work-to-retirement transition. In: Baltes, B.B., Rudolph, C.W. & Zacher, H. (eds.). *Work across the lifespan*. Cambridge: Academic Press, pp. 581–604.

# References

161 Fridriksson, J.F. et al. (2017). *Working environment and work retention.* Copenhagen: The Nordic Council of Ministers. TemaNord 559:1–121.
162 Gyllensten, K., Wentz, K., Håkansson, C. & Nilsson, K. (2018). Older assistant nurses' motivation for a full or extended working life. *Ageing and Society* 39(12):2699–2713.
163 Tunney, O., Henkens, K. & van Solinge, H. (2023). A life of leisure? Investigating the differential impact of retirement on leisure activity. *The Journals of Gerontology* [gbad097] 78:1775–1784.
164 Smith, D.B. & Moen, P. (2011). Retirement satisfaction for retirees and their spouses. Do gender and the retirement decision-making process matter? *Journal of Family Issues* 25(2):262–285.
165 Stimpfel, A.W. et al. (2020). Organization of work factors associated with work ability among aging nurses. *Western Journal of Nursing Research* 42(6):397–404.
166 Bengtsson, E. & Nilsson, K. (2004). *Older employees – a qualitative study on employees' conditions in working life.* Malmö: The Swedish National Institute of Working Life.
167 Hao, Y. (2008). Productive activities and psychological well-being among older adults. *Journals of Gerontology Series B: Psychological Sciences and Social Sciences* 63(2), S64–S72.
168 Ruhose, J., Thomsend, S.L. & Weilage, I. (2023). No mental retirement: estimating voluntary adult education activities of older workers. *Education Economics*, pp. 1–34.
169 Nilsson, K. & Nilsson, E. (2022). Can they stay or will they go? A cross sectional study of managers' attitudes towards their senior employees. *International Journal of Environmental Research and Public Health* 19:1057.
170 Nilsson, E. & Nilsson, K. (2017). The transfer of knowledge between younger and older employees in the health and medical care: an intervention study. *Open Journal of Social Sciences* 5(7):71–96.
171 Eklund, M. et al. (2017). The linkage between patterns of daily occupations and occupational balance: applications within occupational science and occupational therapy practice. *Scandinavian Journal of Occupational Therapy* 24(1):41–56.
172 Hovbrandt, P., Carlsson, G., Nilsson, K., Albin, M. & Håkansson, C. (2019). Occupational balance as described by older workers over the age of 65. *Journal of Occupational Science* 26(1):40–52.
173 Dür, M. et al. (2016). Initial evidence for the link between activities and health: associations between a balance of activities, functioning and serum level of cytokines and C-reactive protein. *Psychoneuroendocrinology* 65:138–148. https://doi.org/10.1016/j.psyneuen.2015.12.015.
174 Håkansson, C. & Ahlborg, G Jr. (2010). Perceptions of employment, domestic work, and leisure as predictors of health among women and men. *Journal of Occupational Science* 17:150–157.
175 Goffman, E. (1959). *The presentation of self in everyday life.* New York: Doubleday & Company Inc.
176 Tennant, M. (2006). *Psychology and adult learning.* 3 uppl. London: Routledge.
177 Fenstermaker, S. & West, C. (2002). *Doing gender, doing difference. Inequality, power, and institutional change.* New York: Routledge.
178 Davies, K. (2001). *Disturbing gender: on the doctor – nurse relationship.* Lund: Department of Sociology, Lund University.
179 Goffman, E. (1963). *Stigma: notes on the management of spoiled identity.* Englewood Cliffs: Prentice-Hall Inc.
180 Oyserman, D., Elmore, K. & Smith, G. (2012). Self, self-concept, and identity. In: Leary, M.R. & Tangey, J.P. (red.). *Handbook of self and identity.* New York, London: The Guilford Press, pp. 64–104.

181 McGoldrick, A.E. & Arrowsmith, J. (2001). Discrimination by age: the organizational response. In: Glover, I. & Branine, M. (red.). *Ageism in work and employment*. England: Ashgate Publishing Ltd, pp. 75–96.
182 French, J. & Raven, B. (1959). The bases of social power. In: Cartwright, E. (red.). *Studies in social power*. Ann Arbor: Institute for Social Research, pp. 150–167.
183 Ås, B. (1978). Hersketeknikker. *Kjerringråd 3*:17–21.
184 Rhodes, C. & Pullen, A. (2010). Editorial: Neophilia and organization. *Culture and Organization 16*(1):1–6.
185 Frankenhaeuser, M. (1993). Kvinnligt, manligt, stressigt. Wiken: Bra Böcker.
186 le Blance, P.M., de Jonge, J. & Schaufeli, W. (2000). Job stress and health. In: Chmiel, N. (red.). *Introduction to work organizational psychology. A European perspective*. Oxford: Blackwell Publishers Ltd., pp. 148–177.
187 Jokela, M. et al. (2010). From midlife to early old age. Health trajectories associated with retirement. *Epidemiology 21*(3):284–290.
188 Maslow, A.H. (1970). *Motivation and personality*. 3 uppl. New York: Addison Wesley Longman.
189 Roethlisberger, F.J. & Dickson, W.J. (1939). *Management and the worker. The early sociology of management and organisation*. Cambridge: Harvard University Press.
190 Araújoa, J. & Pestanab, G. (2017). A framework for social well-being and skills management at the workplace. *International Journal of Information Management 37*(6):718–725.
191 Gao, Q., Woods, O. & Cai, X. (2021). The influence of masculinity and the moderating role of religion on the workplace well-being of factory workers in China. *International Journal of Environmental Research and Public Health 18*(12):6250.
192 White, R. (2023). Intersectionality and discrimination. An examination of the U.S. labor market. London: Springer International Publishing.
193 Davies, K. (2003). The body and doing gender: the relations between doctors and nurses in hospital work. *Sociology of Health and Illness 25*(7):720–742.
194 Esseveld, J. (1997). Seeking to change our thinking about femininity and sex/gender - analysis of three texts. In: Lundqvist, Å. & Mulinari, D. (ed.). *Sociological women's research*. Lund: Studentlitteratur, pp. 27–53.
195 Nilsson, K. et al. (2022). School principals' work participation in an extended working life – are they able to, and do they want to? A quantitative study of the work situation. *International Journal of Environmental Research and Public Health 19*(7):3983.
196 Risberg, G. (2004). *"I am solely a professional – neutral and genderless" on gender bias and gender awareness in the medical profession*. Umeå: Umeå University, Department of Public Health and Clinical Medicine.
197 Glover, I. & Branine, M. (2001). *Ageism in work and employment*. Farnham, Storbritannien: Ashgate Publishing Ltd.
198 Nilsson, K. (2021). Is working life age discriminatory? Reflections on how the working life of the aging workforce is and can be made more sustainable. In: Jönson, H. (ed.). *Perspectives on ageism*. Lund: Social Work Press, vol. 1, pp. 249–269.
199 Thorsen, S. et al. (2012). The association between psychosocial work environment, attitudes towards older workers (ageism) and planned retirement. *International Archives of Occupational and Environmental Health 85*(4):437–445.
200 Nilsson, K. & Nilsson, E. (2022). Managers' attitudes to different action proposals in the direction to extended working life: a cross-sectional study. *Sustainability* (Switzerland) *14*(4):2182.
201 de los Reyes, P. & Mulinari, D. (2005). *Intersectionality: critical reflections on the landscape of (in)equality*. Malmö: Liber.
202 Hörnstedt, K., Nilsson, K., Albin, M. & Håkansson, C. (2017). Managers' perceptions of older workers and an extended working life in Sweden. *International Journal of Gerontology & Geriatric Research 1*(1):14–20.

# References

203 Hersey, P. (1985). *The situational leader*. 4 uppl. New York: Warner Books.
204 Blanchard, K. & Johnsson, S. (2016). *The new one minute manager*. New York: Harper Collins Publicing.
205 Wetherell, M., Taylor, S. & Yates, S.J. (2001). *Discourse theory and practice*. London: Sage.
206 Winter Jørgensen, M. & Phillips, L. (2006). *Discourse analysis as theory and method*. Lund: Studentlitteratur.
207 Foucault, M. (1971, 1993). *The order of discourse*. Installation lecture at the Collège de France on 2 December 1970. Stockholm: B. Östling's publishing house Symposion.
208 Foucault, M. (2008). *The struggle of discourses*. Selected texts. Stockholm: B. Östling book publisher Symposion.
209 Rydén, L. (2015). *Speak up. Come into its own. Come to terms with. About organizational work environment risks and how to manage, prevent and assess them*. Älvängen: EllErr Konsult.
210 House, J.S. (1981). *Work stress and social support*. Readin: Addison-Wesley Publishing Company.
211 Nilsson, K. & Nilsson, E. (2021). Are my employees able to and do they want to work? The baseline investigation in a follow up study regarding managers' attitudes and measures to increase employees' employability in an extended working life. *Springer Nature* 220:10–16.
212 WHO. (1991). *Sundsvall statement on supportive environments for health*. Third International Conference on Health Promotion, Sundsvall, Sweden, 9–15 June 1991. Geneva: World health organisation.
213 Poland, B., Dooris, M. & Haluza Delay, R. (2011). Securing 'supportive environments' for health in the face of ecosystem collapse: meeting the triple threat with a sociology of creative transformation. *Health Promotion International* 26(2):202–215.
214 Wagemakers, A., Vaandrager, L., Koelen, M.A., Saan, H. & Leeuwis, C. (2010). Community health promotion: a framework to facilitate and evaluate supportive social environments for health. *Evaluation and Program Planning* 33(4):428–435.
215 Concordo Harding, S., Baquero Geronimo, M.A. & Bezbaruah, S. (2018). *Financial security of older women: perspectives from Southeast Asia*. Singapore: International Longevity Centre. https://cose.org.ph/wp-content/uploads/2020/12/Financial-Security-of-Older-Women_Perspectives-from-Southeast-Asia.pdf.
216 Blomé, M., Borell, J., Håkansson, C. & Nilsson, K. (2020). Attitudes toward elderly workers and perceptions of integrated age management practices. *International Journal of Occupational Safety and Ergonomics* 26(1):112–120.
217 Bandura, A. (1995). Exercise of personal and collective efficacy in changing societies. In: Bandura, A. (red.). *Self-efficacy in changing societies*. Cambridge: Cambridge University Press, pp. 1–45.
218 Blauner, R. (1973). *Alienation and freedom*. Chicago: University of Chicago Press, 1973.
219 Hirschman, A.O. (1970). *Exit, voice and loyalty. Responses to decline in firms, organizations, and states*. London: Harvard University Press.
220 Antonovsky, A. (1987). *Unraveling the mystery of health: how people manage stress and stay well*. San Francisco: Jossey-Bass Inc.
221 Rutter, M. (1987). Psychosocial resilience and protective mechanisms. *American Journal of Orthopsychiatry* 57(3):316–333.
222 Åhlin, J.K., Peristera, P., Westerlund, H. & Magnusson Hanson, L.L. (2020). Psychosocial working characteristics before retirement and depressive symptoms across the retirement transition: a longitudinal latent class analysis. *Scandinavian Journal of Work, Environment & Health* 46(5):488–497.
223 Bao, M.Z., Gelfand, M.V., Carmi, I. & Goulas, S. (19 November 2018). A space for women's voices. *Developmental Cell* 47(4):393–394.
224 Raihan, M.M.H., Chowdhury, N., Chowdhury, M.Z.I. & Turin, T.C. (2023). Involuntary delayed retirement and mental health of older adults. *Aging & Mental Health*.

225 Clark, R., Milligan, K. & Newhouse, J. (2023). Changing labor market for older workers: short and long-term trends. *Journal of Pension Economics & Finance 22*(4):459–462. https://doi.org/10.1017/S1474747223000094.
226 Yin, R., Xin, Y., Bhura, M., Wang, Z. & Tang, K. (1 April 2022). Bridge employment and longevity: evidence from a 10-year follow-up cohort study in 0.16 million Chinese. *Journal of Gerontology Series B 77*(4):750–758.
227 McCrae, R.R. & Costa, Jr. P.T. (2004). A contemplated revision of the NEO Five-Factor Inventory. *Personality and Individual Differences 36*(3):587–596.
228 Siegrist, J. (2016). Effort-reward imbalance model. In: Fink, G. (red.). *Stress: concepts, cognition, emotion, and behavior.* Handbook of Stress Series Volume 1, Elsevier Europe, Academic Press, ISBN 9780128009512. pp. 81–86.
229 Heilman, M.E. & Caleo, S. (2018). Gender discrimination in the workplace. In: Colella, A.J. & King, E.B. (red.). *Oxford library of psychology. The Oxford handbook of workplace discrimination.* Oxford: Oxford University Press, pp. 73–88.
230 Aronsson, G. (2017). Borderless work, resilience and compensation. In: Vingård, E. (ed.). *Healthy workplaces for men and women of all ages.* Report 2016:8. Stockholm: The Swedish Work Environment Authority, pp. 18–30.
231 Eppler-Hattab, R., Doron, I. & Meshoulam, I. (2023). The role of organizational ageism, inter-age contact, and organizational values in the formation of workplace age-friendliness: a multilevel cross-organizational study. *Journal of Population Ageing.* https://doi.org/10.1007/s12062-023-09424-7.
232 Soja, E. & Soja, P. (2020). Fostering ICT use by older workers. Lessons from perceptions of barriers. *Journal of Enterprise Information Management 33*(2):407–434.
233 Landstad, B.J., Hedlund, M., Tjulin, Å., Nordenmark, M. & Vinberg, S. (2023). Making things work–in spite of a pandemic small scale enterprise managers' approach to business changes and health issues. *PLoS One 18*(7):e0288837.
234 Robinson, O., Demetre, J.D. & Corney, R. (2010). Personality and retirement: exploring the links between the Big Five personality traits, reasons for retirement and the experience of being retired. *Personality and Individual Differences 48*:792–797.
235 Horn, J.L. & Cattells, R. (1967). Age differences in fluid and crystallized intelligence. *Acta Psychologica 26*(2):107–209.
236 Gardner, H. (2006). *Multiple intelligences: new horizons in theory and practice.* New York: Basic Book.
237 Baltes, P.B. (1993). The ageing mind. Potential and limits. *The Gerontologist 33*(5):580–594.
238 Backes-Gellner, U., Schneider, M.R. & Veen, S. (2011). Effect of workforce age on quantitative and qualitative organizational performance: conceptual framwork and case study evidence. *Organization Studies 32*(8):1103–1121.
239 Salthouse, T. (1996). The processing-speed theory of adult age differences in cognition. *Psychological Review 103*(3):403–428.
240 Salthouse, T. (2000). Aging and measures of processing speed. *Biological Psychology 54*(1–3):35–54.
241 Polanyi, M. (1967). *The tacit dimension.* New York: Anchor Books.
242 Bennett, N., Dunne, E. & Carré, C. (1999). Patterns of core and generic skill provision in higher education. *Higher Education 37*(1):71–93.
243 Warr, P. (1994). Age and employment. In: Triandis, H.C., Dunnette, M. & Hough, L. (red.). *Handbook of industrial and organizational psychology.* 2 uppl. Thousand Oaks: Consulting Psychologist Press, pp. 485–550.
244 McAllister, A. et al. (2020). Inequalities in extending working lives beyond age 60 in Canada, Denmark, Sweden and England—by gender, level of education and health. *PLoS One 15*(8):e0234900.

# References

245 Emery, F.E. & Thorsrud, E. (1976). *Democracy at work. The report of the Norwegian industrial democracy program.* Groningen: Wolters-Noordhoff B.V.
246 Hood, C. (1991). A public management for all seasons? *Public Administration* 69(1):3–19.
247 Christensen, T., Lægreid, P. & Røvik, K.A. (2020). *Organization theory and the public sector. Instrument, culture and myth.* New York: Routledge.
248 Lane, J.-E. (2002). *New public management: an introduction.* London: Routledge.
249 Morgan, G. (1997). *Images of organization.* London: SAGE Publications.
250 Pollitt, C. & Bouckaert, G. (2011). *Public management reform. A comparative analysis – new public management, governance, and the neo-weberian state.* 3 uppl. Oxford: Oxford University Press.
251 Schultz, E.M. & McDonald, M. (2014). What is care coordination? *International Journal of Care Coordination* 17:5–24.
252 Schultz, E.M., McDonald, G., Bachmann, K. & Aarseth, T. (2016). Integrated care pathways – a strategy towards better care coordination in municipalities? A qualitative study. *International Journal of Care Coordination* 19:20–28.
253 Kuijpers, R., Joosten, T. & de Natris, D. (2012). Participation in decision- making when designing care programmes and integrated care pathways. *International Journal of Care Pathways* 16(1):25–30.
254 Jonsson, R., Nilsson, K., Björk, L. & Lindegård, A. (2023). Engaging the missing actor: lessons learned from an age-management intervention targeting line managers and their HR partners. *Journal of Workplace Learning* 35(9):177–196.
255 Rydqvist, L.-G. & Winroth, J. (2004). *Sports, wellness, health & health promotion.* Malmö: SISU Idrottsböcker.
256 Argyris, C. (1991). Teaching smart people how to learn. *Harvard Business Review* 69(3):99–109.
257 Jonsson, R., Lindegård, A., Björk, L. & Nilsson, K. (2020). Organizational hindrances to the retention of older healthcare workers. *Nordic Journal of Working Life Studies* 10(1):41–58.
258 Dewey, J. (1997). *Experience and Education.* New York: Simon & Schuster, pp. 1–91.
259 Jovanovic, B. & Nyarko, Y. (1995). The transfer of human capital. *Journal of Economic Dynamics and Control* 19(5–7):1033–1064.
260 Singh-Manoux, A. et al. (2012). Timing of onset of cognitive decline: results from Whitehall II prospective cohort study. *BMJ* 344:d7622. https://doi.org/10.1136/bmj.d7622.
261 Karlsson, P., Thorvaldsson, V., Skoog I., Gudmundsson, P. & Johansson, B. (2015). Birth cohort differences in fluid cognition in old age: comparisons of trends in levels and change trajectories over 30 years in three population- based samples. *Psychology and Ageing* 30(1):83–94.
262 Persson Waye, K. (2016). *Sensory ageing – hearing and occupational noise.* Stockholm: The Swedish Work Environment Authority, pp. 96–109.
263 Blomqvist, S., Xu, T., Paraskevi, P., Låstad, L. & Magnusson Hansson, L.L. (2020). Associations between cognitive and affective job insecurety and incident purchase of psychotropic drugs: a prospective cohort study of Swedish employees. *Journal of Affective Disorders* 266:215–222.
264 Göbel, C. & Zwick, T. (2010). *Which personnel measures are effective in increasing productivity of older workers?* Centre for European Economic Research nr 10-069.
265 Nilsson, K., Jönsson, S. & Nilsson, E. (2022). *Senior workforce.* Systematic review 2022:3, pp. 1–92. The Swedish Agency for Work Environment Expertise.
266 World Health Organization. (2021). *Global report on ageism.* Geneva: World Health Organization. https://www.un.org/development/desa/dspd/wp-content/uploads/sites/22/2021/03/9789240016866-eng.pdf.

267 Odén, B. (2012). *Older through the ages.* Stockholm: Carlssons bokförlag.
268 World Economic Forum. (2022). *The good work framework: a new business agenda for the future of work.* White Paper May 2022. The World Economic Forum's Centre for the New Economy and Society.
269 UNECE. (2023). Policy brief on ageing. *Older persons in vulnerable situations.* No. 28. United Nations Economic Commission for Europe (UNECE). https://unece.org/sites/default/files/2023-06/ECE-WG.1-42-PB28_1.pdf.
270 Laslett, P. (1991). *A fresh map of life: the emergence of the third age.* Cambridge: Harvard University Press.
271 Previtali, F. & Spedale, S. (2021). Doing age in the workplace: exploring age categorisation in performance appraisal. *Journal of Aging Studies 59*:100981.
272 Kuptsch, C. & Charest, E. (2021). *The future of diversity.* Geneva: International Labour Office (ILO).
273 Macnicol, J. (2006). *Age discrimination: an historical and contemporary analysis.* Cambridge: Cambridge University Press.
274 Oakman, J. & Wells, Y. (2013). Retirement intentions: what is the role of push factors in predicting retirement intentions? *Ageing and Society 33*(6):988–1008.
275 Johansson, B. (2016). Cognitive aging. In: Vingård, E. (ed.). *Healthy workplaces for men and women of all ages.* Report 2016:8. Stockholm: The Swedish Work Environment Authority, pp. 50–63.
276 Schaie, K.W. (2005). What can we learn from longitudinal studies of adult development? *Research in Human Development 2*(3):133–158.
277 Rönnlund, M. & Nilsson, L.G. (2009). Flynn effects on sub-factors of episodic and semantic memory: parallel gains over time and the same set of determining factors. *Neuropsychologica 47*(11):2174–2180.
278 Nilsson, K. (2020). The situation during the COVID-19 pandemic for 7,781 healthcare employees: survey responses to the follow-up study Sustainable working life in healthcare. Occupational and Environmental Medicine Lund University vol. 2020. nr. 14, pp. 1–45.
279 Dahler-Larsen, P. (2013). *The evaluation society.* Stanford: Stanford University Press.
280 Pollitt, C. (2003). *Essential public manager.* Oxford: Oxford University Press.
281 Vedung, E. (2009). *Evaluation in public administration.* Lund: Studentlitteratur.
282 Drummond, M.F., Sculpher, M.J., Claxton, K., Stoddart, G.L. & Torrance, G.W. (2015). *Methods fort the economic evaluation of health care programmes.* 4 uppl. Oxford: Oxford University Press.
283 Svensson, M. (2019). *Health economic evaluation. Method and application.* Lund: Studentlitteratur.
284 Ferraz-Nunes, J. & Karlberg, I. (2012). *Health economics – concepts and applications.* Lund: Studentlitteratur.

# Index

Note: **Bold** page numbers refer to tables and *italic* page numbers refer to figures.

acceptable daily intake (ADI) 34
acknowledgement 60, 99, 107, 113, 127, 133–135, 150, 151, 153, 166, 173–175, 177, 178, 195, 199, 210, 256
action plan 74, 79, 155, 261–265
  sustainable working life 262, 265, **266–278**
active work situation 52
activity balance
  recuperation 119
  stimulation 119
acute crisis 46
adaptability 198
adaptedness 15
adaptive learning 204
adrenaline 41
age 4, 5–7, 14, 15, 18, 20, 21, 23–26, 29, 31–33, 36, 37, 39, 40, 48, 50, 56, 59–64, 72–74, 80–83, 86, 88–92, 95, 97, 98, 100, 102, 103, 105–107, 110–113, 115–118, 120, 122–124, 127, 131–133, 135–144, 146, 148, 154, 157, 159–161, 167–171, 178–180, 182, 183, 187, 190, 191, 193, 197, 198, 203, 207, 210–219, *218,* 236–239, 241, 244, 245–249, 258, 279–282, 285
  attitudes 29, 106, 110, 121, 123, 127, 132, 212, 219, 231, 237, 239
  biological 9, 11, 17, 20, 25–27, 29, 31, 33, 36, 39, 43, 53, 55, 58, 61, 73, 75, 76, 79, 83, 88, 91, 99, 106, 112, 131, 140, 218, 221–224, 226, 236–239, 258
  categorisation 137, 231, 236
  chronological 15, 26, 95, 97, 112, 131, 140, 181, 218, 225–227, 229, 231, 236–239, 258
  classification based on normal values 226
  cognitive 61, 100, 112, 163, 180–182, 187, 194, 197, 207, 210, 211, 213, 218, 233–239, 281
  definitions 127, 132, 140, 148, 159, *218,* 218–219, 226, 229, 236
  different cultures 110, 132
  discourses 228, 237
  elite 159, 160, 228, 231
  limit 33, 61, 116, 225–226, 234, 237–238
  management 73, 105, 132, 197, 265, **267–278**
  non-discrimination laws 139, 159
  organisational culture 132, 237
  segregation 137–139
  social 62, 102, 105, 110, 112, 113, 115–118, 120–124, 131, 133, 140, 142, 143, 148, 159–161, 215, 218, 226, 228–233, 236–239, 280
  strive for youth and renewal 132
  sustainable working life 253–255, 261–265, **266–278,** 281, 283, 284, 286, 287
age-conscious and situational leadership 100, 131–132, 146–149, *147,* 199
  considering age in accordance with the SwAge™ model 131
age discrimination 140
  women 142
ageing in working life 160, 218
  lower status 231
  stigmatisation 231
ageing population 216, 227
ageism 140, 229, 231
age-related changes 29, 192, 222, 233
alienation 48, 49, 166–167, 196
anabolism 22
anti-discrimination laws 139–140, 159
anxiety 11, 24, 39, 43, 45, 46, 48, 52, 53, 55, 58, 66, 77, 116, 126, 133, 136, 167, 174, 247
a posteriori knowledge 187
approaches 66, 69, 81, 129, 149, 157, 188, 189, 206, 213, 219, 280
a priori knowledge 187
areas of responsibility 100, 178, 197, 213, 251
authoritative leader 129
authorities 4, 13, 25, 50, 56, 65, 66, 79, 88, 98, 125, 196, 214, 247, 284
  supervision 88
authority in group hierarchy 125
autonomy 45, 50–53, 67, 69, 70, 102, 131, 146, 147, 152, 166, 167, 204

back pain 31
bad condition 126
bad work 101, 126, 135
balance 14, 20–22, 43, 44, 83, 124, 126, 143, 170, 210, 223, 251, 279, 281
  demands and control in work situation 50–53
  effort-reward balance 44, 174–175
  society affecting 160–161
  work-life balance 70, 71, 106, 119, 259

301

becomings, social ages 230
behavioural conflicts 126
behavioural patterns 124
beings, social ages 230
biological age 9, 11, 17, 20, 25–27, 29, 31, 33, 36, 39, 43, 53, 55, 58, 61, 73, 75, 76, 79, 83, 88, 91, 99, 106, 112, 131, 140, 218, 221–224, 226, 236–239, 258
   risk of occupational injury 33, 36, 79, 223
   work capacity 16
   work environment laws 9, 15, 20, 25–27, 29, 33, 39, 53, 76, 88, 89, 99, 223, 224
   working life 221–224
biological ageing 11, 15, 17, 20, 25–27, 29, 31, 33, 36, 39, 43, 53, 55, 58, 61, 73, 76, 79, 88, 99, 106, 221–224
biological diversity 15, 19
   age and gender 258
biostatistical definition of health 65, 66
bodily functions 15–23, 31, 44, 71, 184, 221, 223, 225
   ageing 15, 31, 221, 223, 225
   the circulatory system 15–16
   muscles 18–20
   the nervous system 21–23
   oxygen uptake capacity 15–17
   the pulmonary system 16, 17
   sensory cells 20
   sensory organs 20–21
   skeleton 15, 17–18
   time for recuperation 15, 19
the Bologna model 188
brain 16, 18, 20–23, 30, 31, 33, 41, 42, 55, 60–62, 78, 136, 137, 141, 145, 172, 180, 181–183, 185, 192, 198, 199, 211, 221, 233–235
breathing issues 16–17, 34
bridge employment 90, 213
bridging 90, 213
the buffering hypothesis 133
building windmills 168–171, 198
bullying 53–54, 116, 126, 136, 152, 153, 170, 235
bureaucratic organisational model 69
   hierarchy and power 69
business administration 97

cardiac muscle cells 19
career 13, 75, 83, 89, 100, 105, 108, 118, 123, 127, 138, 140, 160, 167, 168, 175, 184, 207, 210, 212–214, 230, 231, 241, 245, 252, 281, 282
   change 89
career development conversations 256–260, *257, 257*–259
   age 258
   gender 258
case method 157, 208, 280, 285
   transfer of knowledge 207–210, 282
catabolism 23

categorisation 137, 229, 231, 236
   age discrimination 137
   of groups 127, 137, 141, 225, 228, 230, 236, 237
   homogeneity 127, 137
   stereotypes 113, 127, 137, 229
cell death 26, 181
cell renewal 26, 222, 223
central nervous system (CNS) 21–23, 43, 198, 199
change of career and profession 89
chemoreceptors 20
Christmas holidays, for home 4–5
chronic stress 39, 40, 77, 79
chronological age 15, 26, 95, 97, 112, 131, 140, 181, 218, 225–227, 229, 231, 236–239, 258
   discrimination 140
   economy 97, 225, 238
   inclusion and exclusion 226
   normal value 226
   public pension 226, 227
   shift 236
   working life 225–227
circulatory system 15–16, 19, 21, 22, 27, 31, 221, 222
   organs 23, 71
   physical activity 16–17
clear roles 78
climate 2, 6, 30, 35, 37, 73, 76, 144, 196, 200, 204, 250–252, 254, 259, 279
coaching leader 129, 130
cognition 58, 181–182, 209, 233
   changes 187, 234
   difficulties 180, 191, 233
   functions 58, 60, 63, 181, 194, 210, 211, 233, 234, 238, 281
   impairment 191–192, 234
cognitive ability 40, 100, 181–184, 188, 192, 214, 233, 234
   crystallized 182
   decreased 233, 234
   fluid 182
cognitive age 61, 100, 112, 163, 180–182, 187, 194, 197, 207, 210, 211, 213, 218, 233–239, 281
   ability to execute tasks 163
   accidents related to fatigue 61
   effects on productivity 234, 237
   professional skills 182, 234
   tacit knowledge 207, 234
   working life 233–235
cognitive ageing 181, 192, 194, 233–235, 237, 238
   changes in the brain and nervous system 61, 181, 192, 210, 213, 233–235
   cognitive impairment 192, 234
   fatigue syndrome 235
   functional ability 192
   health 112
   stress 233, 235

# Index

comfort zone 35, 171
commanding leader 129
communication 13, 21, 22, 40, 50, 54, 60, 74, 90, 95, 121, 124, 129, 133, 144, 146, 149–152, 166, 189, 191, 201, 202, 205, 206, 251, 256, 263, 264
competence 2, 3, 7, 39, 44, 49, 70, 72, 81, 89, 100, 102, 104, 105, 143, 146, *147,* 150, 154, 159, 163–165, 167, 172, 175, 178, 180, 183, 184, 188–194, 198, 203, 204, 209–214, 231, 233, 239, 245, 246, 252, 255, 257–261, 281, 282
competence development 2, 7, 39, 89, 100, 105–107, 109, 123, 127, 133, 140, 159, 163, *164,* 165, 175, 180–194, 200, 206, 207–210, 212–214, 231, 233, 234, 237, 239, 245, 252, 257, 258, 260, 280–282, 285, 287
    discrimination 105, 140, 159, 231
    goals 257, 260
    recuperation 2, 133, 165, 191
    seniors 140, 187, 191, 231, 233, 234, 282
competitiveness 69, 72, 225, 246
    senior and ill individuals 191
comprehensibility 44, 52, 170–172, 198
the concept of health 11, 23
    classified health 11
    emotional health 12
    financial health 12
    health in relation to time and pace 12
    intellectual health 12
    mental health 11
    physical health 11
    social health 12
    spiritual health 12
conditioned reflex 189
conflict phase 125–126
conflicts 44, 54, 78, 79, 106, 124–126, 130, 144, 148, 152, 154–156, 158, 160, 169, 190, 205, 227
    the determinant areas of the SwAge™ model 155
    exercise 155
    management 154–156
conflicts of interest 126
contact-based professions 55, 56, 89, 91, 203
    recruitment 91
    reduced working hours 91
context 35, 48, 67, 69, 117, 119, 121–133, 140, 149, 151, 161, 167, 170, 171, 186, 188, 199–201, 207, 228, 229, 231, 233, 258
control groups 66, 126, 129, 195
control of one's situation 167, 218
control of one's work situation 80, 150
co-operation 78, 143, 144, 152, 154, 159, 200, 201, 264
coping 40, 47–51, 54, 56, 60, 78, 80, 85, 91, 157, 161, 171, 183, 198, 199

gender 48
    with stress 44–53, 78–79, 133
coping phase, crisis 46
coping strategy 44, 47–51, 54, 56, 60, 78, 198, 199
    frame of reference 49, 50, 199
cost-benefit analysis 101, 286, 287
cost-benefit principle 101
cost-effectiveness analysis 286, 287
cost efficiency 196, 286, 287
cost minimisation analysis 286, 287
cost-utility analysis 286, 287
creativity 12, 30, 63, 124, 129, 130, 150, 154, 171, 173, 178, 182, 190, 197, 204, 205, 211, 213, 214
crisis 44, 45, 54, 78, 107, 129
    acute crisis 46
    coping phase 46
    development 45–47
    orientation towards the future 46–47
    reaction phase 46
    reactions 45, 47, 78
cyber bullying 54

degeneration 224, 234, 238
demand and control 2, 73, 131
Demand-Control Model *51,* 51–52
democratic leader 129, 130, 147
demographic old-age to working-age ratio 216, **217**
the demographic situation 215–218, **217**
    average life expectancy 215
    living conditions 69
    older individuals 221
determinant areas 1–7, *12,* 39, 43, 49, 56, 67, 71, 73, 75, 81–83, 86, 88, *96,* 97, 108, *114,* 132, 133, 142, 144, 148, 149, 155, 156, 158, 160, 165, 199, 211, 237, 241, 253, 256, *257,* 259, 261, 263, 265, 283, 285, 286
    sustainable working life 5–7, *6*
    the SwAgeTM model *10,* 39, 43, 49, 56, 73, 75, 81–83, 86, 88, 96, 108, *114,* 133, 142, 148, 155, 156, 158, 160, *164,* 165, 199, 211, 237, 241, 253, 256, *257,* 261, *262,* 263, 265, 283, 285
    work ability 1–3, *2*
deutero-learning 206
development 2, 7, 13, 14, 21, 27, 39, 44–48, 54, 61, 69, 71, 72, 75, 78, 83, 86, 89–91, 93, 99, 100, 102, 105–109, 112, 122, 123, 125, *125,* 127, 129, 132, 133, 136, 140, 144–146, *147,* 152–154, 159, 161, 163, *164,* 165–167, 173, 175, 180–195, 198, 200, 202, 204–216, 219, 221, 224, 231, 233, 234, 237, 239, 241, 245, 248, 252, 256–260, 280–283, 285, 287
developmental learning 204
diagnosis 24, 27, 47, 65, 66, 168, 261

dialogue 40, 50, 83, 121, 155, 200, 208, 209, 253, 256, 257, *257*
   tool for career development conversations 257
diet and physical activity 67, 87, 223
digitisation 133, 180, 190, 205
directives 82, 87–88, 93, 111, 112, 139, 140, 151, 159, 229, 230, 279
   decreased risk of occupational injuries and ill health 88
   preventive work environment management 73, 87, 111
   work environment management 73
disability 24, 27, 59, 65, 139, 159
discourses 66–68, 70, 85, 86, 121, 123, 157, 228, 237
   social work environment 122, 149, 150
discursive resource 149
   social age 121, 122, 128
discrimination 3, 7, 102, 105, 106, *114*, 121, 136–142, 149–153, 157–160, 170, 205, 219, 231, 248, 251, 259, 280
   age 140, 142
   direct 139
   grounds of 139, 140, 142, 159, 160, 228, 231
   indirect 139
   non-discrimination laws 139, 159
   working life 159–160
disease 13, 14, 23, 24, 32, 34, 40, 43, 51, 61, 65, 85, 87, 89, 107, 111, 174, 192, 222, 227, 233, 236
   objectification 66
disregard 3, 7, 24, 78, 100, 102, 105, 106, 136–142, 149–153, 157, 159–160, 172, 205, 219, 226, 229, 251, 259, 280
dissatisfaction 52, 53, 107, 126, 195, 197
diversified working life 42
diversity 1, 3, 6, 7, 9, *10*, 11–28, 32, 36, 67, 74, 75, 77, 83, 85, 86, 89, 93, 99, 105, 110, 116, 131, 133, 137, 154, 159, 160, 171, 173, 198, 219, 223, 236, 238, 250, 259, 261, 263, 265, 279
division of power 149
division of responsibility 74, 159, 265
division of work 70, 90, 152, 166, 167, 196, 231
   personal life 70
double-loop learning 206, 207
driving force 167, 188, 195
dual continuum model of health 23–24, *24,* 28

economy 86, 97–103, 107, 110–112, 122, 176, 198, 216, 218, 219, 224–226, 232, 238, 283, 285
   evaluations 283, 285–288, **289–290**
   occupational injuries 111
   retirement 112, 216, 218, 232
   social security systems 111–112, 232
   sustainability 216

education 36, 44, 53, 55, 59, 79, 98–101, 105, 110, 112, 118, 122, 123, 159, 160, 186, 188–193, 201, 205, 207–210, 212–214, 218, 224, 233, 238, 247, 265, 284
effort-reward-imbalance (ERI) 174, 175
efforts 13, 22, 24, 44, 51, 52, 65, 66, 78, 83, 90, 99, 109, 112, 118, 119, 127, 134, 142, 144, 151, 166, 167, 169, 170, 174, 175, 177, 186, 190, 195–197, 199, 201, 202, 210–212, 230, 251, 279–281, 283
e-learning organisational culture 206
electroreceptors 20
emotional support 130, 132, 151
empathetic laissez-faire leader 129, 130
empathy 131, 151, 155–156
empiric inductive knowledge 187
employability 1–7, 9, 11, 71, 84, 92, 95, 98–102, 104–113, 115, 123, 131, 141–143, 149, 156, 157, 161, 163, 165, 172, 175, 179, 180, 182, 183, 190–194, 197–200, 203, 206, 207, 209–211, 214, 218, 221, 228, 233, 237, 241, 251, 253–255, 280, 281, 285, 287
   age-conscious and situational leadership 100, 131
   age definitions 218, *218*
   attitudes 99, 100, 106–107
   biological ageing 11, 99, 106, 218, 221
   changes in working life 197–199
   cognitive age 163, 180, 194, 197, 218, 233
   competence 100, 102, 106, 107, 109, 123, 143, 163, 172, 175, 180, 182, 183, 190, 192, 193, 200, 203, 209, 233, 280, 285, 287
   competence development 100, 105–107, 109, 123, 180, 200, 280, 285, 287
   desired competencies 105
   discrimination 102, 105, 106
   education 100, 123, 190, 193, 207
   emotional intelligence 100, 183
   health effects of the work environment 9, 92, 98, 99, 123, 251
   human capital 100
   ill-health 11, 99, 108, 110, 111, 143, 149
   knowledge 99, 100, 102, 105, 107, 109, 112, 163, 180, 182, 190, 209, 233, 280
   leadership 100, 105, 131
   learning 106, 180, 190–192, 200, 203, 207, 233
   long-term sickness absence 141
   norms and attitudes 106
   personal development planning 207
   personal social environment 102, 115, 123, 143, 157
   premature retirement 105
   reorganisations 214, 280
   resources 99
   social support 99–100, 102, 106, 131, 251

# Index

societal support 110–111
technological aids 106, 107
unemployment 101, 141, 280
work ability 99, 100, 253
work life balance 197–199
workplace culture 106
employeeship 144–146, 200–202
   co-operation and teamwork 200
   evaluate 202
   goals 201–202
   implementation phase 202
   leadership 144–146
   put the plan into action 202
   reflection on needs and development 200
   reorient and continue development 202
employer 4, 13, 30, 37, 56, 65, 70, 71, 75, 76, 79–81, 87, 88, 91–93, 98, 104, 105, 107, 108, 112, 131, 132, 155, 159–161, 189, 194, 199, 203, 207, 213, 214, 224, 226, 247, 256–260, 264, 265
employment development 112
   recession 284
   urbanisation 112
   women's participation in working life 86, 127
employment rate 216
empowerment 67, 68, 70, 71, 170, 171, 175–178, 196, 199, 202
   healthy choices 68, 71
enduring organisations 70
engagement 88–90, 135, 146, 147, *147,* 150, 167, 169, 171, 175, 195
entering the labour market 216
   age 216
   education 100, 191
   social welfare 105
environment 4–7, 9–10, *10,* 11, 15, 18, 20–22, 24–60, 63, 65–93, 98, 99, 101, 102, 104–107, 111, 113–120, 143, 145, 147–153, *153,* 154–161, 165, 170, 171, 177, 178, 181, 182, 184, 190, 192, 195, 198, 200, 203, 205, 206, 209, 218, 219, 222–226, 233, 237, 239, 241, 244, 247, 248, 250–256, 259, 261–265, 279–281, 285, 287
   organisation 250–252
   social work environment 3, 7, 39, 51, 53, 55, 102, 113, 114, 116, 121–143, 147–152, 154–158, 160, 161, 165, 170, 171, 237, 239, 251, 259, 261, 280
   supportive 28, 86, 87, 107, 111, *153,* 158–159
episodic memory 181, 184, 186–187, 213
episteme 183
equality 139, 152, 157, 159, 231, 285
equilibrium model of health 14
equilibrium theory 15
ergonomic improvements that contribute to new problems 76

ethical stress 83, 90, 279
ethnic origin, segregation 139
exclusion 7, 44, 53, 54, 102, 113, *114,* 121, 122, 134–136, *137,* 138, 139, 141, 152, 158–160, 170, 205, 226, 230
execution of work tasks 100, *164,* 194–196, 211–214
exercising power 70, 71
explicit long-term memory 186

family situation 2, 36
farmer 50, 51, 121
fatal and non-fatal occupational injuries 24–25, *26,* 104
   senior and younger individuals 25
fatality 25, *26,* 35, 50, 104
   work-related 25
fatigue syndrome 40, 43, 55, 168, 235
feedback 21, 51, 100, 133, 151, 153, 156, 175, 199, 204, 205, 207, 251, 256, 258
female-dominated 77, 91, 137, 138
fight/flight response 21, 185
financial compensation systems 95–96, *96,* 99, 101, 110
   choosing to retire financial consequences 95
   policy instruments in the societal economy 176
financial difficulties 101
financial incentives 2, 3, 5, 6, 81, *96,* 97–99, 101–102, 110, 112, 118, 132, 171, 237, 251, 259, 261, 280
financial security systems 101
the five factor model of personality 173
   agreeableness 174
   conscientiousness 173–174
   extraversion 174
   neuroticism 174
   openness 173
formal groups 124, 135, 136, 152, 177, 195
formal organisation 126, 135
frame of reference 49, 50, 182, 199
friendship 124, 133, 151
functional age 75, 112, 236
functional diversity 1, 3, 6, 7, 9, *10,* 11–28, 36, 67, 74, 75, 77, 83, 85, 86, 89, 93, 99, 116, 131, 133, 137, 159, 171, 198, 236, 238, 250, 259, 261, 263, 265, 279
   segregation 137, 139

gender 15, 18, 25, 31, 37, 48, 63, 77, 121, 122, 124, 137–142, 145, 154, 159, 160, 228, 258, 265
   differences 31, 37, 55–56
   identity/gender expression 137, 139
   segregation 37, 138
gender-dominated professions 138
gender segregation 138
   working life 37
general adaption syndrome (GAS) 40

generational change 81
generic abilities 189
generic knowledge 191, 207–209, 212
goals 14, 15, 45, 52, 53, 58, 74, 78, 85, 100, 104, 116, 123, 124, 126, 127, 129–131, 144, 146, 148, 149, 157, 167, 172–174, 183, 184, 190, 196, 197, 199–202, 204, 205, 208, 210, 212, 225, 226, 256–258, 260, 264, 265, 281
group dynamics 123–126, *125,* 142, 195
   categorisation 127, 137, 141, 225, 228, 230, 231, 236, 237
   conflicts 44, 124, 125–126, 154
   disturbances in the work flow 126
   empathic ability 124
   ill health 67, 126
   loyalty 124
   norms 122, 124, 125, 140, 141, 152, 158, 160, 195, 219, 228, 230, 248
   phases 124–126, *125*
   re-organisations 219, 284
group norms 141, 158
group processes 152, 205

harassment 54, 139, 152, 154, 159, 160
has beens, social ages 229–231
Hawthorn effect 69
health 1–3, 5–7, 9–10, *10,* 11–31, 33–40, 42–53, 55–60, 63–93, 95, 98, 99, 101, 102, 104–108, 110–112, 118–123, 129, 131, 133, 134, 136, 138, 143, 149–151, 153, 154, 158, 159, 165, 170–173, 175, 178, 191, 192, 198, 212, 215, 216, 218, 219, 221, 223, 224, 237, 239, 248–251, 253, 259, 261, 279, 280, 286, 287
   biomedical definition 65, 66
   biostatistical definition 65, 66
   definition levels 13
   dimensions 11–13, *12*
   dual continuum model 23–24, *24*
   empowerment 67, 68, 71, 175
   factors 65, 153, 250
   functional diversity and illness 250
   the individual perspective 14
   individual responsibility 67
   level 14–15
   macro level 13, 85–86
   mental work environment 3, 5, 7, 9, 10, *10,* 11, 30, 39–40, 42, 50, 51, 53, 56, 57, 73, 77, 84, 86, 90, 93, 111, 237, 239
   meso level 13, 65–66
   micro level 13, 219
   motivation 52, 67, 105, 172, 173
   physical work environment 3, 5–7, 10, *10,* 30, 38, 50, 71, 75, 76, 84, 86, 93, 237, 239, 250

recuperation 3, 5, 7, 9, 10, *10,* 58, 60, 63–65, 73, 79, 84, 86, 92, 107, 237, 239
societal norms 160
subjective experience 14, 23
supportive environments 28, 86, 87, 158, 159
working life 27
health effects of the work environment 3, 5, 9–10, *10,* 30, 65–86, 92, 93, 98, 99, 105, 123, 237, 251, 261, 279–280
   action proposals, the societal level 85–93
   for employability 3, 9, 92, 93, 98, 105, 251
   laws and directives 93
   lifestyle diseases 85, 86
   the societal level 85–93
   the SwAge™ model 9, *10,* 86, 92, 98, 237, 261, 279
health prevention 66–69, 71–72, 74, 75, 84, 87, 88, 108
   person-centred 67
   process 71–72
   rehabilitation 67, 108
health promotion 25, *26,* 39, 51, 66–72, 74, 75, 81, 82, 84–87, 92, 93, 111, 154, 158, 159
   employer's control of lifestyle and habits 70
   holistic approach 68
   *The Ottawa Charter for Health Promotion* 85
   productivity 39, 69–71
   societal level 13, 86
   WHO 66, 67, 85, 86, 158
   work environment 39, 51, 71, 72, 75, 82, 84, 93, 111
   workplaces 39, 51, 68–70, 72, 75, 81, 84, 87, 111
healthy and sustainable workplaces 74, 160, 218
healthy choices 67, 71, 86–87, 93
healthy work environment 32, 65, 93
healthy worker effect 26, 36, 123
hearing loss 33, 222
heart 15–19, 31, 41, 42, 46, 222
heavy lifting 29, 30–32, 76, 243, 254
hierarchical memory models 186
hierarchical organisation models 182, 191
hierarchy 69, 125, 135, 137, 141, 149, 172, 173, 177, 189, 191, 210–211, 281
hierarchy of needs 172–173
   situational factors 173
hippocampus 21, 23, 43, 185
homeostasis 11, 12, 21, 22, 62, 63, 119
homogenous groups 127, 137
honeymoon phase 124
horizontal segregation 137
hormones 21, 22, 41, 43, 62, 71, 185, 192
human beings and technology 36
human relations (HR) 69, 90, 160, 195, 279
the human relations school 195

# Index

human resource management (HRM) 69
hypothalamus 22

ice berg metaphor 25, *26*
identity 45, 71, 115, 121–132, 136–139, 149, 177–178, 197, 228, 229
   own identity 100
ill-health 2, 11, *12,* 13, 14, 22, 24–27, 32, 34, 35, 39, 40, 46, 50, 51, 53, 55, 60, 61, 65–68, *68,* 70–72, 75, 81, 82, 84, 85, 91, 98, 99, 102, 105, 106, 108, 110, 111, 117, 126, 129, 136, 143, 149, 154, 161, 166, 168, 175, 195, 198, 218, 221, 224, 250, 256, 265, 280
   disease-centred approach 66
   person-centred approach 67
   preventive measures 68, 108
   working life 27
   work-related 24–26, 35, 39, 40, 53, 55, 61, 82, 221, 256
ill mental health 46, 136
   stress-related 55
illness 11, 13–15, 20–25, 27, 29, 33, 34, 36, 38, 40, 41, 43–45, 50, 51, 55–68, 70–73, 76, 77, 81, 86–88, 93, 98, 99, 101, 102, 104, 105, 107, 108, 110, 116, 142, 153, 158, 169–171, 173, 175, 181, 182, 187, 191, 192, 198, 199, 203, 213, 215, 221–224, 226, 227, 229, 233–236, 238, 248, 250, 261, 264
immaterial resources 99
impact evaluations 283, 285
implicit long-term memory 186
incident 25, *26,* 104, 285
including atmosphere 153, 205
inclusion in groups 3, 113, 137, *137,* 145, 152, 158, 177
incremental cost-effectiveness ratio (ICER) 287, 288, *288*
indirect discrimination 139
inequality 159, 216, 228
   age 159, 228
influence 1–5, 9, 15, 30, 31, 39, 48, 50, 55, 66–68, 78, 85, *96,* 102, 106, 110, 112, 113, *114,* 115–119, 121, 133, *134,* 135, 152, 166, 167, 170, 176, 177, 196, 198–201, 206, 211, 214, 238, *257*
the influence of society 4, 68, 110, 112, 176, 211, 214, 238
informal 24, 54, 71, 77, 124, 128, 131, 134–136, 152, 157, 158, 177, 195, 249, 251, 259, 281
   groups 124, 135, 136, 152, 177, 195
   leaders 24, 54, 131, 152, 157, 158, 259, 281
   organisations 126, 134–136
   support 134, 158

information processing 182, 183, 187
initiatory phase 124
instrumental support 69, 132–134, 142, 151, 202
insufficient accessibility 139
intake (PTWI) 34
integration phase 124, 125
intelligence 47, 100, 180, 182, 183–184
   types of 183–184
international economics 97, 112
intersectionality 141, 142, 228, 229
interviews 73, 156, 169, 200, 229, 234, 263, 284

job lock 143

key competence 189
key figures 195
knowledge 2, 3, 7, 12, 21, 32, 35, 40, 44, 47–49, 55, 67, 68, 77, 79, 82, 87–88, 93, 99, 100, 102, 105–107, 109, 111, 112, 124, 125, 127, 128, 130, 132, 133, 135, 136, 140, 145, 147, 148, 150, 154, 157, 160, 161, 163, 164, *164,* 165–168, 170, 173, 176–178, 180–195, 198, 201, 202, 204, 206–214, 229, 230, 233, 234, 237–239, 245, 247, 250, 252, 255, 259, 261, 265, 280–282, 284
   development 99, 190, 214
   gap 206
   promotion 67, 68, 82, 87, 99, 105, 106, 183, 186, 209, 211–213, 250, 252, 280–282
   transfer between generations 210

labour division 32, 69, 90, 194
labour force 29, 34, 36, 59, 80, 86, 89, 93, 97, 99, 105, 110, 111, 115, 117–119, 136, 159, 178, 211, 216, 218, 225, 226, 229, 233, 237, 238
   participation 80, 89, 93, 117–118, 136, 178, 225, 226, 229, 237
   younger individuals 216, 229
labour market 37, 90, 100, 105, 118, 140, 143, 172, 191, 213, 216
   less prioritisation of seniors 105
   promote employability 105, 111
   working until an older age 11, 26, 32, 72, 80, 82, 102, 123, 193, 198, 212, 218, 234, 246
lack of boundaries between work and personal life 54, 70, 116–117
leader 24, 54, 106, 129–132, 141, 145–148, 152, 157–158, 160, 176, 199, 205, 209, 243, 249, 259, 281
leadership 3, 7, 32, 70, 100, 102, 105–106, *114,* 129–133, 142, 144–149, 152, 157, 160, 161, 170, 195, 199, 201, 205, 219, 237, 243, 251, 259, 280

age conscious 131–132, 219
  bad 131, 132
  employeeship 144–146
  insufficient support 132
  productivity 105, 195
  situational and age-conscious 146–149, 160, 280
leadership styles 129–131, 142, 146–148, 152
  authoritative 129, 147, 148
  coaching 129, 130, 147
  delegative 147
  democratic 129, 130, 147
  empathetic 129, 131
  pacesetting 129
  supportive 130, 146–148
  visionary 129, 130
lean production system 195, 203
learned helplessness 51, 167, 170, 171, 175, 176
learned reflexes 22, 41, 47, 186
the learning organisation 200, 202, 204–207, 210
  adaptive learning 204
  developmental learning 204
  manager 204, 205, 207
learnings 35, 45, 58, 78, 105, 106, 132, 145, 146, 152, 167, 169, 180–183, 186–188, 190–192, 200, 202, 203–207, 209, 210, 212–214, 219, 233, 234, 248, 282, 285
  difficulties 191, 207, 248
  by doing 188, 209
  in the organisation 146, 204, 206, 285
  process 206–207
  through self-reflection 206
legislation against age discrimination 140
leisure activities 81, 95, 114, *114,* 115–119, 157, 160, 237, 247
leisure time 7, 34, 53, 54, 60, 87, 90, 115, 116, 118, 119, 135, 192, 206, 210, 247, 281, 282
life-long learning 105, 190, 191, 213, 214
life phases 89, 91, 92, 122, 123, 228–230
life satisfaction 134, 172, 178, 229
life span 26, 221, 228
lifestyle 2, 12, 13, 25, 27, 32, 34, 63, 67, 68, 70, 71, 85–87, 92, 107, 181, 187, 221–224, 233, 236
  choices 32, 67, 70, 71, 86, 87, 223
  diseases 32, 87, 222, 236
lighting 30, 79, 223
limited work tasks 90
living conditions 69
locus of control 44, 48–50, 78, 170, 171, 198, 199
logical thinking (logos) 183
long-term memory 182, 184–187, 207, 209
long term sickness absence 15, 40, 51, 53, 89, 141, 168, 213, 250, 285
  younger ages 40
loyalty 106, 107, 124, 167, 202
  culture 107

The Luxembourg Declaration on Workplace Health Promotion 87

macroeconomics 97
macro level 13, 85–86, 203, 211, 219, 238
making a living 69, 98, 99, 227, 238
maladies 14
manageability 44, 49, 50, 52, 170–172, 198
management 21, 35, 38, 54, 65, 67, 69–75, 79, 82, 93, 97, 99, 104, 105, 112, 114, 126, 127, 129, 132, 133, 135, 142, 143, 148, 153–157, 161, 166, 175, 194, 195, 197, 201, 203–207, 214, 241, 246–248, 255, 261–265, **266–278,** 279
marginalisation 232
Maslow's hierarchy of needs 172–173
mastering situations 204
master suppression techniques 127–129, 142, 152
material resources 99
maximum expected utility 101
meaningfulness 118, 163–172, 174, 177–179, 194–199, 237
measures 4, 7, 13, 15, 28, 30, 36, 40, 41, 60, 63, 67, *68,* 69–77, 79–86, 91–93, 98, 105, 106, 108, 111, 112, 121, 127, 131, 132, 134, 142, 145, 146, 148, 156, 157, 159–161, 166, 172, 196, 198, 201–203, 205, 207, 209–211, 213, 214, 216, 218, 237, 246, 250, 253, 261, 262, 264, 265, 279–288, **289–290**
  evaluations 63, 81, 202, 283–288, **289–290**
measuring life quality 287
mechanoreceptors 20
member groups 124
memory 12, 21, 23, 34, 42, 43, 46–48, 58, 89, 136, 141, 180–182, 184–187, 194, 207, 209, 210, 213, 233, 234
  cognitive age-related changes 233
  episodic 181, 184, 186, 187, 213
  issues 89
  long-term 182, 184, 185–187, 207, 209
  semantic 182, 184, 186, 187
  sensory 184–185
  visual and motor short-term 185
mental fatigue 44, 60, 61, 91, 192, 221
mental health 11, 36, 39, 40, 46, 53, 55, 56, 77, 88, 89, 91, 111, 118, 119, 121–123, 134, 136, 178, 191
  senior individuals 40
mentally demanding work environment 32, 53, 60, 77, 88, 223
  contact-based professions 55, 56, 89, 91, 203
mental work environment 2, 3, 5–7, 9, 10, *10,* 11, 22, 30, 39–58, 71, 73, 74, 84, 86, 90, 93, 99, 102, 111, 133, 136, 165, 171, 198, 203, 237, 239, 250, 254, 259, 261
  alienation 48
  bullying 53–54, 136

# Index

chronic stress 39, 40, 77
coping with stress 44–53
design of 77–79
fatigue syndrome 40, 43, 55
high work demands 250
ill health 11, 39
informal conversations 251
lost functional ability 44
organisation 39, 48, 49, 51, 52, 54–56
recuperation 2, 3, 5–7, 9, 10, *10,* 11, 58, 71, 73, 84, 86, 93, 99, 102, 133, 165, 237, 239, 250, 261
risk assessment 56
stress 2, 6, 22, 39–44, *43,* 44–53, 55, 73, 77, 198, 250
stressors related working life *43,* 43–44
threats and violence 2, 6, 54–55, 73, 250
work-related sick leave 77
meso level 13, 65–66, 203, 211, 219, 237–238
microeconomics 97, 101
micro level 13, 159, 203, 211, 219, 236–237
motivation 3, 7, 44, 47, 48, 50, 52, 67, 70, 95, 100–102, 105, 108, 112, 124, 130, 132–135, 138, 151, 154, 163, 164, *164,* 165–179, 188, 190, 194, 196, 197, 199–201, 203, 209–211, 214, 230, 233, 237, 246, 251, 259, 261, 281
motor functions 181
muscles 15–21, 23, 26, 27, 30–32, 36, 37, 41, 42, 63, 71, 75, 78, 88, 185, 189, 222, 223, 226, 246
  recuperation 19, 63, 78
  strength 15, 18, 20, 31, 32, 37, 88, 222, 226
  work 19, 30–32, 222
  workload 30
musculoskeletal problems 29, 80

nature *vs.* nurture 86
neophiles 132, 231
nervous system 20, 21–23, 30, 40–44, 75, 78, 79, 180, 192, 198
  autonomic 21
  functions 21–23
  somatic 21, 198
neurons 21, 41, 181, 186
new public management (NPM) 195, 203
night shift work 59
noise 30, 33, 61, 79, 250
  exposure 39
normality 45
norm-breaking 141, 155
norms 4, 13, 67, 70, 85, 106, 107, 110, 122, 124, 125, 129, 134, 136, 140–142, 144, 150, 152, 155, 158–160, 175, 182, 195, 206, 209, 211, 213, 219, 228–231, 236–238, 248, 284
  adapting to 106
  breaking 129, 140–142, 150, 155, 158

criticism 229
  role conflict 106
nudging 86, 175–177

objectification 66, 71
observations 40, 68, 73, 200, 205, 214, 263, 283, 284
occupational accidents 24, 25, 35, 36, 38, 56–61, 64, 72, 73, 79, 87, 102, 110, 142
occupational exposure limit values 33, 34, 75, 76, 82, 83, 88
  International Labour Organisation (ILO) 34
  risk groups 34
occupational health care 72, 81–82, 84, 92, 250, 279, 280
  determinant areas of the SwAge™ model 72, 81, 92, 279
  measures of promotion and prevention 81, 280
  systematic work environment management 279
  work environment laws 84, 92, 250, 279
occupational identity 138, 177–178
occupational illness 36, 38, 56, 57, 60, 64, 72, 73, 76, 86, 87, 98, 102, 110, 142
  organisational causes 72–73, 98
  toxic substances 33, 34
occupational injury 18, 25, *26,* 29, 33–39, 56, 57, 64, 65, 71, 72, 76, 79, 83, 87–89, 98, 102–106, 111, 142, 223, 265, 279
  financial consequences 35, 104–105
  insurance 103
  risk of 25, *26,* 29, 33, 35–38, 57, 64, 65, 72, 76, 79, 83, 87, 89, 98, 104–105, 142, 223, 265, 279
  safety 35–37
  sickness benefit 107
occupational pension 80
occupational role 7, 106, 121, 177, 195
old age 172, 216, 227, 228, 236, 245, 249
on-call time 59, 200
open plan offices 30
organisation
  climate 144
  coordination 204
  culture 83, 107, 132, 204–206, 237, 279, 281
  instability 144
  models 69, 191, 195, 198, 203
  of work 3, 34, 39, 71, 77, 79–81, 90, 119, 126, 149, 166–168, 175, 194, 203–204, 237
organisational structure 213
  flat 213
organisational system 11, 13, 65, 75, 106, 110, 191, 195, 198, 225, 226, 251, 252, 284
  informal 71, 124, 126, 134–136, 157
Ottawa charter for health 85
own responsibility 68, 148
oxygen uptake capacity 15–17, 27, 31, 32, 34, 222
  fatigue 79–80
  static workload 19, 31, 32, 222

pacesetting leader 129
parenthood 159
participation 7, 39, 50, 53, 67, 80, 89, 93, 113, *114,* 115, 117–118, 132, 135, 136, *137,* 141, 148, 149, 158–160, 165, 167, 178, 196, 200, 225, 226, 229, 237, 256
passive work situation 51, 52
pauses 58–62, 64, 78, 92, 107, 128, 151, 157, 280
    working time regulations 59, 91
pension 27, 36, 80, 81, 97, 98, 101, 103, 212, 216, 225–227, 232
pension system 110–112, 216, 226, 227, 238, 284
    chronological age 97, 112, 225, 226, 238
    pension 27, 36, 80, 81, 97, 98, 101, 103, 212, 216, 225–227, 232
    retirement age 112, 216
perception 33, 45, 50, 65–67, 71, 82, 85, 99, 100, 104, 112, 122, 123, 125, 127, 136, 137, 140, 145, 160, 170, 180–182, 184, 186–188, 195, 208, 209, 219, 232, 258, 284
    age-related differences 222
    cell death 181
    cognitive age 112, 181
performance-based self-esteem 174, 175
performance levels 74, 149, 190
personal finance 3, 6, 7, 44, 95, *96,* 97–103, 110, 112, 118, 123, 133, 148, 165, 236, 237, 251, 261
personality traits 48, 165, 174
    situational 48, 165, 174
personal social environment 3, 7, 102, 113, 114, *114,* 115–120, 123, 132, 143, 148, 150, 157, 158, 160, 161, 165, 170, 237, 239, 251, 259, 261, 281
    activities 114, *114,* 115, 116, 118, 237, 281
    recuperation 115, 165
    senior employees 115, 148
photoreceptors 20
physical ability 15, 27, 32, 88, 223, 233
physical activity 16–17, 19, 21, 22, 51, 63, 64, 66, 67, 78, 82, 83, 87, 91, 118
    circulatory system 16–17
physical overload, measures 36
physical work demands 20, 26, 37, 131, 222–224
    increasing age 20, 37, 222, 223
physical work environment 2, 3, 5–7, 10, *10,* 15, 18–21, 25, 27, 29–38, 50, 71, 73, 80, 84, 86, 88, 93, 126, 133, 165, 195, 237, 239, 250, 254, 259, 261
    in the beauty industry 34
    chemical health risks 33–35, 250, 259
    climate 2, 6, 30, 35, 73, 76, 250, 254, 259
    COVID-19 pandemic 76
    design of 75–77
    health issues 29, 30
    physically demanding workloads 26, 29–32, 75, 88

    risks 6, 21, 30, 35, 38, 71, 76, 80, 250
    safety and risk of occupational injuries 35
    toxic substances 2
physiological degeneration
    hearing 222, 223
    reaction ability 223
    sight 222, 223
population policy 216
positive social interactions 135, 167
post-bureaucratic enduring organisation 70, 71
    health promotion as a policy instrument 70
potential for improvement 265
power 4, 49, 54, 65, 69–71, 85, 121, 125–129, 140, 141, 149, 152, 155, 158, 159, 175, 188, 189, 216, 225, 228–230, 237
    forms 127–128
    master suppression techniques 128
    over younger people 159, 189, 216, 228, 230
power structure 228, 229
prejudice, managers 140
premature retirement 53, 98, 101, 105, 136, 140
    decreased possibilities for sickness benefits 101
preventing ill-health 11, *12,* 32, 35, 65, 68, 72, 81, 82, 84, 91, 98, 108, 250, 280
    costs for rehabilitation 98
    financial profits 111
prevention *26,* 37, 66–69, *68,* 69, 71, 74, 75, 84, 87, 88, 108, 111, 264, 280
    primary 67, *68*
prevention process 71–72
    make clear for everyone involved 72, 155
preventive health measures 36, 67, *68,* 69, 72, 74, 108, 111, 280
preventive work environment management 35, 71, 75
    increasing age 35
primary group 124
the principle of cost-effectiveness 286, 287, *288*
the principle of human dignity 288
the principle of need and solidarity 288
problem-based learning (PBL) 209
problem solving 48
processes in the brain 181, 199
process evaluations 283–285
production assets 211, 212
productivity 11, 29, 30, 63, 69–71, 73, 81, 92, 98, 104–106, 108, 110, 118, 120, 130, 134, 135, 154, 157, 166, 167, 194–197, 210, 212, 225, 226, 230, 231, 234, 237, 264, 281, 285
    resources 99
professional skills 100, 168, 170, 182, 190, 226, 234
promotion 127, 159, 175
    activities 111
    discrimination 159

# Index

health promotion 25, *26,* 39, 51, 66–72, 74, 75, 81, 82, 84–87, 92, 93, 111, 154, 158, 159
protective equipment 34, 37, 72, 75, 76, 249
provisional tolerable weekly 34
pseudo conflicts 126
psychosocial work environment 39
public organisations 91
    working time models 91
public pension 81, 226, 227
pull theories 101
push theories 101

quality-adjusted life years (QALY) 287
quality of life 34, 39, 171, 286

rationalisation 4, 47, 101, 123, 166, 176, 195
    decreased time waste 195
reaction patterns 42, 45, 47
reaction phase, crisis 46
readjustments 22
    diversity 198
    lack of qualified labour force 198
    senior labour force 198
receptors 20
recruitment 39, 69, 73, 81, 91, 104, 105, 107, 108, 159, 173
    discrimination 105, 159
recuperation 2, 3, 5–7, 9, 10, *10,* 11, 23, 36, 58–65, 71, 73, 79–81, 83, 84, 86, 89–93, 99, 102, 107, 115, 119, 129, 133, 135, 148, 165, 237, 239, 243, 246, 250, 251, 259, 279
    activity balance 119
    design of work 251
    increasing age 15, 63, 118
    physical activity 63, 64
    risks when lack of 60, 90
    stress reaction 28, 44, 62, 89, 129
    time for 61–63, *62*
reduced working hours 91, 148
    physical activity 91
reemployment, after retirement 172
reference groups 100, 124
reflection 3, 9, *10,* 23, 55, 73, 77, 86, 90, *96,* 112, *114,* 142, 149, 155, 156, 163, *164,* 188, 189, 200, 201, 203, 206–208, 210, 243–249, 251, 253–256, *257,* 258, 262, *262,* 265, 282, 287
rehabilitation 27, 31, 39, 51, 66–68, 73, 90, 107, 108, 213, 250, 258
    policy 108
    responsibility 107–108
relationships 2, 5, 12, 53, 78, 80, 90, 98, 99, 113, 115, 118, 123, 124, 126, 130, 132, 133, 145, 150, 151, 153, 157, 183, 184, 190, 281
relaxed work situation 52
religious beliefs, segregation 139

re-organisations 56, 123, 126, 130, 167, 170, 197, 214, 280, 281
seniors 56
repetitive work tasks 37, 79
resilience 44, 107, 170, 197–199, 223
resource use 71, 238
respiratory organs 16, 71
responsibility for the work environment 71
retirement 3, 9, 25, 27, 29, 36, 45, 53, 56, 61, 63, 77, 81, 88, 90, 92, 98, 101, 102, 105, 107, 112, 113, 115–118, 122, 123, 132, 134, 136, 140, 142, 146, 160, 163, 167, 170–172, 178, 190, 191, 211–213, 216, 218, 224–226, 229, 232, 238, 244, 246, 248, 249
    flexible 53
    normative expectations 238
    postponed 172, 223, 224, 246
    workplace 3–5
retirement age 25, 29, 56, 63, 81, 112, 116, 117, 216, 218, 224, 232, 238, 246
reward 69, 119, 127, 142, 169, 170, 174, 175, 196, 197, 210, 256, 281
risk factors 31, 73–75, 170, 264, 265
risk of accidents 6, 21, 33, 36, 58, 61
role conflicts 106, 125, 148, 160
roles 5, 7, 48, 54, 78, 87, 89, 106, 113, 115, 121, 122, 124–126, 128, 138, 139, 143, 144, 148, 149, 152, 153, 155, 159, 160, 168, 170, 177, 188, 195, 197, 199, 202, 204, 205, 208, 213, 228, 233, 237, 247
rotation of work tasks 76, 83, 192, 210, 279, 281
    manageability 195
    prioritisation 251
routines 46, 51, 63, 72, 76, 79, 108, 118, 119, 122, 159, 163, 164, 169, 176, 190, 192, 200, 202, 204, 243, 248, 250–252, 264, 265

safety representative 104
salary negotiations 256, 259
salutogenesis 50, 170–171
satisfaction in working life 196–197
    quick 172
    work conditions for senior individuals 88
satisfying needs 172, 175
scapegoat mentality 46, 244
schemata 182, 183, 187
scientific management 69, 166, 194
secondary groups 124
secondary prevention 67, *68*
security 6, 45, 47, 49, 54, 61, 67, 78, 98, 101, 102, 107, 108, 110–112, 124, 129, 131, 133, 145, 147, 148, 150–152, 154, 156, 158, 161, 173, 195, 199, 200, 205, 224, 226, 232, 238, 251, 256, 259, 280
    measures 105, 111
    risks 54, 98, 111, 117, 152, 158, 224, 256
    routines 200

security systems 101, 110, 111–112, 224, 226, 232, 238
segregation 37, 137–139
self-actualization 12, 167, 172, 173
self-confidence 51, 122, 174, 182, 190
self-esteem 50, 122, 123, 143, 157, 173–176, 182, 190, 233, 281
self-identity 121–123
self-monitoring groups 195
self-rated health 1, 6, 9, 11–28, 30–32, 56, 58, 66, 92, 101, 119, 170, 171, 191, 198
  finance 101
semantic memories 182, 184, 186, 187
senior employees 15, 21, 29, 32, 33, 36, 40, 56, 63, 80, 81, 88, 92, 100, 105, 115, 132, 135, 140, 148, 171, 177, 187, 189, 191, 208, 210, 211, 213, 215, 216, 221, 226, 229–231, 233, 234, 237, 281, 282
  competence development 100, 140, 187, 191, 231, 233, 234, 282
  demographic situation 216
  employability 100, 105
  estimation of the old age support rate 216
  learning connected to previous knowledge 187, 191
  occupational role 29, 132, 177
  retirement 56, 63, 92, 105, 115, 132, 171, 191, 211, 213, 229, 234
sense of coherence (SOC) 40, 44, 90, 152, 171, 198, 199
  the determinant areas of the SwAge™ model 199
sense of community 3, 5, 6, 39, 80, 98, 99–100, 105, 113–114, *114*, 115, 120, 121, 135, 136, 141–161, 166, 167, 196, 205, 237, 251, 255, 259, 261, 280, 281
  in a group 152
sense of meaningfulness 118, 198
sense of security 49, 54, 67, 78, 107, 108, 111, 129, 131, 133, 147, 150, 154, 158, 161, 205, 251, 256
senses 5, 6, 14, 19, 20, 32, 44–46, 49, 50, 52, 53, 79, 88, 118, 121, 122, 133, 136, 145, 148, 152, 158, 167, 172, 180, 183–186, 198, 202, 209, 210, 216, 222, 223, 234, 279
sensory impressions 180, 181, 185, 186
sensory memory 184–185
sensory organs 20, 21, 27, 31, 33, 71, 221
  occupational injuries among senior individuals 33
sensory system 20–21, 32–33, 38, 180, 185
sexual harassment 54, 139, 159
sexual orientation 122, 139, 159
shift work 31, 44, 59, 61
short-term memory 185–187
short-term sickness absence 53, 250

sickness 13
  benefit 101, 111, 168
  compensation 97
sickness absence 15, 25, 29, 34, 36, 37, 39, 40, 51, 53, 55, 63, 66, 70, 73, 77, 81, 88, 89, 91, 92, 102, 104, 105, 122, 129, 135, 141, 146, 160, 168, 178, 213, 250, 263, 280, 285, 286
  costs 107–108
  decreased, the Hawthorn effect 135
  statistics 25, 91, 263
sick pay 29, 31, 34, 81, 97, 101, 102, 108, 111, 280
single loop learning 206
skeleton and joints 17–18
  biological ageing 17
skills 3, 7, 15, 21, 99, 100, 106, 107, 123, 130, 132, 136, 146, 150, 157, 163, 164, *164*, 165, 166, 168, 170, 177, 180–194, 197, 200, 207, 210–212, 214, 226, 233, 234, 243, 245, 246, 248, 252, 259, 261, 280–282
sleep disturbances 46, 58, 60–61
smooth muscle cells 19
social age 62, 102, 105, 110, 112, 113, 115–118, 120–124, 131, 133, 140, 142, 143, 148, 159–161, 215, 218, 226, 228–233, 236–239, 280
  categorisation 113, 127, 229–231
  groups 3, 114, 125, *137*, 140, 141
  social constructions 237
  societal perspectives 231–232
  stereotypes 139
  working life 228–232
social alliances 77, 79
social contexts 117, 121–133, 151, 161, 228, 229
  inclusion and exclusion 121, 122, 161
social control 195
social group 3, 114, 125, 137, *137*, 140, 141
social interaction 115, 118, 121, 124, 134, 135, 152, 167
social isolation 48, 53, 135, 136
social learning theory (SLT) 78
socially ineffective structure 135
  counteracted recuperation 135
  decreased work ethics 135
  increased staff turnover 135
social networks 105, 107, 115
social roles, mental impact when losing 233
social security systems 101, 110–112, 226, 232, 238
social service professions *see* contact-based professions
social status 134
  coping with stress 44, 79
  exercise 155–156
  improved health 123
social welfare 13, 81, 97, 99, 110, 111

# Index

social work environment 3, 7, 39, 51, 53, 55, 102, 113, 114, 116, 121–143, 147–152, 154–158, 160, 161, 165, 170, 171, 237, 239, 251, 259, 261, 280
    symptoms of mental illness 229
societal culture 93, 112, 159, 161, 211, 214
    sustainable working life 159
societal economy 86, 99, 101, 110–112, 216, 218, 219, 224, 225, 238
    GDP 111
    health 85–86
    ill-health 85, 91, 111
    occupational accidents 87, 110
societal ideals 184
societal norms 136, 160
    senior people's knowledge and experience 160
socio-cultural ideals 184
sociotechnical systems theory 11, 195
Socratic dialogue (thoughtful dialogue) 208, 209
sole proprietors 50
source of power 188
spatial abilities 181, 182
sphere for action 3, 5–7, *6,* 7, 86, 92, 93, 98, 110, 112, 131, 132, 177, 201, 211, 214, 237, 261, 279, 283
    determinant areas 3, 5–7, 86, 92, 93, 110, 112, 132, 211, 214, 261
spread of infection 246
static work tasks 203, 243
staying in working life until an older age 138
    welfare systems 81
stereotype 139
    attitudes 29
    characteristics 229
    expectations 122, 123
stigmatisation 122
stimulation 3, 7, 44, 47, 48, 50, 52, 102, 105, 108, 119, 133, 134, 152, 163, 164, *164,* 165–179, 194, 196, 197, 199–201, 209–212, 214, 237, 247, 251, 255, 259, 261, 281
    activity balance 119
strain injuries 20, 37
streamlining 195, 203
stress 2, 6, 11, 16, 22–24, 28, 36, 39, 40–43, 55, 56, 58, 60–62, 70, 73, 77, 83, 88–91, 113, 117, 129, 133, *134,* 136, 150, 151, 154, 160, 167, 168, 174, 175, 182, 189, 192, 196, 205, 213, 230, 233, 235, 245, 246, 250, 251, 254, 259, 279, 284
    chronic 39, 40, 77, 79
    cognitive ageing 233, 235
    cognitive impairment 192
    coping strategy 44, 47, 48, 78
    coping with 44–53, 78–79
    hormones 41, 62
    ill-health 40, 53, 61, 117, 129, 175

mental work environment 2, 6, 22, 36, 39, 40, 43, 53, 73, 250, 254, 259
physical activity 22, 91
senior women 56
social support 44, 77, 79, 113, 133, 151, 205
societal problem 88–89
stressors 22, 23, 27, 30, 40–43, *43,* 44, 45, 47–51, 54, 60, 78, 133, *134,* 170, 171, 180, 185, 198, 199
stress reactions 22–24, 28, 42–44, 46, 62, 78, 83, 89, 91, 113, 129, 133, *134,* 167, 175, 279
    coping strategy 78
    recuperation 28, 44, 62, 89, 129
    risks with lasting stress reactions 28, 60
stress-related ill-health 55, 91, 117
    sickness absence 55, 91
stress-related mental illness 51, 91, 235
striated skeletal muscle cells 19
subgroups 124, 152, 158
subjective perception of oneself 71, 122
supervision 88, 247, 248
support 3, 5–7, 17–19, 25, 31, 39, 44, 47, 48, 55, 60, 68–70, 72, 74, 77–79, 81–84, 86–93, 98–100, 102, 105–107, 110–114, *114,* 116, 119, 120, 130, 131, 136, 141, 142, 170, 175, 176, 192, 194, 195, 199, 202, 204–207, 209, 212, 216, 227, 244–246, 249–251, 255, 258, 259, 280
    the determinant areas of the SwAge™ model 7, 39, 72, 86, *114,* 133, 156–158, 165, 243–249, 280
    instrumental 69, 132–134, 142, 151, 202
    sense of community 3, 5, 6, 39, 98, 99–100, 105–114, *114,* 120, 142–161, 166, 167, 196, 205, 237, 251, 255, 259, 261, 280 281
    social 3, 5–7, 25, 39, 44, 48, 77, 79, 99–100, 102, 106, 113, 114, 119, 120, 131–134, 142, 150–153, 155–156, 170, 202, 205, 251, 259, 281
supportive environment 28, 86, 87, 107, 111, *153,* 158–159
    age 107, 111, 159
    knowledge 107
    recuperation 107
    stress 107
    the SwAge™ model 86, 158
supportive functions 90
supportive resources in work 99, 105, 113, 119, 132, 133, *134,* 147, 202, 204–206
surveys 7, 73, 127, 253–255, 262, 263, 284
sustainable working life 1–7, 9, *12,* 17, 26, 31, 33, 36, 39, 40, 59, 71, 73, 81–83, 85, 87, 89, 92, 98, 116, 123, 132, 133, 138, 144, 146, 160, 197, 199, 211, 215–219, *218,* 241, 249, 253–257, *257,* 259, 261–265, **266–278**, 279, 284–286

# Index

action plan 262, 265, **266–278**
age 253–255, 261–265, **266–278,** 281, 283, 284, 286, 287
  determinant areas 1–7, *6,* 11, *12,* 39, 71, 73, 81–83, 133, 144, 160, 199, 241, 256, *257,* 259, 261, 262, *262,* 285, 286
  determinant areas of the SwAge™ model 1–7, 39, 73, 81–83, 98, 133, 199, 201, 241, 253, 256, *257,* 261, *262,* 265, **266–278,** 279, 285
  knowledge, creativity and development 12, 211
  process evaluations 283–285
  work motivation and work satisfaction 178, 197
the SwAge™ model 1–7, 9, *10,* 30, 39, 49, 56, 72, 73, 75, 77, 81–83, 86, 88, 92, *96,* 98, 108, 110, *114,* 131, 133, 142, 144, 146, 148, 155–158, 160, *164,* 165, 198–199, 201, 202, 207, 211, 237, 241, 243–249, 253, 256, *257,* 261, 262, *262,* 263–265, **266–278,** 279, 280, 283, 285
  assessment tool 253
  cases 243–249
  determinant areas *10,* 39, 43, 49, 56, 73, 75, 81–83, 86, 88, 96, 108, *114,* 133, 142, 148, 155, 156, 158, 160, *164,* 165, 199, 211, 237, 241, 253, 256, *257,* 261, *262,* 263, 265, 283, 285
  dialogue tool 83, 256, 257, *257*
  employability 98, 108, 156, 237, 241, 253, 285
  financial security 110
  'individuals' financial security 110
  nine determinant areas 7, 12–13, 43, 49, 56, 81–83, 86, 88, 108, 133, 142, 148, 155, 156, 160, 199, 237, 241, 253, 256, 265, 283, 285
  risk factors 264
  sphere of action 9, *10*
  workplace analysis 73, 82, 201, 262, 265, 279, 285
systematic work environment management 21, 71, 75, 82, 241, 261, 279
  information 71, 72, 241
  work-related injuries 24, 50, 107, 111

tacit knowledge 55, 157, 177, 178, 188–190, 206, 207, 210, 214, 234, 247, 282
  transfer of 207–209
Taylorism 166, 195
teamwork 144, 195, 200, 206, 208
technologies 31, 35, 36, 54, 73, 75, 106, 107, 123, 126, 132, 133, 140, 150, 152, 167, 168, 171, 177, 189–191, 194, 197, 198, 201, 203–207, 209, 229, 247, 250, 259, 263
  aids 35, 106, 107, 259
  development 54, 191, 207
  system 152
telomeres 26, 221
tense work situation 51, 60

terms of employment 159
  discrimination 159, 231
tertiary prevention 67, 68
theoretical deduction 187
Theory X 166
Theory Y 167
thermoreceptors 20
threats and violence 2, 6, 40, 44, 54–55, 73, 83, 88, 250, 279
  measures 79
thyroid hormones 22, 41, 43, 185
time 2–7, 9, *10,* 12, 15, 16, 19, 22–26, 34, 36, 39, 40, 42, 44–48, 53–56, 58–65, 71, 75, 77–79, 81, 83, 86–93, 98, 99, 102, 106–108, 115–120, 122, 123, 125, 132–136, 138, 141–143, 146, 148, 150–152, 154–156, 160, 161, 164–166, 168, 169, 172, 175, 177, 178, 181, 182, 185–187, 189–192, 194–197, 199–210, 212, 213, 215, 216, 218, 221–230, 233, 234, 237, 243–250, 254, 256–259, 264, 279, 281, 282, 284–286
  constraints 106, 151, 154, 168
  plan 92, 117, 189
  recuperation 61–63, *62*
  studies 63, 166, 194, 195
  waste 195
time for rest and recuperation 58, 61–63, 237
  gender 63
  increasing age 63
tiredness 19, 44, 53, 58, 60, 79, 80, 168
  accidents 33, 53, 58, 79
  biological ageing 53, 79
  studies of age groups 19
tolerable daily intake (TDI) 34
toxic substances 34, 40
  occupational illness 33, 44
Toyota production system 195
transforming resources 99
treatment as usual (TAU) 286
trust 40, 50, 113, 144, 147, 148, 151, 152, 158, 257

unconscious processes 180
unemployment 97, 101, 102, 108, 111, 118, 141, 216, 238, 280
  benefit 102, 216
unilateral movements 30–32, 83, 254, 259, 279

value conflicts 126
values 4, 13, 33, 34, 61, 65–66, 70, 71, 75, 76, 82, 83, 88, 107, 112, 119, 124, 126, 132, 138, 154, 155, 169, 177, 178, 182, 184, 188, 195, 201, 206, 208, 219, 222, 225–226, 230, 251, 279, 285, 286, 288
Vattenfall 81
  decreased sickness absence 81
  increased productivity and profitability 81
  promotion 81
  secured competence supply 81

# Index

ventilation 30, 79, 249
vertical segregation 137–139
vibrations 20, 29, 31, 250, 254, 259
victimisation 152, 153, 212
visionary leader 129, 130
vitality resources 99

waking hours 117, 119, 196
welfare system 13, 81, 85, 97, 99, 111, 226
well-being 2–5, 7, 9–14, 22, 24, 27, 28, 30, 38, 39, 42, 48, 50, 52, 57, 59, 63–72, 77, 81, 85–87, 98, 99, 102, 110, 114, 126, 131, 133–136, 142, 149, 150, 152, 154, 158, 163, 169–174, 191, 196, 197, 200, 223, 250, 251, 259, 280, 286
wellness allowance 71, 86
whistle blower 150
willingness to work 1–3, 5, 9, 52, 54, 95, 111–113, 115, 117, 121, 131, 137, 138, 143, 147, *147*, 148, 163, 165, 170, 192, 194, 196, 203, 221, 228, 233, 237, 261, 286
wisdom 47, 126, 182, 187, 208, 212, 234
work ability 7, 15, 27, 29, 30, 32, 33, 35, 45, 65, 82, 99, 100, 112, 118, 131, 132, 134, 146, 147, 225, 226, 237–239, 250, 253–255
    biological, social and cognitive age 112
    determinant areas 1–3, *2*
    functional age 112
    leisure activities 118
    personal social life 118
    sense of meaningfulness 118
    social support 131, 134
    survey 7, 253
    work satisfaction 118
work ability index (WAI) 29
work climate 196, 200, 250–252
    creative 252
work content 31, 78, 91, 100, 150, 167, 178, 213–214, 226
work environment
    chemical reactions 33–35
    COVID-19 76
    effects on the body 71
    factors that cause ill-health 65, 195
    factors with positive or negative impacts 73
    health promotion measures 71, 92
    insufficient ergonomics 30, 150
    insufficient risk assessments 25, 26, 32, 53, 56, 60, 132
    issues 72–75
    legislation 72, 87, 160
    management 21, 35, 71, 73, 75, 82, 241, 261, 265, 279
    mental 2, 3, 5–7, 9, 10, *10*, 11, 22, 30, 39–58, 71, 73, 74, 84, 86, 90, 93, 99, 102, 111, 133, 136, 165, 171, 198, 203, 237, 239, 250, 254, 259, 261
    mentally demanding 32, 53, 60, 77, 88, 223
    physical 2, 3, 5–7, 10, 15, 18–21, 25, 27, 29–38, 50, 71, 73, 75–76, 80, 84, 86, 88, 93, 126, 133, 165, 195, 237, 239, 250, 254, 259, 261
    predictor of ill-health 24–26
    priority list of positive and negative factors 263–264
    promotion 39, 51, 71, 72, 75, 84, 93, 111
    psychosocial 39
    report accidents 25
    risk assessment based on the SwAge™ model 73–74, 88
    work content 31, 78, 91, 100, 150, 167, 178, 213–214, 226
    work postures 31, 75, 76, 263
    work tasks 3, 5, 7, 15, 16, 18–21, 24, 27, 30–33, 35–37, 39, 44, 52, 53, 56, 58–61, 64, 69, 70, 73–76, 78–80, 83, 90–92, 95, 98, 99, 102, 105, 106, 108, 109, 116, 117, 119, 122, 123, 126, 128–131, 135, 136, 143, 146–148, 150, 151, 163–172, 174–181, 188, 193, 197–200, 202–210, 222, 224–226, 231, 233, 237–239, 243, 245–247, 250–252, 254, 255, 258–261, 263, 279–282
work environment laws 83, 87
psychosocial work environment 39
work flow 30, 69, 126, 135, 167
working hours 2, 3, 5–7, 9, 10, *10*, 34, 58–65, 70, 73, 78, 80, 83, 84, 86, 90–93, 99, 102, 103, 106, 116–120, 122, 131, 135, 148, 150, 154, 160, 165, 174, 200, 205, 221, 224, 232, 237, 239, 250, 251, 254, 259, 279
    adjustment to social life 118
    breaks 59–60
    pace and recuperation 58–64
    parental leave 91, 122
    work environment laws 83, 87
    working time regulations 59
working life 1–7, 9–13, 15, 19, 20, 25, 26, *26,* 27, 29, 32–34, 36, 39, 40, 45, 52–54, 58, 59, 61, 63–65, 69–71, 73, 75, 79, 82, 83, 85–92, 95, 97–102, 105, 107–112, 114, 116, 118–120, 122, 123, 127, 132, 133, 136, *137,* 138–142, 144, 146, 148, 149, 156, 158, 165, 168, 171, 176–178, 182, 183, 187–192, 203, 205, 207, 208, 211–213, 215–219, 236–238, 241, 246–249, 251, 253–256, *257,* 259, 261–265, **266–278,** 279, 280, 283–287
    age (*see* age)
    age definitions *218,* 218–219
    biological age 221–224
    chronological age 225–227
    cognitive age 233–235
    discrimination 159–160
    employability changes 197–199
    extended 80–81

gender differences between physical demands 55–56
gender segregation 37
ill-health 27
  the individual's possibility of participating in 110, 213, 236
  measures of health/ill-health 27
  overview of efficiency and meaningfulness 194–196
  satisfaction 196–197
  social age 228–232
  social effects of informal organisations 134–136
  social inclusion 115–116
  stressors *43*, 43–44
working time model 80-90-100 80, 81, 92, 148
  Vattenfall 81
working time models 80–81, 91–92, 148
  80-90-100 80, 81, 92, 148
  generational change 81
  pauses 91, 92
  possibility of decreased working hours 91, 92
  recuperation 92, 148
  senior employees 80, 81, 92
  work pace 80, 92
working time regulations 59, 79, 88, 161
  breaks 59
  the determinant areas of the SwAge™ -model 88
  night shift work 59
  occupational health care 81–82
  on-call time 59
  overtime 107, 174, 246
  pauses 59
  preventive measures 79
  recuperation 59, 79
work-life balance 70, 71, 106, 119, 142, 160–161, 259
  employability 197–199
workload 2, 6, 14, 15, 17, 26, 30–32, 37, 75–77, 119, 152, 196, 254, 259, 264, 279
work motivation 70, 165–167, 175, 176, 178, 196, 197
  satisfaction 172–174
work pace 2–3, 5–7, 9–10, *10*, 58–60, 63, 64, 71, 73, 78, 80, 83, 84, 86, 91–93, 99, 102, 135, 150–152, 165, 174, 203, 221, 224, 237, 239, 245, 250, 251, 254, 259, 261, 263, 279
  recuperation 63
workplace *2*, 3, 7, 9–11, 18, 19, 25, 28, 32, 34–40, 44, 45, 48, 50, 51, 53–54, 56, 60, 68–70, 72, 73–77, 79, 80–84, 86–88, 90, 91, 95, 100, 101, 104–109, 111, 113–117, 121–124, 126, 127, 129–132, 135, 136, 138–140, 142–145, 148–151, 155, 157, 159–161, 163, 164, 166–168, 171, 175–178, 189, 190, 192–197, 199–201, 206–208, 210, 212, 214, 215, 218, 229, 231, 243–246, 248, 250, 251, 253, 279–287
  analysis 73, 82, 201, 262, 263, 265, 285
  back pain 31
  bullying 53–54
  climate 35, 76, 279
  culture 106, 152–154, *153*
  design of 30
  examinations 73, 253, 261, 263–264
  inclusion 132–134
  management 261–265, **266–278**
  retirement 3–5
  risks 77–78
  social activities 154
work postures 31, 75, 76, 263
work-related ill-health 82, 256
  societal support 110
work satisfaction 3, 7, 31, 44, 47, 48, 50, 52, 56, 118, 133–135, *164,* 165–179, 195, 196–197, 199, 209, 214, 261, 281, 287
work schedule 2, 58–59, 83, 157, 161, 250, 259, 261, 279, 281
work situation 4, 7, 9, 19, 24, 25, 27–29, 31–33, 36, 39, 40, 42, 44, 45, 48, 49, 54–58, 60, 61, 64, 70, 72, 73, 77, 78, 80–84, 86–88, 90, 91, 93, 95, 98, 101, 102, 109, 123, 131, 133, 135, 136, 142, 143, 145, 146, 149, 150, 156, 160, 165–168, 170–172, 176, 178, 194, 197, 198, 200, 201, 208, 210, 213, 224, 225, 237, 243, 246, 250, 251, 253–260, 263–265, 279–281
  balance between demands and control 50–53
  societal perspectives 88–89
work tasks 3, 5, 7, 15, 16, 18–21, 24, 27, 30–33, 35–37, 39, 44, 52, 53, 56, 58–61, 64, 69, 70, 73–76, 78–80, 83, 90–92, 95, 98, 99, 102, 105, 106, 108, 109, 116, 117, 119, 122, 123, 126, 128–131, 135, 136, 143, 146–148, 150, 151, 163–172, 174–181, 188, 193–200, 202–210, 222, 224–226, 231, 233, 237–239, 243, 245–247, 250–252, 254, 255, 258–261, 263, 279–282
  execution 100, 163–164, *164,* 194–196, 211–214
work unit 24, 70, 73, 74, 77, 107, 121, 124, 126, 127, 130, 134, 135, 146, 153–157, 167, 170, 178, 196, 199–202, 204–206, 208, 243, 259, 263, 264, 280
  age diversity 105
  differences 154
  relationship 130
World Health Organization (WHO) 159
  health 66, 67, 85, 86, 110, 158
  Ottawa charter for health promotion 85, 158
  physical activity 66, 67

Milton Keynes UK
Ingram Content Group UK Ltd.
UKHW031328071224
451979UK00004B/33